McGraw Hill

Chemistry
Review and Workbook

McGraw Hill

Chemistry
Review and Workbook

John Moore
Richard Langley

New York Chicago San Francisco Athens London Madrid
Mexico City Milan New Delhi Singapore Sydney Toronto

1 2 3 4 5 6 7 8 9   LHS   27 26 25 24 23 22

ISBN   978-1-264-25904-5
MHID   1-264-25904-2

e-ISBN   978-1-264-25905-2
e-MHID   1-264-25905-0

McGraw Hill products are available at special quantity discounts to
use as premiums and sales promotions or for use in corporate training
programs. To contact a representative, please visit the Contact Us pages
at www.mhprofessional.com.

Contents

·24·

Solutions II

·25·

Kinetics

·26·

Introduction to equilibrium

·27·

Acids and bases

·28·

Other equilibria

·29·

Buffers and titrations

·30·

Thermodynamics

·31·

Electrochemistry

·32·

Nuclear chemistry 247

·A·

Constants and equations 257

·B·

SI units and conversions 259

·C·

Some properties of water 261

Periodic table 287

Introduction

To The Student

Learning chemistry, like learning to play a violin, is not a spectator sport. Mastering chemistry, like most sciences, involves active participation. This workbook is designed for the high school or college student who is taking an introductory chemistry course or someone who wants to brush up on their chemistry background in preparation for taking some type of standardized exam, such as the MCAT. We designed this workbook to give you a review of the basic chemistry concepts and to provide you with problems to test your understanding. In addition, these specific tips will help you in your study of chemistry:

- Strive for understanding, not just memorization.
- Study some chemistry every day—long study sessions right before an exam are not nearly as effective as shorter, regular study sessions, which usually take less overall time.
- Work many, many problems, but again strive for understanding—it is a waste of time to simply memorize how to do a particular problem; and it is also a waste of time to simply look over the solution for a problem without striving to understand why a certain procedure was followed.
- Nomenclature, the naming of chemical compounds, is extremely important. When the time comes, learn the rules, and apply them. Calling a chemical compound by the wrong name is certainly not the way to impress your chemistry teacher or a potential boss.
- Practice, practice, practice.

You will be doing many problems in your study of chemistry. Here are some specific suggestions to help you in your problem solving:

- Identify and write down what quantity you wish to find.
- Extract and write down just the pertinent information from the problem, especially the numbers *and* units—this is especially important for long word problems.
- Identify and write down any equations or relationships that might be useful.
- Look for relationships among the information from the problem, the equations, and the quantity you wish to find.
- Use the unit conversion method (illustrated in Chapter 2) in solving for the desired quantity.
- Round *only* your final answer—to the correct number of significant figures.

- When you look at a worked example, try covering up the solution and slowly uncover it after you guess what the next step will be, especially when reviewing the material.
- Practice, practice, practice and practice some more.

Note, unlike your textbook where you are supposed to begin at the beginning and work your way through, topics in this workbook may be used in any order. This means if what you are covering in an early chapter of your textbook appears in a later chapter here, you can skip straight to it without using the earlier chapters.

In addition, not all books use the same values for all numerical quantities, so do not be surprised by minor differences; they will work the same. If there appears to be a conflict, your teacher will tell you what to do. Also, some equations may appear in different forms in different books. You must understand that these are not different equations, but the same old equations rearranged.

McGraw Hill

Chemistry
Review and Workbook

What is chemistry?

Chemistry is the study of matter. Matter is anything you can touch or that can touch you. Matter has mass and occupies space. Matter is the chair you are seated on or the breeze blowing against your face. For matter to do anything, energy must be involved.

To understand how matter and energy undergo chemical or physical processes, it is necessary to measure various properties. In some cases, the amounts of matter and energy may be extremely small; therefore, the measurements must be made very carefully to ensure that the results are very precise and very accurate.

When making these measurements, there are certain terms that apply and certain ideas that must not be confused.

The properties of matter fall into two categories, some properties are extensive (measurable) and other properties are intensive (observable).

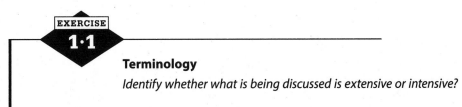

Terminology

Identify whether what is being discussed is extensive or intensive?

1. The book has a mass of 2.5 kg.

2. The book has a blue cover.

3. Under standard conditions, water boils at 100°C.

4. Under standard conditions, water has a density of 1.00 g/cm³.

5. The volume of the water sample is 100 mL.

Precision and accuracy

A precise measurement may be repeated to give the same value. An accurate value is how close a value is to the "true" value.

1. Two students measure the height of a book. Student A measures the height five times and gets 27.2 cm, 27.4 cm, 27.2 cm, 27.4 cm, and 27.3 cm. Student B also measures the height five times and obtains the following values; 27.2 cm, 27.5 cm, 27.1 cm, 27.4 cm, and 27.0 cm. Which student's measurements were more precise?

2. Student A averages their five measurements and determines the height of the book is 27.3 cm and student B gets 27.2 cm as the height. Which student's value was more accurate?

Some seemingly simple concepts may lead nonscientists to misconceptions. Examples of these misconceptions are weight/mass and heat/temperature. An object has a mass (quantity of matter); however, it does not have weight unless gravity acts upon it. Since gravity may vary, so may weight. Temperature is an intensive property, while heat is an extensive property.

Basic concepts

The gravity on Mars is slightly less than 40% of the gravity on Earth. An astronaut in a space suit weighs 200 lbs on Earth. This same astronaut and suit has a mass of 90 kg.

1. What is the weight of the astronaut and space suit on Mars?

2. What is the mass of the astronaut and space suit on Mars?

There are two pans, A and B, of water on a stove. Each contains water initially at 25°C. Pan A contains 1 qt of water and pan B contains 1 gal of water. The stove heats both pans at the same rate.

3. Eventually the water in each pan boils. When the water begins to boil, how does the temperature in pan A compare to the temperature in pan B?

4. Once heating begins, which pan boils first?

5. Which question, 3 or 4, involves an extensive property?

6. Which question, 3 or 4, involves an intensive property?

7. Explain your choice for question 5.

8. Explain your choice for question 6.

Chemistry and numbers

Many things in chemistry are measured. It should be remembered that in chemistry, no measurement is of any use if it does not contain a number *and* a unit. For example a measurement of 4 is meaningless unless there is a unit, as it could mean 4 dollars, 4 miles, 4 days, or 4.... Make sure you always express the appropriate units for any value you report.

It is also important to remember that the numbers 4, 4.0, 4.00, and 4.000 mean different things to scientists. Make sure you use the appropriate number of digits when expressing your answer.

Units

In most cases, chemistry employs the SI (Le Système International d'Unités) system, which relies on seven base units. These seven units were chosen because anything a scientist can measure may be expressed by one of these units or a combination of two or more base units. To simplify some combinations, other names are used. For example, energy is expressed as mass times distance squared divided by time squared. However, these combined units may always be broken down to the base units (a process commonly used to simplify problems).

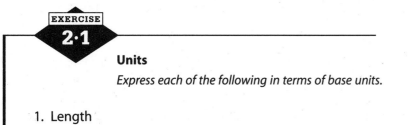

EXERCISE 2·1

Units

Express each of the following in terms of base units.

1. Length

2. Time

3. Electric current

Modifying units

Units may be modified by combining the units or by using SI prefixes. The following exercise contains some examples.

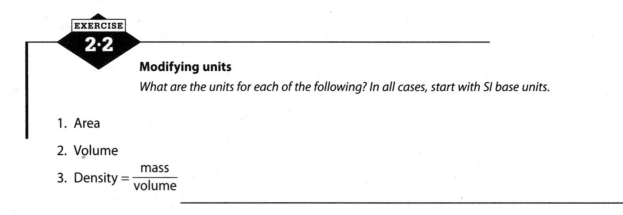

EXERCISE 2·2

Modifying units

What are the units for each of the following? In all cases, start with SI base units.

1. Area

2. Volume

3. Density $= \dfrac{\text{mass}}{\text{volume}}$

Types of numbers

Scientist deal with two types of numbers. Some numbers are measured (directly or indirectly), and it is imperative to report such values to the appropriate number of significant figures. The other numbers are "exact" numbers. These are numbers that are exactly what you see, and the rules of significant figures do not apply to these. Exact numbers should not be rounded. Examples of exact numbers include exactly 12 eggs in a dozen eggs and exactly 2.54 cm in 1 in.

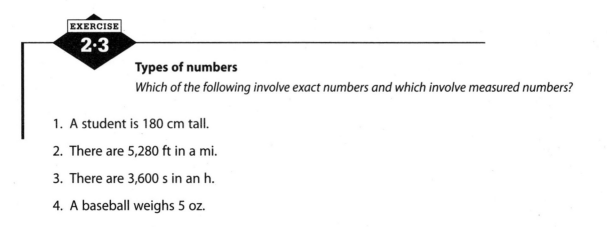

EXERCISE 2·3

Types of numbers

Which of the following involve exact numbers and which involve measured numbers?

1. A student is 180 cm tall.

2. There are 5,280 ft in a mi.

3. There are 3,600 s in an h.

4. A baseball weighs 5 oz.

Significant figures

When reporting any measured number, it is necessary to use the appropriate number of significant figures. The correct number of significant figures must be followed when different measured numbers are combined.

Significant figures may be any digit from 1 through 9, and in some cases, 0. A 0 is not significant if it is used only as a placeholder to express a multiple of 10. In many cases, to clarify if a 0 is only a placeholder or not, it will help to express a number in scientific notation. Placeholding zeroes are between the decimal point and a digit to the right of the decimal point (for example, the zeroes in 0.054) or to the right of the last non-zero digit (for example, the zeroes in 1,200).

There are a few cases where a zero to the right may be significant; to eliminate ambiguity express the number in scientific notation. For example, to indicate that the first zero in the 1,200 example is significant, it should be written as 1.20×10^3.

If a number is in scientific notation, all digits shown, even zeroes, are significant as all place-holding zeroes have been separated and given as a power of 10. For example, the number 100 may be expressed in scientific notation in a number of ways, including 1×10^2, 1.0×10^2, and 1.00×10^2. These three choices have 1, 2, and 3 significant figures, respectively. Another example is writing 0.0032 as 3.2×10^{-3}, which has two significant figures in both cases.

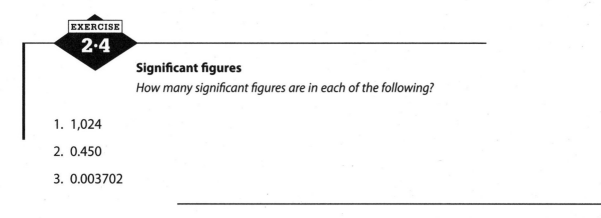

EXERCISE
2·4

Significant figures

How many significant figures are in each of the following?

1. 1,024

2. 0.450

3. 0.003702

Significant figures in mathematical operations

Converting units may be as simple as multiplying the length of a rectangle by its width to determine the area. It may also involve changing the length and width of the rectangle from inches to centimeters and then to area.

When performing the conversions, it is necessary to maintain the correct number of significant figures. There are various rules for maintaining the correct number of significant figures. These rules depend on the mathematical operation being performed. The two rules discussed here deal with (1) multiplication and division and (2) addition and subtraction.

Under both rules, it is necessary to determine how many significant figures each of the numbers used in the calculation has. When dealing with multiplication and division, either alone or in combination, the final answer will have the same number of significant figures as the least number of significant figures in the numbers used in the calculation. In the case of addition and subtraction, either alone or in combination, the final answer will have the same number of digits past the decimal point as the number used in the calculation with the least number of digits past the decimal point.

Only measured numbers are important when determining the significant figures resulting from a calculation. The number of digits in an exact number are not used to determine the significant figures in an answer.

Significant figures in mathematical operations

Perform the following unit conversions and report the answers to the correct significant figures with the correct units.

1. 53.21 in. + 27.0 in. + 22.321 in. + 22.013 in. =

2. $\dfrac{0.533 \text{ kg}}{0.100 \text{ m} \times 0.100 \text{ m} \times 0.10 \text{ m}} =$

3. $(3.42 \text{ in})(4.0 \text{ in})\left(\dfrac{2.54 \text{ cm}}{1 \text{ in.}}\right)^2 =$

4. $(25 \text{ mL})\left(\dfrac{1 \text{ cm}^3}{1 \text{ mL}}\right)\left(\dfrac{1 \text{ mm}}{0.1 \text{ cm}}\right)^3 =$

5. $\dfrac{(5.42 - 5.00) \text{ g}}{(2.30 \text{ cm})(3.27 \text{ cm})(1.88 \text{ cm})} =$

What is matter?

Atoms are the fundamental building blocks of matter. Under normal conditions, all matter consists of atoms. Atoms are built from yet smaller particles. All atoms contain electrons and protons, and all atoms except the simplest (normal hydrogen) contain neutrons. The electrons are located outside the core or nucleus of the atom (where the protons and neutrons are found). Compared to the size of the atom, the nucleus is extremely small. In later chapters, we will see that the electrons may be subdivided based upon how much energy they have.

Atomic theory

The identity of an atom may be defined by the number of protons present. For the atoms of an element, the number of neutrons may vary to produce different isotopes (atoms containing the same number of protons but differing numbers of neutrons). An atom is neutral because there are equal numbers of electrons and protons. If the numbers of electrons and protons are not equal, what you have is no longer an atom, but an ion. Ions may form by adding or removing electrons from an atom.

The name *neutron* alludes to the fact that these particles are neutral. Neither electrons nor protons are neutral. Electrons and protons have charges of –1 and +1, respectively. A neutron is slightly heavier than a proton, and both are significantly heavier than an electron. Many times, it is possible to ignore the mass of electrons when talking about all three particles. For simple calculations, the masses of protons and neutrons are often taken as being 1 amu (atomic mass unit) and electrons have a mass of 0 amu. Assuming the mass of an electron is 0 amu introduces a very small error in mass calculations.

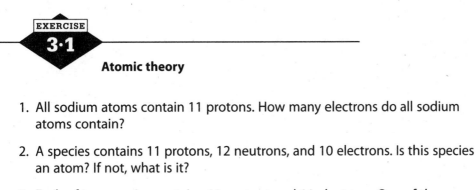

EXERCISE

3·1

Atomic theory

1. All sodium atoms contain 11 protons. How many electrons do all sodium atoms contain?

2. A species contains 11 protons, 12 neutrons, and 10 electrons. Is this species an atom? If not, what is it?

3. Each of two species contains 11 protons and 11 electrons. One of the species contains 12 neutrons and the other contains 11 neutrons. How are these two species related?

Locating the components of atoms

1. The number of electrons in an atom may be altered to form an ion. How may the location of the electrons in an atom facilitate the formation of ions?

2. Ions form when an atom gains or loses electrons. Why is there no equivalent particle formed by the gain or loss of protons?

Modern atoms

Before the 1920s, atoms were described as having a small central nucleus with electrons orbiting. This was the Bohr model of the atom. The electron orbits depended on how much energy the electrons have. More precise experiments indicated discrepancies with this simple description. The development of quantum mechanics, beginning in the 1920s, was directed toward explaining these discrepancies. The quantum mechanical description of atoms will be covered in a later chapter.

Modern atoms

1. In the Bohr model of the atom, where are the protons located in an atom?

2. In the Bohr model of the atom, where are the electrons located in an atom?

Atomic weight (mass)

Every atom has a mass expressed as its atomic mass (weight). A related term is the mass number, which is the sum of the number of protons and the number of neutrons. The mass number of an atom will be close to the atomic mass. Thus, a cesium atom with 55 protons and 78 neutrons has a mass number of 133. The atomic mass of this cesium atom is 132.9 amu. (Note: mass numbers are exact numbers, while atomic masses are measured numbers.)

The presence of isotopes complicates the determination of the atomic mass. For example, in a sample of an element, if half of the isotopes of an atom have a mass of 20 amu and the other half have a mass of 21 amu, then the atomic mass of the element would be 20.5 amu. For real elements, there are often more than two isotopes, and it is unlikely that the amount of the different isotopes will be equal, which means the calculation is more involved. For these cases, it is necessary to determine the contribution of each isotope and finally add all the contributions.

EXAMPLE 1

Let's use chlorine as an example. Natural chlorine consists of two isotopes. One chlorine isotope has a mass number of 35 (= chlorine-35) and the other has a mass number of 37 (= chlorine 37). Approximately 76.0% of the chlorine atoms have a mass number of 35 and the remainder (24.0%) have a mass number of 37. The contribution to the atomic mass for chlorine-35 is:

$$(35)\left(\frac{76.0\%}{100\%}\right) = 26.6$$

The contribution to the atomic mass for chlorine 37 is:

$$(37)\left(\frac{24.0\%}{100\%}\right) = 8.88$$

If there were more isotopes, this procedure would be repeated for each of them. The final step in the calculation is to sum the individual contributions of the isotopes and round to the appropriate number of significant figures.

$$26.6 + 8.88 = 35.5$$

Using mass numbers simplifies the numbers used. If atomic masses and better percentages were used, the numbers used for chlorine-35 would be 34.96885 amu and 75.78%, and for chlorine-37 the numbers would be 36.96590 amu and 24.22%.

EXERCISE

3·4

Atomic weight (mass)

1. Determine the atomic mass of chlorine, which consists of two isotopes, chlorine-35 and chlorine-37. Information on these two isotopes is in the following table:

ISOTOPE	ATOMIC MASS	PERCENT ABUNDANCE
Chlorine-35	34.96885 amu	75.78%
Chlorine-37	36.96590 amu	24.22%

2. Determine the atomic mass of magnesium, which consists of three isotopes, magnesium-24, magnesium-25, and magnesium-26. Information on these three isotopes is in the following table:

ISOTOPE	ATOMIC MASS	PERCENT ABUNDANCE
Magnesium-24	23.9850 amu	78.990%
Magnesium-25	24.9858 amu	10.000%
Magnesium-26	25.9826 amu	11.010%

3. Determine the atomic mass of zinc, which consists of five isotopes, zinc-64, zinc-66, zinc-67, zinc-68, and zinc-70. Information on these three isotopes is in the following table:

ISOTOPE	ATOMIC MASS	PERCENT ABUNDANCE
Zinc-64	63.929146 amu	48.63%
Zinc-66	65.926036 amu	27.90%
Zinc-67	66.927131 amu	4.10%
Zinc-68	67.924847 amu	18.75%
Zinc-70	69.925325 amu	0.62%

Introduction to the periodic table

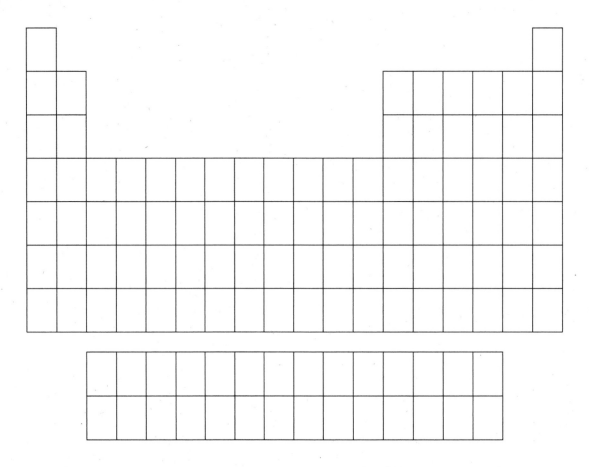

A blank periodic table appears above. This is one of many different forms of the periodic table. Each box represents an element. The elements are organized from left to right in order of increasing atomic number, and as with reading a book, your eyes scan left to right to the end of a lines, then you look back to the left and down one row. The elements in any column form a group or family. Like human families, members of a chemical family are similar but not identical. Each row, from left to right is a period. The current periodic table has seven periods. The two rows below the main body of the periodic table are parts of periods 5 and 6, respectively, in the main part of the periodic table. The two columns on the left plus the six columns on the right are the representative elements. The remaining 10 columns in the main part of the periodic table are the transition elements. The two rows below the main part of the periodic table are the inner transition elements. Other area of the periodic table may have names also. There may be additional information above each column or to the left or right of the periods.

The contents of each box vary with the manufacturer of the periodic table. The amount of information is limited by the dimensions of the table. In addition, the planned use of the table may dictate what information is present. Most tables contain the symbol of the element along with its atomic number and its atomic mass.

It is extremely important for you to understand the structure of the periodic table and be able to navigate around the table to extract the information you are seeking. As you learn about a new element, locate it on your periodic table, and practice finding it quickly.

For the following questions you will need access to a periodic table with, at a minimum, the chemical symbols, the atomic numbers, and the atomic masses of the elements.

EXERCISE
3·5

Introduction to the periodic table

1. What is the atomic mass of element 64?

2. What is the atomic number of the element with the symbol Tl?

3. What are the chemical symbols of the elements most like element number 35?

4. What is unusual about the relative atomic masses of the elements with the atomic numbers 52 and 53?

5. What is the only element that has an atomic mass close to its atomic number?

Chemical substances

As stated previously, all matter contains atoms under normal conditions. Those atoms may be all of the same elements, such as in a diamond, which is considered an element because all atoms are atoms of carbon. It is also possible for a substance to contain atoms of more than one element. If the atoms of the different elements consist in a fixed ratio, such as in H_2O, the substance is a compound. If the ratio is not fixed, the substance is most likely a mixture, such as salt dissolved in water. In a salt-water mixture, the ratio may vary depending on the relative amounts of salt and water present. Both elements and compounds are sometimes grouped together and called a pure substance.

EXERCISE
3·6

Chemical substances

Classify each of the following as a pure substance or a mixture. If it is a pure substance, is it an element or a compound?

1. Ammonia gas

2. A soft drink

3. Metallic iron

Nomenclature

While it is usually possible to look up the names or formulas of compounds, it is much faster if you do not need to. Learning at least the most common names will speed up your work and save you time, especially in situations such as during an exam.

During this course some of these rules will be modified to use new information you have learned.

The International Union of Pure and Applied Chemistry (IUPAC) sets the "rules" for naming compounds. The following examples follow the IUPAC rules for inorganic substances. Note that IUPAC has a different set of rules for organic compounds; however, since organic chemistry is not normally covered in this course, these rules will not be covered here, except in the limited situation where the compounds fit into both categories.

Most of the compounds named here will fall into one of two categories. There will be either binary or ternary compounds. A binary compound consists of only two elements, and a ternary compound consists of three different elements. Compounds containing more than three elements normally follow the rules for ternary compounds.

In some cases, additional information may be added to the formula. Such information may include the state of matter for the substance (this does not alter the name). Water, for example, may exist as a solid (ice), a liquid, or as a gas (steam). These are indicated as $H_2O(s)$, $H_2O(l)$, and $H_2O(g)$. If this information is given to you, you will find that the state will simplify some problem later in this course.

> Note that while there are a number of rules in chemistry, not just for nomenclature, hydrogen will often be an exception.

There is a nomenclature list in the Appendix.

Special names

A few compounds, such as water (H_2O), have special names. According to IUPAC these special names apply regardless of what any other rules may say. There are very few names that fall into this category. These are among the few one-word names for compounds.

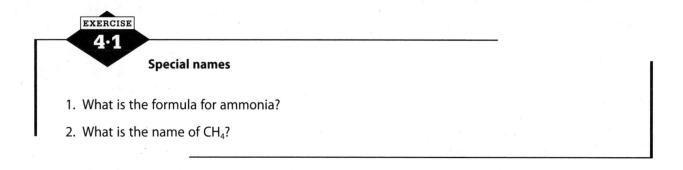

Special names

1. What is the formula for ammonia?

2. What is the name of CH_4?

Nonmetal-nonmetal

Compounds consisting of two nonmetals have two-word names. The first name refers to one of the nonmetals and the second name is that of the other nonmetal modified by changing the ending to *-ide*. The order is important; however, the rule for ordering involves concepts not covered yet. For the present, other than the noble gases (the elements to the far right of the periodic table), the nonmetal farther to the right and/or higher on the periodic table will usually be last in both the name and formula. If there is more than one atom of either one or both of the nonmetals, a multiplying prefix is used. (Hydrogen is an exception here, as it does not get a prefix in binary compounds.) The first 10 multiplying prefixes used here are:

mono-	1	hexa-	6
di-	2	hepta-	7
tri-	3	octa-	8
tetra-	4	nona-	9
penta-	5	deca-	10

Mono- is becoming less common; one of the few situations where it is still used is for carbon monoxide, CO. When using these prefixes, the terminal *-o* or *-a* may be dropped if the name of the element begins with a vowel.

Nonmetal-nonmetal

Name or give the formula for each of the following.

1. ClF_3

2. $H_2S(g)$

3. Cl_2O_4

4. Carbon dioxide

5. Xenon tetrafluoride

6. Tetraphosphorus decoxide

Metal-nonmetal

Like compounds consisting only of nonmetals, metal-nonmetal compounds consist of two names. The name of the metal always goes first. While prefixes are not supposed to be used, this was not always the rule. It is possible to see names like manganese dioxide even though it is technically no longer correct.

The question is how you determine the number of atoms of each element present in the formula without prefixes. To solve this problem, it is necessary to recognize if the metal is a representative element or a transition metal. The key is that metal-nonmetal compounds may be considered to consist entirely of ions and that the total charge for all the positive ions (cations) in the formula must exactly cancel to total charge for all the negative ions (anions) in the formula.

Why some elements form certain ions and others do not will be discussed later in this book. For the representative elements, the metals are on the left side of the periodic table and nonmetals are on the right side. Metals form positive ions (cations), and nonmetals form negative ions (anions). Beginning with the metals, the metals in the first column form +1 ions, the second column form +2 ions (skip the transition metals), the next column forms +3 ions, then +4, and so on. Exceptions are treated like transition metals. Some examples of this are K^+, Ba^{2+}, and Al^{3+}. A similar procedure works for the nonmetals, except that since nonmetals are the opposite of metals, it is necessary to count in the opposite direction. So nonmetal (on the right) start with the last column being 0 (no ions form), the next column to the left –1, moving one more column to the left gives –2, and then –3. Some examples are I^-, S^{2-}, P^{3-}. There are exceptions for the less common elements. You will not be expected to understand these exceptions; just use the name/formula given to you.

If the metal is a transition metal, the rule is slightly different because most transition metals can form ions with different charges. For example, iron commonly forms both Fe^{2+} and Fe^{3+} ions. Due to this variability, iron chloride might be either $FeCl_2$ or $FeCl_3$; therefore, something must be added to iron chloride to differentiate which compound the name represents. The solution is to use what is known as Stock notation. Stock notation represents the charge of an ion with Roman numerals; thus, Fe^{2+} become iron(II) and Fe^{3+} becomes iron(III). Note that Stock notation alters the name of the metal (as iron(III)) and does not follow it (as iron (III)). Prior to the use of Stock notation, every ion had a "new" name. For example, Fe^{2+} was the ferrous ion and Fe^{3+} was the ferric ion. While these alternate names are no longer preferred, they do appear from time to time. If the charge on the transition metal ion is not given to you in the name, then it must be determined from the fact that the compound must be neutral and the charges of the other ions in the formula.

EXAMPLE 4.1

Here are some examples:

Sodium chloride   Sodium should be +1 and chloride (from chlorine) should be –1.
The compound is NaCl.

Calcium chloride   Calcium should be +2 and chloride should be –1.
The compound is $CaCl_2$ (If it were CaCl, the changes would not cancel).

Magnesium nitride   Magnesium should be +2 and nitride (from nitrogen) should be –3.
The compound is Mg_3N_2 (using the charges 3(+2) + 2(–3) = 0).

In most cases, it is possible to swap the charge of the counter ion with the subscript; however, use care when doing this, as the subscripts must be reduced by dividing by the lowest common denominator. For example, using this approach for Sn^{4+} and O^{2-} would give Sn_2O_4, which *must* be reduced to SnO_2.

BaS   This compound contains Ba^{2+} and S^{2-} and is named barium sulfide.

Al_2O_3 This compound contains Al^{3+} and O^{2-} and is named aluminum oxide.

EXAMPLE 4.2

Here are some transition metal examples:

Copper(II) oxide   Copper is given as Cu^{2+} and oxide is O^{2-}.
                        The compound is CuO (in older texts, this was cupric oxide).

Manganese(IV) oxide   Manganese is given as Mn^{4+} and oxide is O^{2-}.
                        The compound is MnO_2 (not Mn_2O_4).

CuCl   Since chlorine is Cl^-, the copper must be Cu^+.
                        The compound is copper(I) chloride.

Fe_2O_3 Since oxide is O^{2-}, three oxygen must yield –6; so two irons must = +6.
                        This gives +3 for each iron. The compound is iron(III) oxide.

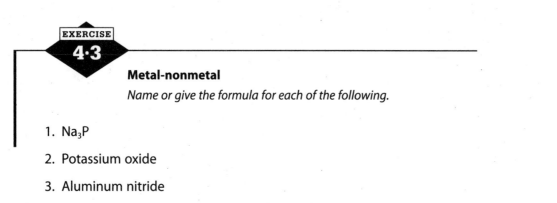

EXERCISE
4·3

Metal-nonmetal
Name or give the formula for each of the following.

1. Na_3P

2. Potassium oxide

3. Aluminum nitride

4. Titanium(IV) oxide

5. $CrCl_2$

6. Co_2O_3

Compounds containing polyatomic ions

Some compounds contain polyatomic ions. Polyatomic ions are like other ions in that they have a charge; however, they consist of more than one atom. An example is the sulfate ion, SO_4^{2-}. Polyatomic ions behave as a unit; thus, the sulfate ion is a single entity, not separate sulfur and four oxygens. In formulas, polyatomic ions behave just like other ions. For example, the oxide ions, O^{2-}, in aluminum oxide, Al_2O_3, may be replaced by sulfate ions to yield $Al_2(SO_4)_3$ to give aluminum sulfate. Note, the *-ite* prefixes are reserved for polyatomic ions with fewer oxygen atoms than those with an *-ate* suffix.

EXERCISE
4·4

Compounds containing polyatomic ions

Name or give the formula for each of the following.

1. Potassium phosphate

2. $Ca(NO_3)_2$

3. Ammonium sulfate

4. Iron(II) sulfite

5. $Cr_2(SO_4)_3$

6. Nickel(II) nitrate

Acids

We will see later that an important subcategory of compounds are the acids. Two important features of acids are first, their names always contain the word *acid* and second, their formulas always contain hydrogen (normally listed first).

Binary acids contain hydrogen and one other element.

Ternary acids contain hydrogen and two other elements. The two other elements normally constitute a polyatomic ion.

There are not very many binary acids. Using $H_2S(aq)$ as an example, the hydrogen is accounted for by the word *acid* and a *hydro-* prefix. (The (aq) indicates that the substance is dissolved in water.) The sulfur is modified by adding the *hydro-* prefix and adding an *-ic* suffix. So $H_2S(aq)$ becomes hydrosulfuric acid. This also works for the few polyatomic ions that have an *-ide* suffix such as the cyanide ion. For this reason HCN(aq) is hydrocyanic acid. Note the formulas of these two acids, $H_2S(aq)$ and HCN(aq), both contain "(aq)" with the formula. The presence of the (aq) is important because if the substance is not dissolved in water, the substance is no longer named as an acid; therefore, $H_2S(g)$ and HCN(l) are named hydrogen sulfide and hydrogen cyanide, respectively.

When naming ternary acids containing polyatomic ions, the procedure depends upon whether the polyatomic ion has an *-ite* suffix or an *-ate* suffix. An *-ite* suffix always changes to *-ous*, while an *-ate* suffix always changes to *-ic*. (In some cases there may be some additional minor name changes.) Using HNO_2 and HNO_3 as examples, the *nitrite* becomes *nitrous* and the *nitrate* becomes *nitric*, while in both cases, the hydrogen is incorporated into the word *acid*. So HNO_2 and HNO_3 are nitrous acid and nitric acid, respectively.

In a few cases, the procedure discussed previously needs to be expanded. For example, using the polyatomic ions containing chlorine and oxygen:

	Less oxygen	Even less oxygen	Change to acid
HOCl	Chlorite	Hypochlorite	Hypochlorous acid
$HClO_2$	Chlorite	Chlorite	Chlorous acid
	More oxygen	Even more oxygen	
$HClO_3$	Chlorate	Chlorate	Chloric acid
$HClO_4$	Chlorate	Perchlorate	Perchloric acid

This procedure also works for bromine and iodine, and to a lesser extent for fluorine, which only forms HOF, hypofluorous acid.

A few other elements also behave in this manner; however, it is rare to see them in courses at this level other than the permanganate ion.

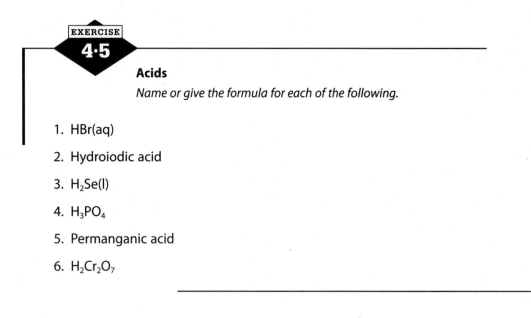

EXERCISE
4·5

Acids

Name or give the formula for each of the following.

1. HBr(aq)

2. Hydroiodic acid

3. $H_2Se(l)$

4. H_3PO_4

5. Permanganic acid

6. $H_2Cr_2O_7$

Chemical equations

Terminology

Chemical equations are important in chemistry. A chemical equation is of little use unless it is balanced. A balanced chemical equation follows the law of conservation of mass, which means that each side of the reaction arrow ($\rightarrow$) equals the mass of the opposite side. This equivalence of mass is achieved by making sure there are the same number of each type of atom on each side of the reaction arrow.

EXAMPLE 1

Here is an example of a balanced chemical equation:

$$2\,H_3PO_4(aq) + 3\,Sr(OH)_2(aq) \rightarrow 1\,Sr_3(PO_4)_2(s) + 6\,H_2O(l)$$

The "1" in front of the $Sr_3(PO_4)_2(s)$ is entirely optional as are any other 1s. To check to see if the reaction is balanced, check the numbers of each type of atom on each side of the reaction arrow. The numbers in front of the chemical formulas are the coefficients, which in normal chemical equations are integers. In this case:

ATOM	REACTANT SIDE	PRODUCT SIDE
H	12	12
P	2	2
O	14	14
Sr	3	3

If any of the numbers for an element do not match, there is an error, and the equation is not balanced

The reaction arrow separates the reactants from the products. All substances to the left of the reaction arrows are reactants, and all substances to the right are products. The states given in parentheses, in this case, (aq), (s), and (l), do not affect the balancing, but are extremely important in some of the later chapters (at this point do not worry about where these come from).

Terminology

Using the following balanced chemical equation, answer the following questions.

$$2\ Al(s) + 3\ H_2SO_4(aq) \rightarrow Al_2(SO_4)_3(aq) + 3\ H_2(g)$$

1. List the products of this reaction.

2. List the reactants of this reaction.

Balancing equations

There are many methods for balancing equations. The following method will work for any type of reaction seen in any general chemistry course except, sometimes, one type. You will be shown how to deal with this other type later.

EXAMPLE 2

Begin with the following unbalanced chemical equation:

$$\_\_\_\ C_6H_{14}(l) + \_\_\_\ O_2(g) \rightarrow \_\_\_\ CO_2(g) + \_\_\_\ H_2O(l)$$

From this point on, this is a "fill in the blank" problem, which is finished when the blanks are filled with integers.

Balancing goes one element at a time. It is possible to start with any element; however, it helps not to choose an element that appears more than once on either side of the reaction arrow. In this case, we will begin with C.

$$\_1\_\ C_6H_{14}(l) + \_\_\_\ O_2(g) \rightarrow \_\_\_\ CO_2(g) + \_\_\_\ H_2O(l)$$

The "1" is not necessary in the final answer and is only present here to indicate where we started.

The 1 indicates that there are 6 carbon atoms on the left side, so it is necessary to make sure there are 6 carbon atoms on the right side. This is done as:

$$\_1\_\ C_6H_{14}(l) + \_\_\_\ O_2(g) \rightarrow \_6\_\ CO_2(g) + \_\_\_\ H_2O(l)$$

The "1" we entered earlier also affects the hydrogen, so we need to make sure there are 14 hydrogen atoms on the right side:

$$\_1\_\ C_6H_{14}(l) + \_\_\_\ O_2(g) \rightarrow \_6\_\ CO_2(g) + \_7\_\ H_2O(l)$$

This procedure is continued for each element that appears once on each side of the reaction arrow. We have finished all the elements appearing once on a side, and in this case, we are left with oxygen. With the current coefficients, there are 19 oxygen atoms on the product side; therefore, we need 19 oxygen atoms on the reactant side. The easiest way to achieve this is:

$$\_1\_\ C_6H_{14}(l) + \_\frac{19}{2}\_\ O_2(g) \rightarrow \_6\_\ CO_2(g) + \_7\_\ H_2O(l)$$

The presence of the fraction may be disconcerting; however, we are not done yet. We have filled all the blanks, so we can move on.

The next thing to do is to clear the fractions by multiplying all the coefficients by the smallest value that will clear all fractions present, leaving us with only whole numbers. In this case, the smallest value is 2. Doing this gives:

$$\_2\_\ C_6H_{14}(l) + \_19\_\ O_2(g) \rightarrow \_12\_\ CO_2(g) + \_14\_\ H_2O(l)$$

To finish the problem, it is important to check to make sure each type of atom is balanced.

ATOM	REACTANT SIDE	PRODUCT SIDE
C	12	12
H	28	28
O	38	$24 + 14 = 38$

If everything is equal, the equation is balanced. If something does not match, do a quick scan to see if you made a simple mistake, such as not multiplying one of the coefficients by 2. If you did not make a simple mistake, the fastest way to finish the problem is to start over again. You may wish to begin by balancing a different element first.

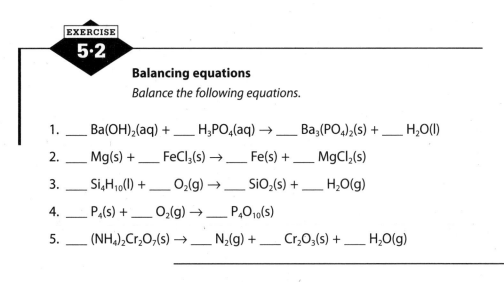

EXERCISE
5·2

Balancing equations

Balance the following equations.

1. ___ $Ba(OH)_2(aq)$ + ___ $H_3PO_4(aq) \rightarrow$ ___ $Ba_3(PO_4)_2(s)$ + ___ $H_2O(l)$

2. ___ $Mg(s)$ + ___ $FeCl_3(s) \rightarrow$ ___ $Fe(s)$ + ___ $MgCl_2(s)$

3. ___ $Si_4H_{10}(l)$ + ___ $O_2(g) \rightarrow$ ___ $SiO_2(s)$ + ___ $H_2O(g)$

4. ___ $P_4(s)$ + ___ $O_2(g) \rightarrow$ ___ $P_4O_{10}(s)$

5. ___ $(NH_4)_2Cr_2O_7(s) \rightarrow$ ___ $N_2(g)$ + ___ $Cr_2O_3(s)$ + ___ $H_2O(g)$

Light and matter

Light

Advancements in understanding the nature of light, also known as electromagnetic radiation, began near the beginning of the twentieth century. Before that time, physicists believed that light was well understood and that it consisted of waves. However, as science advanced, so did the ability to make observations. Better observations led to the discovery of discrepancies that could not be explained by light as waves. From these observations, many significant advancements in the understanding of both light and matter were made in the early decades of the twentieth century. Improvements on these advances continue to this day.

Before examining these advances, it is important to understand what scientists knew about light before the advances were made.

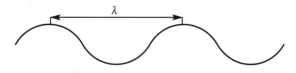

Light, as waves, had the properties of waves. Waves have, among other properties, a wavelength (λ), a frequency (v), and a velocity (v). A wave is a continuous series of crests and troughs with the wavelength being the distance between one point and the next equivalent point (see diagram). The frequency is how many peaks (or troughs) pass a point in a given period of time (usually a second). The velocity is how fast the waves are moving, in general, this is v; however, in the case of light, it is called the speed of light, c (3.00×10^8 m/s). The variables λ and v, are related to c comes from the relationship $c = \lambda v$.

Max Planck discovered that for light the energy, E, could be related to the frequency by the relationship $E = hv$. In this equation, h is Planck's constant, which equals 6.626×10^{-34} J s. To simplify some problems, it is possible to combine this equation with the preceding equation to give $E = \dfrac{hc}{\lambda}$.

The two constants c and h imply that the wavelength needs to be expressed in meters, the frequency in 1/s or s^{-1}, and the energy in joules. If you are given or need different units, it will be necessary to make some type of conversion.

Light

Answer the following questions. Do not forget that the answers need the correct significant figures and units.

1. Determine the frequency for light with a wavelength of 5.00×10^{-9} m.

2. Determine the wavelength for light with a frequency of 3.25×10^{17} s^{-1}.

3. Determine the energy for light with a frequency of 3.25×10^{17} s^{-1}.

4. Determine the energy for light with a wavelength of 5.00×10^{-9} m.

5. Determine the energy for light with a wavelength of 1.75 nm.

Quantized energy and the Bohr atom

One of the major advancements in the twentieth century was made by Neil Bohr. Bohr was studying the spectrum of hydrogen. Hydrogen, being the simplest element, had the simplest spectrum. Hydrogen's spectrum consists of a series of lines, and depending upon the type of spectrum being examined, these lines corresponded to the gain or loss of a certain amount of energy, no more or no less. According to classical physics, there should be variation with the possibility of all wavelengths being gained or lost. Bohr's explanation was that the electrons in an atom could have only certain amounts of energy indicated by the lines in the spectrum. Since only certain amounts of energy were allowed, energy was quantized. The amount of energy lost or absorbed were labeled quanta. Bohr came up with the following picture of an atom:

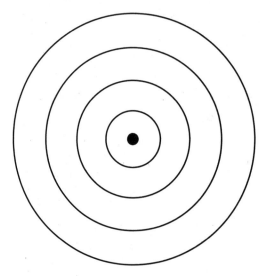

The nucleus is in the center. Electrons in one "circle," or shell, could absorb energy and move (excite) to a shell farther out, and electrons in an outer shell could lose energy and move to a shell nearer the nucleus. Since the same two shells have the same separation, the amounts of energy lost or gained are the same. Note, this diagram is in terms of energy and not physical appearance. Moving away from the nucleus by less than the separation between the shell would be equivalent to climbing a stairway by half steps. This model, now known as the Bohr atom, explains the spectrum of hydrogen; however, for other elements the explanation is not as good. It was up to later scientists to explain why the Bohr model only partly explained the spectra of other atoms.

Quantized energy and the Bohr atom

1. Light with a frequency of 3.085×10^{15} s^{-1} will excite a hydrogen electron from the first shell (nearest the nucleus) to the fourth shell. How much energy was absorbed?

2. Light with a wavelength of 121 nm will excite a hydrogen electron from the first shell (nearest the nucleus) to the second shell. How much energy was absorbed?

Wave behavior of matter

Another important step in the understanding of electrons was proposed by Louis de Broglie. This step became the equation that now bears his name. The equation is $\lambda = \dfrac{h}{mv}$. The equation allows one to calculate the wavelength, λ, of matter. It is very important for electrons and some other subatomic particles, but the values are too short for other particles to be useful. The terms on the right side of the equation are Planck's constant (h), the mass of the particles (m), and the velocity (v, not ν) of the particle.

Note, in some problems involving the de Broglie relationship, the joules will need to be expressed as $\dfrac{\text{kg m}^2}{\text{s}^2}$. Masses, if not in kilograms, will need to be converted to kilograms, and velocities not in meters/second will need to be converted. These conversions may be performed separately or, to save time, incorporated into the equation.

When considering electrons, the wavelength of an electron may be considered to be equal to the circumference of the first shell, two wavelengths equal to the circumference of the second shell, and so on.

> We use analogies for many of the things discussed in this chapter because these things are very complicated. For example, many times we use diagrams, such as the one shown for a wave and for the Bohr atom, but those diagrams are two-dimensional, and the world is three-dimensional.

Wave behavior of matter

1. Determine the wavelength of an electron traveling at 1.38×10^7 m/s. The mass of an electron is 9.109×10^{-31} kg.

2. Determine the wavelength of an electron traveling at 7.67×10^4 km/h. The mass of an electron is 9.109×10^{-31} kg.

3. A very fast, 195 lb ice skater can skate with a velocity of 33.3 mi/h. What is the wavelength of this ice skater?

Quantum chemistry

Quantum numbers

Once it was accepted that electrons had both particle and wave properties, the path to the next major advance was open. This advance was made by Erwin Schrödinger, and in his honor, this advance is known as the Schrödinger equation. This equation is the basis of quantum mechanics. As waves, electrons have no particular location is space. This ties to the Heisenberg uncertainty principle, which states that it is impossible to simultaneously determine both the position and the momentum (mass × velocity) of an electron.

When solving this equation for a particular electron, Schrödinger found that three integers were necessary to explain the behavior of the electrons in an atom. Later work determined that a fourth quantum number was needed to complete the understanding of electron behavior.

The first quantum number, known as the principal quantum number, designates which shell the electron occupies. (These are the same shells as in the Bohr atom.) This quantum number has the symbol n. The shell nearest the nucleus has $n = 1$, and the shells are numbered outward from there so that n may have the values 1, 2, 3, 4 . . . . The principal quantum number indicates how much energy the electron has. This quantum number directly controls the second quantum number and indirectly controls the third quantum number.

The second quantum number is the angular momentum quantum number, which is symbolized by an l. Like the principal quantum number, this quantum number may assume a variety of values. Unlike n, which has a limitless range of values (positive integers), l is limited. The values of l are limited by the value of n for the electron under consideration. The values of the angular momentum quantum number range from 0 to $(n-1)$. This means that if, for example, $n = 4$, then l may be 0, 1, 2, or 3. For a hydrogen atom, all values of l for a particular n have the same energy; however, for all other atoms, energy increases with increasing l. This quantum number separates the shells into subshells. In some cases, it is useful to substitute letters for the l values. These substitutions are s for $l = 0$, p for $l = 1$, d for $l = 2$, and f for $l = 3$. For example, an electron with $n = 3$ and $l = 2$ is in the $3d$ subshell.

The third quantum number is the magnetic quantum number, which is symbolized by an m_l. This quantum number depends upon the angular momentum quantum number. The values of m_l depend directly on the l value for that electron (and indirectly on the n value). For an electron, m_l may assume any value from $-l$ to $+l$. Thus, if $l = 3$ ($n \geq 4$), m_l may assume any of the following -3, -2, -1, 0, $+1$, $+2$, or $+3$. This quantum number divides the subshell into orbitals.

The final quantum number is the electron spin quantum number, which is symbolized by a m_s. This quantum number has only two values. It may be either

+1/2 or −1/2. These two values relate to the electron being able to spin either clockwise or counterclockwise.

The "rules" discussed previously for the four quantum numbers cover the possible values for these quantum numbers. Each individual electron in an atom can have only one value for each quantum number at a time. For example, an electron may have the following set of quantum numbers: $n = 2$, $l = 1$, $m_l = −1$, and $m_s = +1/2$. Other sets are possible provided the members of the alternate set do not break any of the rules.

Wolfgang Pauli tied these sets of quantum numbers together with what is now known as the Pauli exclusion principle. According to this principle, no two electrons in an atom may have identical sets of the four quantum numbers. Thus the set of quantum numbers given earlier ($n = 2$, $l = 1$, $m_l = −1$, and $m_s = +1/2$) is unique among the electrons in that atom. Another electron in the same atom must have at least one quantum number that is different. This principle limits the number of electrons in an atom.

Finally, how do the quantum numbers relate to the arrangement of the electrons in an atom? There is a tendency for the electrons to seek the lowest energy state. This state of lowest energy is known as the ground state. If all the electrons in the atom are not in the lowest possible energy state, the arrangement is an excited state. Atomic spectra, such as that of hydrogen used by Bohr, involve electrons leaving or returning to the ground state. An atom may have any number of excited states, but there is only one ground state. The first shell can hold only two electrons (one with $n = 1$, $l = 0$, $m_l = 0$, and $m_s = +1/2$ and one with $n = 1$, $l = 0$, $m_l = 0$, and $m_s = −1/2$). Once the first shell is full, electrons begin to fill the second shell until the second shell is filled. This filling proceeds out from the nucleus until all the electrons are accounted for. The order of filling is known as the Aufbau principle. According to this principle, the electrons enter an atom in order of increasing energy.

EXERCISE
7·1

Quantum numbers

1. If $n = 3$, what are the possible values of l?

2. If $l = 2$, what are the possible values of m_l?

3. If $m_l = +2$, what are the possible values of m_s?

4. If $n = 2$, what are the possible values of l and m_l?

5. It was stated in the discussion that the first shell may hold two electrons. How many electrons may fit into the second shell?

6. Write the sets of possible quantum numbers for an electron in the 2p orbital of hydrogen. (This is an electron in an excited state such as may be found from the atomic spectrum of a hydrogen atom.)

7. A boron atom has five electrons. Give a possible set of quantum numbers for these five electrons.

8. Why is it not possible to have three electrons in the first shell?

9. Why is it not possible to have a 2d subshell?

10. Why do electrons enter the 2p subshell before the 3p subshell?

Atomic structure

Atomic orbitals

Both the Bohr theory and quantum mechanics agree that the electrons in an atom occupy shells. The fact that these shells are referred to as energy levels alludes to energy being the key to distinguishing between the shells. The lowest energy electrons are in the first shell ($n = 1$). Electrons with higher energy enter the second shell ($n = 2$), the electrons enter the third shell ($n = 3$) with higher energy still, and so on. To date, ground state electrons are known to enter shells up to $n = 7$, with excited-state electrons entering shells of higher energy.

Every shell contains one or more subshells. The azimuthal quantum number, l, differentiates between subshells. Electrons normally enter subshells with $l = 0, 1, 2$, or 3, which are more commonly referred to by the letters s, p, d, and f. Theoretically, there are other possible quantum numbers for $l \geq 3$, with the next subshell ($l = 4$) being the g subshell. To date, these higher subshells have not been observed in ground-state atoms.

The subshells occupy regions in space that relate to what type of subshell they are. Each of these subshells contains one or more orbitals. For example, the s subshells are, in general, considered to be spherical; again as with many things quantum mechanical, this is a simplification. The Schrödinger equation describes each of these subshells by a wave function (equation) with the square of this function being related to where the electron is located. Since we are dealing with waves, the electrons do not have a specific location in space at any time. This uncertainty in the electron location means that "shell" may be the wrong name as shell implies a defined surface (such as an eggshell). To bypass this possible misconception, scientists often refer to the electron shells as electron clouds. The nebulous nature of a cloud better relates to what an electron is doing in that atom.

Every shell contains an $l = 0$ (s) subshell. For the first shell, the s subshell is the only subshell, and it contains only one orbital (designated 1s). Subshell types may be referred to by the appropriate l quantum number, or the letter designation. Normally, when referring to a specific subshell, it is important to designate both the n and l quantum numbers. For example, the subshell with $n = 1$ and $l = 0$ is labeled the 1s subshell instead of 10. The 1s electron cloud is spherical in shape (circular in two dimensions as on a sheet of paper). The 2s electron cloud looks like the 1s except that it is larger, while the 3s electron cloud is still larger. These three examples are illustrated as follows. There are internal differences between the different *s* electron clouds, which do not affect this discussion or introductory chemistry courses.

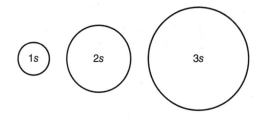

Beginning with the second shell ($n = 2$), it is possible to have *p* subshells ($l = 1$). The shells with $n \geq 2$ have *p* subshells. The *p* subshells contain three orbitals each (one for each of the following: $m_l = -1$, 0, or +1). The "shapes" of *p* orbitals are more complicated than the shape of the *s* orbitals. One simple way of representing *p* orbitals is:

This shape is often referred to as a dumbbell. The color coding (one end is shaded) is for situations beyond this book; suffice it to say that this shading is used to distinguish one end (lobe) from the other. Every *p* subshell contains a set of three different *p* orbitals. The three *p* orbitals in the set look identical; however, they have different orientations in space. One member of the set, if necessary labeled p_x, is aligned along the *x*-axis. The other two members of the set are the p_y, along the *y*-axis, and the p_z, along the *z*-axis. (Note the *x*, *y*, and *z* subscripts are usually not necessary unless specifically asked for.) The 3*p* subshell also contains a set of three different *p* orbitals; however, each of these is larger than the orbitals in the 2*p* set. As with the *s* orbitals in the different shells, the *p* orbitals look the same externally, except for size, but differ internally.

Beginning with the third shell ($n = 3$), there is a *d* subshell containing a set of *d* orbitals. The *d* orbitals occur in sets of five. The shapes of the *d* orbitals are more complicated than the *p* orbitals; however, these shapes are not important here. Similarly, beginning with the fourth shell ($n = 4$), there is an *f* subshell present. The *f* subshell contains seven different *f* orbitals with shapes that are more complicated than the *d* orbitals. In general, all you need to know about the *d* and *f* subshells is in which shells they appear (for *d*: $n \geq 3$ and for *f*: $n \geq 4$), and the number of orbitals present in the subshell (for *d* subshells: 5 orbitals, and for *f* subshells: 7 orbitals).

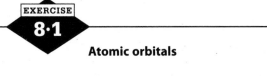

EXERCISE 8·1

Atomic orbitals

1. What subshells are present in the fourth ($n = 4$) shell?

2. How many orbitals are present in the third ($n = 3$) shell?

3. List the *n*, *l*, and m_l quantum numbers for one of the *p* orbitals in the 2*p* subshell.

4. How many electrons can fit into the 2*p* orbital you listed in question 3?

5. Based upon your answer to question 2, what is the maximum number of electrons that may fit in the third shell?

6. Based upon your answer to question 1, what is the maximum number of electrons that may fit in the fourth shell?

Many electron atoms

In both the Bohr model and the quantum mechanical model of the atom, electrons are associated with a particular nucleus because the positive charge of the nucleus (which depends upon the number of protons present) attracts the negative charge on the electrons. Recall that the Bohr model works accurately only for hydrogen atoms. Using quantum mechanics, it may be seen why there is a limitation to the Bohr model. This limitation is because the electrons also interact (repel) with each other. For a hydrogen atom, there is only one electron; therefore, there is no electron repulsion.

If we consider an electron in the third shell, the repulsion from electrons in higher shell ($n \geq 4$) is minimal. Repulsion from shells nearer to the nucleus and to a lesser extent, from electrons in the same shell shields or screens the attractive charge from the nucleus. For example, if we consider a chlorine atom, the atomic number is 17, which means there are 17 protons in the nucleus. For a chlorine atom, the nuclear charge is +17. The distance between the nucleus and the third shell reduces the attraction to a degree; however, a more important factor is that the 10 electrons in the first two shells (core electrons) significantly reduce the charge attracting the third shell electrons through shielding (not necessarily equal to the number of electrons as the electrons are continually moving and not always in position to effectively shield). The attraction between the third shell electrons and the nuclear charge becomes the nuclear charge—the shielding = the effective nuclear charge. For all atoms where the third electron shell is being filled, the shielding remains approximately constant. For this reason, the third shell electrons experience an increasing effective nuclear charge as the nuclear charge increases. Many of the consequences of the effective nuclear charge are beyond a course at this level; however, consequences that apply to the periodic table are important.

If one compares sodium and potassium, the outermost electron for sodium is in the third shell, while the outermost electron is in the fourth shell. Potassium has 8 more protons in its nucleus than sodium. This means the nuclear change for potassium is +8 higher than for sodium, which would imply that potassium has a greater attraction for the electron in its fourth shell than sodium has for the electron in its third shell; however, this does not consider that the core of potassium has 8 more electrons than the core of sodium. For this reason the effective nuclear change of the two elements is similar except for the fact that electrons in the fourth shell average farther from the nucleus than the electrons in the third shell. This greater distance reduces the attraction of the electrons to the potassium nucleus also.

EXERCISE

8·2

Many electron atoms

1. Why does the Bohr model not work for atoms other than hydrogen?

2. What is the expected effective nuclear charge for a hydrogen atom?

3. How do the effective nuclear charges of boron and aluminum compare to each other?

4. How many electrons are in the core of a magnesium atom?

Electronic structure of atoms

Electron configurations

To understand the behavior of atoms, that is, their chemistry, it is necessary to know something about the arrangement of the electrons within the atom. Most important, what electrons are in the outermost shell? The arrangement may be expressed in a number of ways. One of these ways is to write the electron configuration.

To determine the electron configuration for an atom, it is necessary to know the number of electrons present, which for an atom is the atomic number. (Do not forget atoms that have gained or lost electrons become ions, where the number of electrons no longer matches the atomic number.) The electrons present are then entered into the electron configuration according to certain rules. The rules are the Pauli exclusion principle (no two electrons in an atom may have identical sets of the four quantum numbers) and the Aufbau principle (filling of the electron configuration progresses from the lowest energy level available toward higher energy levels). Later, we will see that Hund's rule may also be necessary.

To save time it is useful to remember that all s subshells may hold a maximum of 2 electrons, all p subshells may hold a maximum of 6 electrons, d subshells may hold a maximum of 10, and f orbitals may hold a maximum of 14 electrons. These are fixed limits governed by the Pauli exclusion principle. A subshell may contain fewer than the maximum. Though not always needed, it will help to remember that each orbital may contain no more than 2 electrons. The number of electrons present in a subshell is indicated with a superscript after the orbital designation. For example, a filled 1s subshell is indicated as $1s^2$, a partly filled 1s subshell (1 electron) would be $1s^1$ (1 superscripts are necessary).

According to the Aufbau principle, the 1s subshell should be filled first. Once the 1s is filled, we move to the 2s and then the 2p. However, complications arise starting with the third shell. The third shell does fill in the order $3s \rightarrow 3p \rightarrow 3d$; however, before the 3d subshell is reached, the fourth shell begins to fill (the 4s subshell is lower energy than the 3d subshell). A similar situation occurs for all the d subshells. Another compilation occurs for the f subshells. The first f subshell to fill is the 4f, which does not begin to fill until the 6s subshell is filled. Similarly, the 5f begins to fill after the 7s. These are not exceptions to the Aufbau principle, these occur because the energy levels overlap. (Note that none of these complications occur for the energy levels in a hydrogen atom.) Keeping all of these complications straight may lead to errors; therefore, some mnemonic device may aid the writing of electron configurations. One such device is pictured as follows.

(There are other diagrams for this also.) As an aid, the maximum number of electrons that may enter a subshell is included.

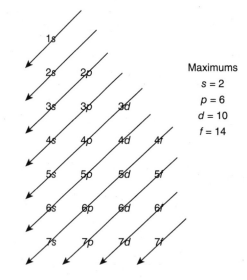

Maximums
$s = 2$
$p = 6$
$d = 10$
$f = 14$

To construct this diagram begin by listing the possible subshells by shell. The first shell contains only the 1s subshell, so only the 1s is shown. The second shell contains both the 2s and 2p subshells, so only these are listed. The third shell contains the 3s, 3p, and 3d subshells, and so on. This diagram may be extended to include other energy levels and possible subshells; however, in the form shown, it can accommodate more atoms than are currently known. At this time, it appears that the last filled subshell is the 7p. The arrows relate to the Aufbau principle. To use this diagram, begin with the first (top) arrow, which passes only through the 1s subshell. Once the 1s is filled, move to the second arrow (passing through the 2s). Once the 2s is filled, move to the third arrow (through the 2p and 3s); the 2p is filled first and then the 3s. After the 3s, move to the next arrow and so on until all the electrons are accounted for.

EXAMPLE 1

To illustrate how to use the diagram, the electron configuration of manganese will be determined. First, Mn has 25 electrons. According to the diagram, you need to take 2 of these electrons and place them in the 1s subshell to give $1s^2$ and leave 23 electrons to be accounted for before moving to the next arrow. Moving to the second arrow, take 2 more electrons for the 2s subshell (now you have $1s^2 2s^2$), with 21 electrons remaining. Shifting to the third arrow, begin by using 6 electrons to fill the 2p subshell ($1s^2 2s^2 2p^6$) and leaving 15 electrons. Still using the third arrow, the next subshell is the 3s, which takes 2 more electrons ($1s^2 2s^2 2p^6 3s^2$), leaving 13 electrons. Now comes the fourth arrow, where the 3p subshell is next followed by the 4s. It takes 6 electrons to fill the 3p ($1s^2 2s^2 2p^6 3s^2 3p^6$), which leaves 7 electrons. Then the 4s subshell takes two more ($1s^2 2s^2 2p^6 3s^2 3p^6 4s^2$), leaving 5 electrons. Now comes the fifth arrow (3d, 4p, and 5s). The 3d can hold 10 electrons; however, there are only 5 left. The last 5 electrons enter the 3d subshell, leaving it partially filled. The final electron configuration of manganese is $1s^2 2s^2 2p^6 3s^2 3p^6 4s^2 3d^5$. As a check, the sum of the superscripts must equal the number of electrons present, in this case, 25.

For practice, you might try with uranium (92 electrons): $1s^2 2s^2 2p^6 3s^2 3p^6 4s^2 3d^{10} 4p^6 5s^2 4d^{10} 5p^6$ $6s^2 4f^{14} 5d^{10} 6p^6 7s^2 5f^4$.

There are a number of things happening here, so take it slowly at the beginning.

The diagram is very useful; however, for a variety of reasons, there are a few exceptions. If you see a list of electron configurations, you will easily be able to pick them out. Your instructor may require you to learn one or two of the exceptions: just do so with the understanding that for around 100 of the known elements there are no exceptions.

Hund's rule was mentioned previously. In the type of electron configurations that we have been writing its effect is not obvious. Indeed, it is only apparent when one looks at the electron orbitals instead of the subshells. For the $2p$ subshell, the orbitals are $2p_x$ $2p_y$ and $2p_z$. Adding one electron at a time to the $2p$ orbitals involves Hund's rule. For example, for a $2p^2$ subshell, the orbitals might be $2p_z^1$ $2p_y^1$ $2p_z^0$ (one electron in each of two of the three 2p orbitals). The actual configuration must have one electron in each of two different orbitals (any two). This is part of Hund's rule. The second part of Hund's rule is more apparent if we examine the quantum numbers for these electrons:

n	l	m_l	m_s
2	1	−1	+1/2
2	1	0	+1/2

(The m_l may be any two of the possible values [−1, 0, or +1], and both m_s values could be −1/2. All of these variations are equally correct.)

Now, with this table in mind, Hund's rule says that when dealing with a set of orbitals, the electrons enter the orbitals individually with the same spin (spins parallel). Entering the orbitals individually is why we can use any two of the three $2p$ orbitals. Spins parallel means that all the electrons have $m_s = +1/2$ or all have $m_s = −1/2$. In the case of the $2p$ subshell, this filling continues until each orbital has one electron ($2p_z^1$ $2p_y^1$ $2p_z^1$), after which it is necessary to add a second electron to each of the orbitals (in any order), with these second electrons having the opposite m_s value as the first electron (Pauli exclusion principle). Single electrons in an orbital are unpaired electrons, and two electrons in an orbital are paired electrons (spin paired). For the $2p$ electrons, it is common to differentiate them by the x, y, and z subscripts; however, for the d and f orbital sets the designations are more involved; therefore, we will simply refer to them according to their m_l values (−2, −1, 0, +1, and +2 for the d set and −3, −2, −1, 0, +1, +2, and +3 for the f set). Hund's rule ensures that the lowest energy state (most stable) is filled first (Aufbau principle).

None of the preceding discussions means anything if the predictions could not be confirmed experimentally. If every electron in an atom is spin paired, the atom is diamagnetic, which means that it is weakly repelled by a magnet. If at least one electron is unpaired, the atom is paramagnetic and is attracted to a magnet. The more unpaired electrons, the greater the attraction. The terms *diamagnetic* and *paramagnetic* apply to other substances also.

For negative ions (anions), the diagram still works, all that needs to be done is to add the appropriate number of electrons to the electron configuration of the atom. For example, O = $1s^2 2s^2 2p^4$ and $O^{2-} = 1s^2 2s^2 2p^6$. For positive ions (cations), it is necessary to remove a number of electrons equal to the magnitude of the charge. However, these electrons will come for the outer shell first. This distinction is not significant for elements such as magnesium, Mg = $1s^2 2s^2 2p^6 3s^2$ going to $Mg^{2+} = 1s^2 2s^2 2p^6$. However, for the transition elements (and inner transition elements) it is significant. For this reason, Mn = $1s^2 2s^2 2p^6 3s^2 3p^6 4s^2 3d^5$ and $Mn^{2+} = 1s^2 2s^2 2p^6 3s^2 3p^6 3d^5$.

Later we will see ways to abbreviate (condense) the electron configurations of an atom or an ion. In addition, you will see $s^2 p^6$ (an octet) many times.

Electron configurations

1. How many electrons do each of the following atoms have?
   (a) He, (b) Si, (c) Co, (d) Gd, (e) Bk

2. How many electrons do each of the following ions have?
   (a) Li^+, (b) S^{2-}, (c) Pd^{2+}, (d) As^{3-}, (e) I^-

3. An element utilizes the following subshells: $1s2s2p3s3p4s3d4p5s4d$. What is the maximum number of electrons that may be present?

4. What is the electron configuration of Al?

5. What is the electron configuration of Ni?

6. What is the electron configuration of Sb?

7. Show, with sets of quantum numbers, how nitrogen is expected to follow Hund's rule. The electron configuration of $N = 1s^2 2s^2 2p^3$.

8. Which atoms are always paramagnetic?

9. The manganese ion has the electron configuration $Mn^{2+} = 1s^2 2s^2 2p^6 3s^2 3p^6 3d^5$. How many unpaired electrons are present?

10. What is the maximum number of unpaired electrons that an atom in its ground state may have?

11. What is the electron configuration of Be^{2+}?

12. What is the electron configuration of P^{3-}?

13. What is the electron configuration of V^{2+}?

14. The elements F, Cl, and Br appear on the same column of the periodic table. What do the electron configurations of these elements have in common?

15. The most stable ions for Mg, N, and Ti are Mg^{2+}, N^{3-}, and Ti^{4+}. What do these three ions have in common?

16. Stable ions for Cl, K, and Ca are Cl^-, K^+, and Ca^{2+}. Which element has atoms with the same electron configuration as these three stable ions?

17. Stable ions for C, Na, and Al are C^{4-}, Na^+, and Al^{3+}. Which element has atoms with the same electron configuration as these three stable ions?

18. What do the answers to questions 16 and 17 indicate about the electron arrangements for the noble gases (He, Ne, Ar, Kr, Xe, Rn, Og)? These elements do not easily form ions.

19. What do the electron configurations of the elements Ne, Ar, Kr, and Xe have in common?

Periodic properties

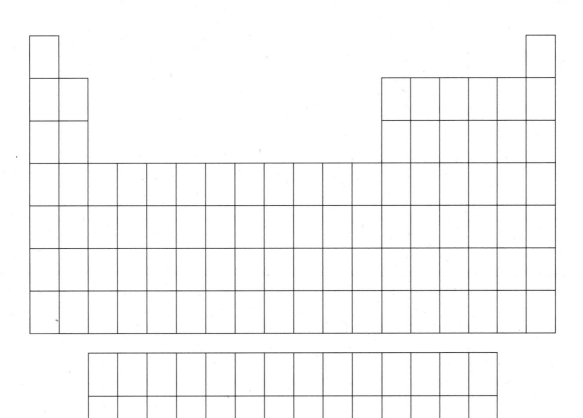

This chapter focuses on the periodic table. Each "box" on the table deals with one element. This table contains many properties of the elements, and by an element's very position on the table, we know many other properties of an element. Most periodic tables contain the symbols, atomic numbers, and atomic masses (weights) of the elements. There are usually various numbers and letters above the tops of each column (these will vary with the age of the table). The columns are the groups or families. Sometimes there are numbers to the left of each row (period) on the table. Additional information may be present for specific uses.

Most periodic tables have 2 columns on the left side, 10 columns in the center, 6 columns on the right side, and 14 columns in two rows beneath the remainder. (This same combination of numbers [2, 6, 10, and 14] also appears in Chapter 9.)

The original arrangement of the elements on the periodic table was determined by Dmitri Mendeleev in 1869. This arrangement was in order of increasing atomic mass. In the early twentieth century, it was found that arranging the elements in order of increasing atomic number gave better results.

The atomic numbers are always integers, while the atomic masses usually are not integers; however, for certain nonnatural radioactive elements, the mass number (integer) of the most stable isotope is used.

Mendeleev's periodic table had a combination of Roman numerals and letters (such as IA, IIA, IIIB, VB, etc.) at the head of each column. (In Mendeleev's day, the 14 columns below the main body of the table were unknown. Later tables substituted Arabic numbers for the Roman numerals. A's and B's were still present with American tables using A's for the representative elements and B's for the transition elements (except VIII, or 8, has no letter). European tables used A's for the 7 columns to the left, then 8, followed by 1B through 7B. More recent tables have the numbers 1 through 18 at the heads of the columns. For students at this level, the American system tends to be more useful. (To convert the 18 numbering system to the American system, alter the numbers of all columns after 10, to 1 through 8, then add B's after each column number for the transition elements and A's after each of the other numbers. The slight difference remaining in columns 8, 9, and 10 does not alter their use.)

The elements in each group or family have similar properties, which is how Mendeleev knew when to start a new period. Today, we know that a new period begins when the electrons begin to enter a new energy level (shell). With the development of quantum mechanics, we now know that the elements in a family have similar properties because they have a similar arrangement of electrons in their outer shells. Considering that Mendeleev did not know about electrons, atomic numbers, or quantum mechanics, his discovery is even more amazing.

In this chapter, we will examine some of the properties of the elements that may be determined by the position of the element on the periodic table. In general, these will be relative properties (for example, larger or smaller). There are other periodic properties that will be discussed in other chapters.

Note that some of the exceptions noted in the following section deal with filled or half-filled subshells. The filled subshells are elements in columns 2, 11 (IB), and 18 (8A). The half-filled subshells are elements in columns 7 (7B) and 15 (5A).

Note that in the questions for this chapter, you can assume there are no exceptions unless stated otherwise. If you actually examine the numbers, you will find some exceptions.

Atomic radii

The atomic radius is basically the "size" of an atom. In general, values are around 10^{-10} m. There are two general trends in the atomic radii of the elements (note, a "general trend" means there are exceptions). The first trend deals with progress down a column. When going down a column, each step down adds a new electron shell, and since each shell is larger than the preceding one, the atomic radius increases down a column. There are very few exceptions to this, for example, the radii of Zr and Hf are identical. Since different periods are of different lengths, the steps are not equal.

The second trend is within a period. Here we are concerned with one shell. If all the elements are using the same shell, they would be expected to be the same size. However, there is a gradual increase in the nuclear charge from left to right across a period. Consider the third period (Na through Ar): for these elements, the nuclear charge is decreased by the 10 electrons closer to the nucleus (shielding). Therefore, the charge is reduced from that expected by the atomic number to the effective nuclear charge. Since for these elements the number of shielding electrons is constant, the effective nuclear charge for these elements increases toward the right. The increasing effective nuclear charge results in a greater attraction for the outer electrons toward the right, which means that the radii decrease toward the right. These steps tend to be smaller than the vertical steps. Within the transition elements, there are a number of exceptions to this trend due to factors beyond the scope of this discussion. There is an additional minor factor in that as the effective nuclear charge pulls the electrons nearer the nucleus, there is an increased attraction due to the decrease in distance between the electrons and the nucleus (a slight increase in the effective nuclear charge).

The atomic radii refer to atoms (neutral and not connected to anything else). There are other types of radii. Older works sometimes mixed different types of radii with erroneous conclusions, some of which still persist.

For the other periodic properties discussed here, certain conclusions may be related back to the atomic radii (and indirectly to the effective nuclear charge). There is a tendency for smaller atoms to have a greater attraction for electrons.

Generally, the actual values are unnecessary, just trends. The actual values, if needed, are given in tables. If you are going to use the values, it is safest to only take numbers from the same table.

EXERCISE 10·1

Atomic radii

1. Explain why atomic radii decrease toward the right on the periodic table.

2. Place the following elements in order of increasing atomic radii (assume there are no exceptions): F, Cl, Br, I, and At.

3. Place the following elements in order of increasing atomic radii (assume there are no exceptions): Na, Mg, Al, Si, and P.

4. Place the following elements in order of increasing atomic radii. (Assume there are no exceptions): Mo, Sb, Be, Al, and Ca.

5. Which of the following is the most likely to be an exception to the general trends in atomic radii: Mg, As, and Ru? Why?

Ionization energy

The **ionization energy** is the energy required to remove an electron from a gaseous atom in its ground state. The reaction is:

$$A(g) \rightarrow A^+(g) + 1\ e^- \qquad \Delta H_1 = \text{(First) ionization energy}$$

It is possible to remove more electrons up to the removal of all electrons. Very few chemical processes involve sufficient energy to remove more than four electrons. The ionization energies for removing more electrons are:

$$A^+(g) \rightarrow A^{2+}(g) + 1\ e^- \qquad \Delta H_2 = \text{Second ionization energy}$$

$$A^{2+}(g) \rightarrow A^{3+}(g) + 1\ e^- \qquad \Delta H_3 = \text{Third ionization energy}$$

$$A^{3+}(g) \rightarrow A^{4+}(g) + 1\ e^- \qquad \Delta H_4 = \text{Fourth ionization energy}$$

The first ionization energy involves the removal of an electron from a neutral atom. The other ionization energies involve the removal of electrons from ions with increasingly positive charges. The greater the positive charge, the more energy necessary to remove the electron. Meaning:

$$\Delta H_1 < \Delta H_2 < \Delta H_3 < \Delta H_4$$

The two general trends for the (first) ionization energy is that the values decrease when going down any column and increase when moving to the right on the periodic table. These trends are related to the effective nuclear charge experienced by the electron about to be removed. The trends are not smooth; especially when crossing the transition and inner transition series. There are also higher than expected values whenever an electron is being removed from a filled or half-filled subshell (indicating that these arrangements are more stable than expected).

Generally, the actual values are unnecessary, just trends. The actual values, if needed, are given in tables.

Ionization energy

1. Explain why ionization energies increase toward the right on the periodic table.

2. Place the following elements in order of increasing ionization energy (assume there are no exceptions): O, S, Se, Te, and Po.

3. Place the following elements in order of increasing ionization energy (assume there are no exceptions): B, C, O, F, and Ne.

4. Place the following elements in order of increasing ionization energy (assume there are no exceptions): Mo, Sb, Be, Al, and Ca.

5. Which of the following is the most likely to be an exception to the general trends in ionization energy: C, N, and O? Why?

Electron affinity

The (first) **electron affinity** refers to the energy change when an electron is added to a gaseous atom in its ground state. This reaction is:

$$A(g) + 1\ e^- \rightarrow A^-(g) \qquad \Delta H_1 = \text{(First) electron affinity}$$

Very few chemical processes involve sufficient energy to add more than four electrons. The electron affinity energies for adding more electrons are:

$$A^-(g) + 1\ e^- \rightarrow A^{2-}(g) \qquad \Delta H_2 = \text{Second electron affinity}$$

$$A^{2-}(g) + 1\ e^- \rightarrow A^{3-}(g) \qquad \Delta H_3 = \text{Third electron affinity}$$

$$A^{3-}(g) + 1\ e^- \rightarrow A^{4-}(g) \qquad \Delta H_4 = \text{Fourth electron affinity}$$

The (first) electron affinity may be positive, negative, or zero. All remaining electron affinities are positive because energy is required to force a negative electron onto a negative ion (anion).

The trends for electron affinities and ionization energies are similar for the same reason (effective nuclear charge). Unfortunately, because of numerous other factors, this trend is far from simple. Elements with filled subshells and half-filled subshells have lower electron affinities than expected (often zero or close to zero).

Generally, the actual values are unnecessary, just trends. The actual values, if needed, are given in tables. Remember, electron affinities are less regular than either atomic radii or ionization energy trends. Part of this is because of all three properties discussed here, the electron affinities are the most difficult to measure accurately.

EXERCISE 10·3

Electron affinity

1. Explain why electron affinities increase toward the top on the periodic table.

2. Place the following elements in order of increasing electron affinity (assume there are no exceptions): N, P, As, Sb, and Bi.

3. Place the following elements in order of increasing electron affinity (assume there are no exceptions): Li, B, C, O, and F.

4. Place the following elements in order of increasing electron affinity (assume there are no exceptions): Mo, Sb, Al, and Ca.

5. Which of the following is the most likely to be an exception to the general trends electron affinity: C, O, and Ne? Why?

Types of chemical reactions

·11·

This chapter covers a number of reaction types, but these are not the only reaction types possible.

Note that it is possible for a reaction to fall into more than one category.

Combination reactions

Combination reactions involve two or more substances reacting to form one product.

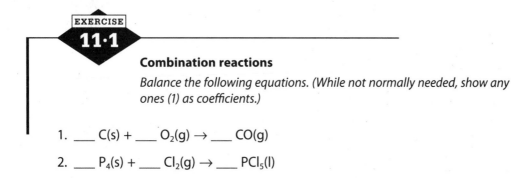

EXERCISE

11·1

Combination reactions

Balance the following equations. (While not normally needed, show any ones (1) as coefficients.)

1. ___ $C(s)$ + ___ $O_2(g)$ → ___ $CO(g)$

2. ___ $P_4(s)$ + ___ $Cl_2(g)$ → ___ $PCl_5(l)$

Decomposition reactions

Decomposition reactions involve one substance reacting to form two or more products.

EXERCISE

11·2

Decomposition reactions

Balance the following equations. (While not normally needed, show any ones [1] as coefficients.)

1. ___ $XeO_3(s)$ → ___ $Xe(g)$ + ___ $O_2(g)$

2. ___ $C_3H_5N_3O_9(l)$ → ___ $CO_2(g)$ + ___ $N_2(g)$ + ___ $H_2O(g)$ + ___ $O_2(g)$

   $C_3H_5N_3O_9(l)$ is nitroglycerin.

Displacement reactions

Displacement reactions involve two elements changing places. In general, one element enters a compound to push (displace) another element out.

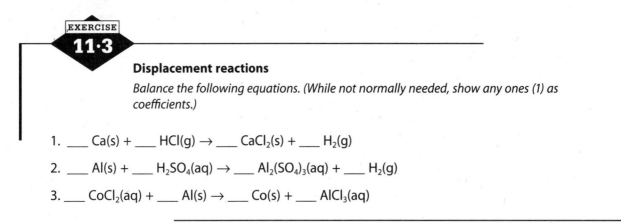

Displacement reactions

Balance the following equations. (While not normally needed, show any ones (1) as coefficients.)

1. ___ Ca(s) + ___ HCl(g) → ___ CaCl$_2$(s) + ___ H$_2$(g)

2. ___ Al(s) + ___ H$_2$SO$_4$(aq) → ___ Al$_2$(SO$_4$)$_3$(aq) + ___ H$_2$(g)

3. ___ CoCl$_2$(aq) + ___ Al(s) → ___ Co(s) + ___ AlCl$_3$(aq)

Metathesis reactions

A **metathesis reaction** is a double displacement, where two compounds are reacting. Two exchanges take place. The formation of a precipitate (solid) when two solutions are mixed is a common example.

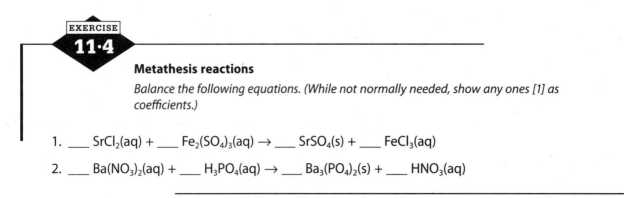

Metathesis reactions

Balance the following equations. (While not normally needed, show any ones [1] as coefficients.)

1. ___ SrCl$_2$(aq) + ___ Fe$_2$(SO$_4$)$_3$(aq) → ___ SrSO$_4$(s) + ___ FeCl$_3$(aq)

2. ___ Ba(NO$_3$)$_2$(aq) + ___ H$_3$PO$_4$(aq) → ___ Ba$_3$(PO$_4$)$_2$(s) + ___ HNO$_3$(aq)

Neutralization reactions

A **neutralization reaction** involves an acid reacting with a base to form a salt and in most cases water. This is a special type of metathesis reaction.

EXERCISE 11·5

Neutralization reactions

Balance the following equations. (While not normally needed, show any ones [1] as coefficients.)

1. ___ HCl(aq) + ___ Ca(OH)$_2$(aq) → ___ CaCl$_2$(aq) + ___ H$_2$O(l)

2. ___ H$_2$SO$_4$(aq) + ___ Al(OH)$_3$(s) → ___ Al$_2$(SO$_4$)$_3$(aq) + ___ H$_2$O(l)

3. ___ H$_3$PO$_4$(aq) + ___ NH$_3$(aq) → ___ (NH$_4$)$_3$PO$_4$(aq)

Combustion reactions

This type of reaction involves a substance "burning" in the presence of another (oxidizer). This is a type of redox reaction. In most cases, the oxidizer is oxygen, O$_2$.

EXERCISE 11·6

Combustion reactions

Balance the following equations. (While not normally needed, show any ones [1] as coefficients.)

1. ___ P4(s) + ___ O2(g) → ___ P4O10(s)

2. ___ C$_4$H$_{10}$(g) + ___ O$_2$(g) → ___ CO$_2$(g) + ___ H$_2$O(g)

3. ___ H$_2$O(l) + ___ F$_2$(g) → ___ HF(g) + ___ O$_2$(g)

Redox reactions

A redox reaction is short for oxidation-reduction reaction. In a redox reaction, the oxidation state of one element is increased while the oxidation state of another element is decreased. The increase (oxidation) involves a loss of electrons, and the decrease (reduction) involves a gain of electrons. The number of electrons lost must equal the number of electrons gained. In general, combination, decomposition, displacement, and combustion reactions are all redox reactions. More involved redox reactions, to be covered in the chapter on electrochemistry, normally require a more involved procedure to balance them.

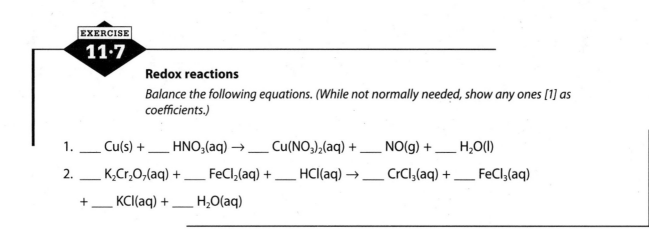

Redox reactions

Balance the following equations. (While not normally needed, show any ones [1] as coefficients.)

1. ___ Cu(s) + ___ HNO₃(aq) → ___ Cu(NO₃)₂(aq) + ___ NO(g) + ___ H₂O(l)

2. ___ K₂Cr₂O₇(aq) + ___ FeCl₂(aq) + ___ HCl(aq) → ___ CrCl₃(aq) + ___ FeCl₃(aq)

   + ___ KCl(aq) + ___ H₂O(aq)

Ionic bonds

When two or more atoms interact, they may form a chemical bond. There is more than one type of chemical bond that may form. The three types of chemical bonds are ionic, covalent, and metallic. Ionic bonds are the subject of this chapter. Covalent bonds are discussed in Chapter 13. We will not be examining metallic bonds until Chapter 21. In general (there are exceptions), ionic bonds form between a metal and a nonmetal, covalent bonds form between nonmetal or metalloid atoms, and metallic bonds form between metal atoms.

While deciding on the type of bond based upon the type of atoms involved is useful, it is sometimes more important to have some theoretical basis to decide on the bond type. The theoretical basis uses the periodic property known as electronegativity. **Electronegativity** is the relative attraction that an atom in a compound has for electrons (and bonding is all about electrons). (Electronegativity is loosely related to ionization energy and electron affinity.) The definition and first table of electronegativity values were first devised by Linus Pauling. Other electronegativity scales have been developed and are useful at times. The following is a periodic table showing the revised Pauling values.

1	2	3	4	5	6	7	8	9	10	11	12	13	14	15	16	17	18
H 2.20																	He
Li 0.98	Be 1.57											B 2.04	C 2.55	N 3.04	O 3.44	F 3.98	Ne
Na 0.93	Mg 1.31											Al 1.61	Si 1.90	P 2.19	S 2.58	Cl 3.16	Ar
K 0.82	Ca 1.00	Sc 1.36	Ti 1.54	V 1.63	Cr 1.66	Mn 1.55	Fe 1.83	Co 1.88	Ni 1.91	Cu 1.90	Zn 1.65	Ga 1.81	Ge 2.01	As 2.18	Se 2.55	Br 2.96	Kr
Rb 0.82	Sr 0.95	Y 1.22	Zr 1.33	Nb 1.6	Mo 2.16	Tc 1.9	Ru 2.2	Rh 2.28	Pd 2.20	Ag 1.93	Cd 1.69	In 1.78	Sn 1.96	Sb 2.05	Te 2.10	I 2.66	Xe
Cs 0.79	Ba 0.89	La 1.1	Hf 1.3	Ta 1.5	W 2.36	Re 1.9	Os 2.2	Ir 2.20	Pt 2.28	Au 2.54	Hg 2.00	Tl 1.62	Pb 2.33	Bi 2.02	Po 2.0	At 2.2	Rn
Fr 0.7	Ra 0.9	Ac 1.1	Rf	Db	Sg	Bh	Hs	Mt	Ds	Rg	Cn	Nh	Fl	Mc	Lv	Ts	Og

Ce 1.12	Pr 1.13	Nd 1.14	Pm 1.13	Sm 1.17	Eu 1.2	Gd 1.20	Tb 1.1	Dy 1.22	Ho 1.23	Er 1.24	Tm 1.25	Yb 1.1	Lu 1.27
Th 1.3	Pa 1.5	U 1.38	Np 1.36	Pu 1.28	Am 1.13	Cm 1.28	Bk 1.3	Cf 1.3	Es 1.3	Fm 1.3	Md 1.3	No 1.3	Lr 1.3

Note that there are no Pauling values for the noble gases or many of the artificial elements. In most cases, you will not need to actually use the values given in this table, but if you do use the values, take the absolute value of the difference in the values of the two interacting elements. A difference of 1.7 or more implies the formation of an ionic bond, and if it is less than 1.7, the bond is covalent (there are subdivisions of covalent bonding discussed in the next chapter). A difference of 1.7 means that the bond is 50% ionic and 50% covalent. This simplified use of electronegativity does not predict metallic bonding; metallic bonds normally form between two atoms of "low" electronegativity.

In a molecule such as F_2, the electronegativity difference is 0; therefore, the bond is covalent. The electronegativity difference between F and Cl is 0.82; therefore, the bond in covalent (actually we will see later that it is polar covalent). The electronegativity difference between Na and Cl is 2.23; therefore, the bond is ionic.

A simpler way of utilizing electronegativity does not involve the numerical electronegativity values. This method begins by locating fluorine on the periodic table; note that fluorine has the highest value (3.98). The values decrease as you move away from fluorine (the transition and inner transition metals are exceptions, and this method will not work for them). The element nearer to fluorine has the higher value. If the two elements are the same distance from fluorine, the one higher on the periodic table has the higher value. The farther apart the elements are, the greater the electronegativity difference.

Bond polarities and electronegativity

As the electronegativity difference between two atoms increases, the bond type shifts from covalent through polar covalent to ionic. This shift corresponds to increasing separation of the elements on the periodic table. As noted earlier, a difference greater than 1.7 is ionic. The separation between covalent and polar covalent is less well defined but is commonly assumed to be in the 0.2 to 0.3 range. The distinction between covalent and polar covalent bonding will be important in later chapters.

Bond polarities and electronegativity

1. Using the electronegativity values from the periodic table in this chapter, what is the electronegativity difference between Si and O?

2. Using the electronegativity values from the periodic table in this chapter, what is the maximum electronegativity of an element that can form an ionic bond to oxygen?

3. Without using the electronegativity values given in the periodic table in this chapter, what type of bond will form between Mg and Cl?

4. Without using the electronegativity values given in the periodic table in this chapter, what type of bond will form between two Br atoms?

5. Without using the electronegativity values given in the periodic table in this chapter, what type of bond will form between Si and Se?

Lewis symbols of atoms

Lewis symbols or Lewis structures are a means of representing how elements interact to form bonds. The concept of Lewis symbols was first conceived by Gilbert N. Lewis. A Lewis symbol contains the symbol for an element plus some indication of the electrons surrounding the atom. Not all electrons are shown, just the valence electrons. For the representative elements, the valence electrons are the electrons in the atom's outer shell. For example, the electron configuration of chlorine is $1s^2 2s^2 2p^6 3s^2 3p^5$, which means that the outer shell is the third shell containing the valence electrons ($3s^2 3p^5$). (All other electrons [$1s^2 2s^2 2p^6$] are the core electrons.) There is no simple means of determining the valence electrons for the transition and inner transition elements; therefore, Lewis symbols are not normally done for these elements. Since chlorine has 7 valence electrons, its Lewis symbol will be the symbol Cl plus an indication of the seven electrons (often seven dots).

The number of valence electrons present for the representative elements may be quickly determined from the position of an element on the periodic table. This avoids writing the electron configuration of the element. First, determine the position of an element on the periodic table. For example, arsenic, As, is in column 15 (5A). For elements in the right-hand six columns, the number of valence electrons is the modern group number (15) minus 10 = 5, and for the older designation (5A) it is simply the number (5). So arsenic has 5 valence electrons, and the Lewis symbol will have the symbol, As, surrounded by some indication of the 5 electrons. For representative elements in columns 1 (1A) and 2 (2A), the number of valence electrons is simply 1 and 2 respectively. Only helium, with 2 valence electrons, is an exception to this.

EXAMPLE 1

If one examines the elements in any particular column on the periodic table, for example some of the elements in Group 17 are F ($1s^2 \mathbf{2s^2 2p^5}$), Cl ($1s^2 2s^2 2p^6 \mathbf{3s^2 3p^5}$), Br ($1s^2 2s^2 2p^6 3s^2 3p^6 \mathbf{4s^2} 3d^{10} \mathbf{4p^5}$), and I ($1s^2 2s^2 2p^6 3s^2 3p^6 4s^2 3d^{10} 4p^6 \mathbf{5s^2} 4d^{10} \mathbf{5p^5}$). In these electron configurations, the outer shell (valence electrons) are in boldface. For these four elements, the valence electrons are always $s^2 p^5$, and if we determined the electron configurations of the remaining elements in the column, we would find that their valence electrons would be $6s^2 6p^5$ and $7s^2 7p^5$, respectively. Recall that

Mendeleev placed elements in a column because they had similar properties. Now, it is apparent that these similar properties are due to the valence electrons being the same for all of the elements in a family. Again, helium is an exception as it has only 2 valence electrons while all other noble gases have 8.

Looking at the Lewis symbols of a few elements, we see

H·   He⁚ (special case)

Li·

·Ḃ·

Using Be as an example, there are a number of correct ways of writing its Lewis symbol, and some of these ways are:

·Be·   ·Be
                    ·

Be   Ḃe·
  ·

The rules for drawing Lewis symbols does not include where the electrons need to be placed around the chemical symbol. However, in later chapters, we will see that it is usually more convenient to place the electrons in the four positions (above, below, left, and right) as shown in these examples. In addition, add the electrons one at a time into these four positions and then place a second electron beside the first until you reach a maximum of 8 electrons (4 pairs). Argon is an example of an atom containing the maximum number of valence electrons:

⁚Är⁚

Note, while it is customary to use dots to represent the valence electrons, it is possible to use other symbols, for example:

8Är8

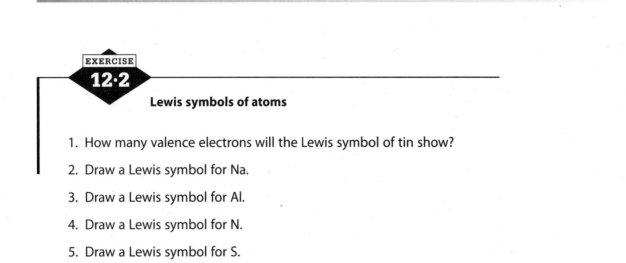

EXERCISE

12·2

Lewis symbols of atoms

1. How many valence electrons will the Lewis symbol of tin show?

2. Draw a Lewis symbol for Na.

3. Draw a Lewis symbol for Al.

4. Draw a Lewis symbol for N.

5. Draw a Lewis symbol for S.

Ionic bonding

Ionic bonding occurs when there is a large electronegativity difference between the interacting atoms. This large difference results in the transfer of electrons from one atom to another. The atom losing electron(s) becomes a cation, and the atom gaining electron(s) becomes an anion. The attraction of the cations to the anions is the ionic bond. The ionic bond is nondirectional. In an ionic solid, such as NaCl, there are a large number of sodium cations and an equal number of chlorine anions. The numbers must be equal so that the overall change is neutral. All of the sodium cations attract all the chloride anions, and all the sodium cations and chlorine anions repel all the like charged ions.

The secret to the number of electrons an atom is likely to need is the octet rule (8-electron rule). (There are various forms of this rule, and we will see a better way of stating it later.) For ions, the octet rule is that atoms tend to gain or lose electrons to give an ion with an octet of electrons in its outer shell. If we use sodium and chlorine as examples, we have the following electron configurations:

Sodium	Na: $1s^2 2s^2 2p^6 3s^1$	Na$^+$ $1s^2 \mathbf{2s^2 2p^6}$
Chlorine	Cl: $1s^2 2s^2 2p^6 3s^2 3p^5$	Cl$^-$ $1s^2 2s^2 2p^6 \mathbf{3s^2 3p^6}$

In the electron configurations of the ions, the octets are in boldface. Note, that in the case of sodium, the formation of the sodium ion involves the elimination of the valence shell ($3s^1$), which leaves the second shell as the outer shell.

Using Lewis symbols to illustrate this same example, with an arrow keeping track of the electron movement:

$$\text{Na}^\circ \; {\overset{\cdot\cdot}{\underset{\cdot\cdot}{\text{Cl}}}}\text{:} \longrightarrow \left[\text{Na}\right]^+ \left[{\overset{\cdot\cdot}{\underset{\cdot\cdot}{\text{\r{C}l}}}}\text{:}\right]^-$$

> Different symbols are being used for the sodium atom and the chlorine atom to help see what has occurred. The use of different electron symbols for different atoms (instead of all dots) makes tracking electrons easier. The square brackets are optional.

EXERCISE

12·3

Ionic bonding

1. Draw the Lewis symbol for the ion that K will most likely form.

2. Draw the Lewis symbol for the ion that Ba will most likely form.

3. Draw the Lewis symbol for the ion that P will most likely form.

4. Draw the Lewis symbol for the ion that O will most likely form.

5. Show how magnesium atoms combine with fluorine atoms to form magnesium fluoride.

Covalent bonds and Lewis structures

While some combinations of atoms result in the formation of ionic bond, this is not always the case. In this chapter, we will examine one of the alternatives to ionic bonding, namely covalent bonding. In general, covalent bonds occur between nonmetal atoms. The other alternative, metallic bonding will be discussed later.

Covalent bonding

Ionic bonds form when there is a relatively large difference in the electronegativity of the atoms involved. Covalent bonds (and metallic bonds) occur when the electronegativity difference is small or zero. The small electronegativity difference results in the atoms sharing electrons instead of being transferred. The sharing of the electrons is accomplished by the outer electron shell of one atom overlapping with the outer electron shell of another atom. Shared electrons count as electrons belonging to both atoms.

The octet rule is also seen to apply to covalent bonding; however, as we will see later, there are exceptions.

EXERCISE
13·1

Covalent bonding

1. Is it likely that a covalent bond will form between atoms of H and C? Why?

2. Is it likely that a covalent bond will form between atoms of Na and F? Why?

Lewis structures

EXAMPLE 1

If we examine the formation of a covalent bond between two fluorine atoms, we see:

$$:\!\ddot{F}\!\cdot \quad \cdot\!\ddot{F}\!: \longrightarrow :\!\ddot{F}\!:\!\ddot{F}\!:$$

As seen previously, different symbols are being used for the electrons on the two fluorine atoms involved to help you visualize what is happening. Any electrons between two atoms are being shared, and in this case, it is apparent that of the two electrons being shared, one is from each fluorine atom. Each fluorine atom in the molecule is surrounded by its 7 original electrons plus 1 electron from the other fluorine atom to make a total of 8 electrons (an octet). The pair of electrons being shared is known as a bonding pair, the remaining electrons (not between the atoms) are nonbonding pairs or lone pairs.

EXAMPLE 2

Replacing the fluorine atoms in Example 1 with oxygen leads to the formation of an oxygen molecule as shown here:

$$:\ddot{O}\cdot \quad \cdot\overset{\circ\circ}{\underset{\circ}{O}}: \longrightarrow :\ddot{O}: \quad \overset{\circ\circ}{\underset{\circ\circ}{O}}: \longrightarrow :\ddot{O}::\overset{\circ\circ}{O}:$$

Each unpaired electron on the oxygen atoms is moved from the "normal" Lewis structure to the center of the arrangement so the unpaired electrons will be between the two atoms (even though they appear to be paired). Finally, the two oxygen atoms are moved together so the electron shells may overlap. For an oxygen molecule, there are four electrons between the two atoms (two pairs each consisting of 1 electron from each oxygen atom). Two pairs of electrons being shared between two atoms is a double bond. Each oxygen atom is the molecule has its original 6 electrons plus 2 electrons from the other atom to give a total of 8 (octet).

EXAMPLE 3

The third example will be the cyanide ion, CN^-. This will be slightly different than the preceding examples because the atoms are not the same and this is an ion instead of a molecule. The carbon atom has 4 valence electrons, and the nitrogen atom has 5 valence electrons. In addition, there is an additional electron from the charge. The Lewis structure must include these 10 electrons $(4 + 5 + 1)$. The Lewis structure of the cyanide ion is:

$$\cdot\dot{C}\cdot \quad \cdot\overset{\circ}{\underset{\circ}{N}}: \longrightarrow \cdot C: \quad \overset{\circ}{\underset{\circ}{N}}: \longrightarrow \left[\overset{\cdot}{C}\overset{\circ}{::}\overset{\circ}{N}\overset{\circ}{}\right]^-$$

In this case, there is a triple bond between the two atoms (three pairs). One of the electrons, indicated by ×, is used to indicate the charge, which is also indicated by the negative sign. As in the preceding examples, each atom has an octet.

A triple bond is the highest observed bond between atoms of two representative elements. So while quadruple bonds have been observed for a few transition elements, they do not appear in any Lewis structures seen at this level of chemistry class.

Commonly, Lewis structures use lines to represent pairs of bonding electrons, and all electrons are symbolized the same way. So the Lewis structures of the three previous examples are:

$$:\ddot{F}—\ddot{F}: \quad \ddot{O}=\ddot{O} \quad \left[:C\equiv N:\right]^-$$

While these structures are correct, it is more difficult to check them to make sure they are correct.

Lewis structures involving more than two atoms form in the same way as those in the diatomic species done so far. One important consideration is the arrangement of the atoms involved.

In general, for up to five atoms, the arrangement has up to four atoms arranged about a central atom with the central atom being the one requiring the most electrons to achieve an octet. (If more than one atom has the same electron requirement, the larger atom tends to be in the center.) Hydrogen will never be the central atom because unlike the other nonmetals, other than helium, it never achieves an octet. The maximum number of electrons for a hydrogen atom is 2.

The following are some general factors to consider when drawing Lewis structures. If we consider hydrogen cyanide, HCN, the arrangement of the atoms may be as in the formula, HCN, HNC, or CHN. The last option is eliminated because H is never the central atom. Of the other two options, the atom needing the most electrons to complete an octet will be the central atom (if two atoms need the same number of electrons, the larger atom will be in the center). For HCN, the C is the central atom with the H and N attached to it. This means that H bonds to C and N bonds to C (bonds mean that the two atoms must be adjacent). The exact arrangement of the atoms is not important as long as the appropriate atoms are adjacent. For example when drawing the Lewis structure of HCN, the arrangement could be HCN or NCH, as either places both the H and N adjacent to the C. It is also allowed to have other than a linear arrangement. For example, for HCN, the H could be above or below the C (either leaves the H adjacent).

Another consideration is that identical atoms generally avoid bonding to each other unless there is no alternative (for example, H_2O_2). Thus, for ammonia, arrangements such as NHHH and HNHH do not occur. For ammonia, all three hydrogen atoms are bonded to a central nitrogen atom.

For oxyacids (such as HNO_3, H_2SO_4, and $HC_2H_3O_2$), each of the acidic hydrogen atoms are attached to an oxygen atom (no more than one H per O) with that O atom attached directly to the central atom.

Finally, do not alter the formula to try to follow some rule. You cannot change the number of atoms or electrons under any circumstances.

It is now time to update the octet rule to consider both ionic and covalent bonds. The **octet rule** is that atoms tend to gain, lose, or share electrons to achieve an outer shell arrangement of 8 electrons. In compounds, the most electronegative atom will get an octet.

EXERCISE
13·2

Lewis structures

1. Draw the Lewis structure for H_2O.

2. Draw the Lewis structure for CF_4.

3. Draw the Lewis structure for HCN.

4. Draw the Lewis structure for H_2O_2.

5. Draw the Lewis structure for HNO_2.

6. Draw the Lewis structure for NO_2.

Resonance

For molecules like SO_3, there is another factor to be considered. Before we tell what the other factor is, let's look at some possible Lewis structures:

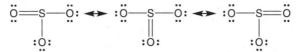

The difference between these structures is the position of the double bond. If you look carefully at the electrons (dots) shown, you will see that most have not moved. In each structure, all four atoms have an octet. Any of these is an acceptable Lewis structure for SO_3. However, there is another question to consider: What does SO_3 do? Sulfur trioxide does all three simultaneously. The entire set is the best indication of what really happens in SO_3. Each of these is a resonance structure, with the structure of SO_3 being the average known as a resonance hybrid. The presence of the double-headed arrows between the structure indicates that the structures are related by resonance.

> When drawing resonance structures, moving bonds (electrons) is the key, and moving atoms is *never* acceptable.

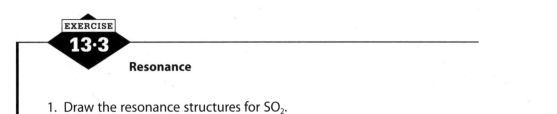

Resonance

1. Draw the resonance structures for SO_2.

2. Draw the resonance structures for NO_2^-.

3. Draw the resonance structures for HNO_3.

Exceptions to the octet rule

While the octet rule is important, it should be remembered that the rule is only a tendency, not a law that must be obeyed. As noted previously, H never has more than 2 electrons because the first electron shell can hold only 2 electrons (helium would follow after H if it formed any stable compounds). Elements with fewer than 3 valence electrons may also be exceptions. The most common example of an element with fewer than 3 valence electrons is B. For example, when B combines with F:

$$:\!\overset{..}{\underset{..}{F}}\!:\!\overset{}{\underset{\circ}{B}}\!\overset{\circ}{:}\!\overset{..}{\underset{..}{F}}\!:$$
$$:\!\overset{..}{\underset{..}{F}}\!:$$

B has only 3 valence electrons (indicated by small circles), and when those are used in bonds, no more F atoms will attach. However, electron-deficient compounds like BF_3 tend to be very reactive. For example, BF_3 will react with ammonia, NH_3, because ammonia has a lone pair of electrons:

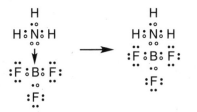

After the two molecules react, the B has an octet (3 electrons of its own, 1 electron from each of three F atoms, and 2 from the N). This type of bond (with 2 electrons from one atom) is a coordinate covalent bond. Once formed, a coordinate covalent bond is identical to any other covalent bond.

Exceptions may also occur when an element has more than an octet of electrons. Such exceptions are reserved to elements in the third period or below on the periodic table. There are never any exceptions of this type for elements in the first two periods. These exceptions fall into two groups. The first group contains elements with more than four atoms bonded to a central atom. The second group is less obvious and contains compounds where the central atom has at least 4 valence electrons and lone pairs of electrons around the central atom. It may help in these two groups of compounds to draw the central atom a little larger than the other atoms to reduce the problem of crowding.

Exceptions are also present when the molecule contains an odd number of electrons as in nitrogen dioxide, NO_2, with 17 electrons. In such cases, one atom will not achieve an octet. However, the more electronegative element always gets an octet. Note, if you are given NO_2, resist changing it to NO_2^-, just so it will not be an exception.

While there are exceptions to the octet rule, in most cases, there is only one atom in the compound (other than H) that is an exception.

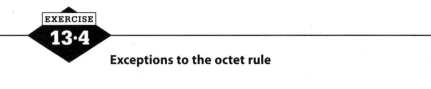

Exceptions to the octet rule

1. One of the few metals that will form covalent bonds is Be. The compound BeI_2 is an example (check the electronegativity difference to see why). Draw a Lewis structure for BeI_2.

2. Draw a Lewis structure for XeF_2.

3. Draw a Lewis structure for SbF_5.

4. Draw a Lewis structure for ClF_3.

5. Draw a Lewis structure for ClF_4^-.

Bond energies

There is a certain amount of energy associated with a covalent bond between two atoms. This energy is known as the bond energy. The **bond energy** is the energy required (endothermic) to break a bond between two atoms. The energy involved depends upon the atoms involved and on the type of bond (single, double, or triple). Here is a typical table of bond energies:

Average Bond Energies in kJ/mol

Br:Br	192.5	C:::N	891	F:N	272	N:O	176
Br:C	276	C:O	351	F:O	185	N::O	607
Br:Cl	218	C::O*	781	F:P	485	N:P	209
Br:F	237	C:::O	1072	F:S	285	O:O	142
Br:H	366.1	C:P	263	F:Si	540.	O::O	498.7
Br:I	180.	C:S	255	H:H	436.4	O:P	350.
Br:N	243	C::S	477	H:I	298.3	O::P	502
Br:P	270.	C:Si	360.	H:N	393	O:S	347
Br:S	215	Cl:Cl	242.7	H:O	464	O::S	469
Br:Si	290.	Cl:F	253	H:P	326	O:Si	370.
C:C	347	Cl:H	431.9	H:S	340.	P:P	215
C::C	615	Cl:I	210.	H:Si	395	P::P	489
C:::C	812	Cl:N	200.	I:I	151.0	P:S	230.
C:Cl	331	Cl:O	205	I:O	200.	P:Si	215
C:F	439	Cl:P	330.	I:P	215	S:S	215
C:H	414	Cl:S	250.	I:Si	215	S::S	352
C:I	240.	Cl:Si	359	N:N	159	S:Si	225
C:N	293	F:F	150.6	N::N	418	Si:Si	230.
C::N	615	F:H	568.2	N:::N	941.4		

*C::O   799 in CO_2

To utilize this table, it is important to know what bonds are present in a compound. This normally requires a correct Lewis structure. Care should be taken when using a table like this. The values in the table are average values, which means that a particular bond may be above or below average. If resonance is involved, the stability of the molecule is increased, which leads to higher than average values for the bond energies.

One use of bond energies is to determine the heat of reaction. This is especially useful if other thermodynamic information on one or more of the substances involved is lacking.

In order to determine the heat of reaction, it is necessary to determine the sum of the bond energies for all reactants and the sum of the bond energies for all of the products. The heat of reaction is then equal to the sum for the reactants minus the sum for the products.

EXAMPLE 4

Determine the heat of reaction for the reaction of oxygen gas with hydrogen gas to form water.
   First, a balanced chemical equation is needed.

$$2 H_2(g) + O_2(g) \rightarrow 2 H_2O(g)$$

Next, it is necessary to know what bonds are present (in this case Lewis structures are needed).

Now use the table to determine the bond energies involved.

$$H–H = 436.4 \text{ kJ/mol} \qquad O=O = 498.7 \text{ kJ/mol} \qquad H–O = 464 \text{ kJ/mol}$$

Combining this information with the balanced chemical equation gives 2 mol H_2 = 2(436.4 kJ) and 1 mol O_2 (498.7 kJ), or 1,371.5 kJ for the reactants. For the products, there are 4 H–O bonds, giving 4 mol HO = 4(464 kJ) = 1,856 kJ. The heat of reaction is 1,371.5 kJ – 1,856 kJ = –484 kJ.

EXERCISE

13·5

Bond energies

1. Determine the heat of reaction for the following reaction:

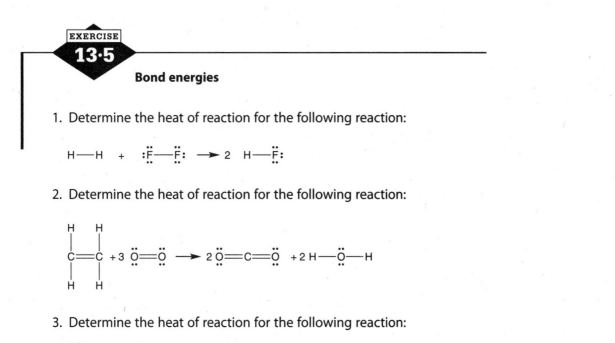

2. Determine the heat of reaction for the following reaction:

3. Determine the heat of reaction for the following reaction:

$$N_2(g) + 3 H_2(g) \rightarrow 2 NH_3(g)$$

Intermolecular forces

In general, intermolecular forces are any type of interaction between molecules (or individual atoms). Intermolecular forces are weaker than chemical bonds (covalent, ionic, and metallic); however, they are important to the behavior of many substances. The primary emphasis here is with the intermolecular forces between identical molecules. When different types of substances are mixed as in a solution (homogeneous mixture), additional types of intermolecular forces may be present. These additional types will be discussed in the chapter on solutions. Some sources group the intermolecular forces discussed here into the general term *van der Waals' forces*, which generally means any interaction between molecules other than chemical bonds.

Types of intermolecular forces

London dispersion forces are present in all substances; however, they are normally so weak that they may be ignored, especially if some other type of intermolecular force is present.

If we consider atoms of helium, we have a nucleus "surrounded" by two electrons. These two electrons are spread around the nucleus so the atom appears neutral. As the electrons move around the nucleus, there will be occasions where for an instant they will be on the same side of the atom. For this instant, one side of the atom will be negative, leaving a net positive charge on the opposite side. This uneven distribution of charge is known as an instantaneous dipole. For the instant that the dipole exists, the positive side will attract electrons on nearby helium atoms. This attraction is an intermolecular force known as a London dispersion force.

The presence of London dispersion forces between helium atoms allows helium to be liquified at very low temperatures. For helium atoms, this intermolecular force is so weak that it is necessary to cool the helium to about 4 K for it to liquify. Another noble gas, neon, has 10 electrons around its nucleus; thus with five times as many electrons as helium, there is a greater likelihood of an instantaneous dipole. The increased likelihood of an instantaneous dipole means that the London dispersion forces are greater than those between helium atoms. The greater attraction between neon atoms leads to neon liquifying at a higher temperature (about 27 K), though there is a slight contribution due to the greater mass of neon atoms. This increase in the strength of attraction with increased number of electrons applies to all substances. But what about a substance like a hydrogen molecule? A hydrogen molecule, like a helium atom, contains 2 electrons; however, a hydrogen molecule has two nuclei instead of one. The additional nucleus increases the likelihood of an instantaneous molecule as illustrated by the fact that hydrogen gas liquifies at about 20 K.

The next strongest intermolecular force is a dipole-dipole force. Dipole-dipole forces occur between polar molecules (molecules with a dipole). Unlike instantaneous dipoles, polar molecules have a permanent dipole (a molecule with a positive end [pole] and a negative end [pole]). The presence of a permanent dipole means that the molecules involved always attract each other instead of an instantaneous attraction that quickly disperses. Permanent dipoles occur when there is an uneven distribution of the electrons. Just how uneven the distribution is, is measured as the dipole moment. This uneven distribution of electrons occurs whenever two atoms with differing electronegativities bond together. For a hydrogen molecule, the electronegativity difference is zero and H_2 is nonpolar. For a molecule of hydrogen chloride, the electronegativity difference is 0.96; therefore, there is a permanent dipole. This, plus an increased London dispersion force, leads to hydrogen chloride liquifying at about 188 K. The strength of the dipole increases with an increasing electronegativity difference until the difference is greater than 1.7. Then it is better to describe the situation as an ionic bond and no longer treat it as a molecule. Note that some textbooks use a different cutoff than the 1.7 mentioned here. In addition, some sources consider a difference of less than 0.2 to be too small to be significant, which means that it is acceptable to consider substances with such a small difference to be nonpolar.

The presence of a permanent dipole is often indicated as follows:

The arrow points toward the more electronegative atom, while the cross at the opposite end indicates a partial positive charge. (This is also known as a polar covalent bond.) The δ's indicate a partial change (positive or negative) as opposed to the full charge present for an ion. The δ+ from one molecule attracts the δ− of another molecule. The greater the partial charges are, the greater the strength of the dipole-dipole force.

The previous discussion works well for diatomic molecules (molecules with only two atoms); however, when more than two atoms are present, it is necessary to also consider the structure of the molecule. If we look at two ways of drawing the correct Lewis structure of water:

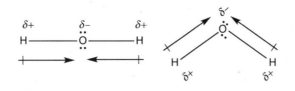

The diagram on the left shows a molecule with a partial positive charge on each end and a partial negative charge in the center. Since there are no positive and no negative ends, there is no dipole (the molecule is nonpolar). The diagram on the right shows a molecule with a partially negative top end and a partially positive bottom end, which means there is a dipole present. If water molecules had the structure indicated on the left, they would be attracted by weak London dispersion forces, while the structure on the right would involve stronger dipole-dipole forces (plus another interaction seen later). The properties of water indicate that the structure of the right is the correct structure. It is estimated that if water had the structure on the left, it would liquify at 170 K instead of the observed 373 K.

The chapters on molecular molecules discuss means of determining the correct structures of molecules like H_2O.

The final type of intermolecular force to be discussed here is hydrogen bonding. The name is unfortunate because hydrogen bonding does not involve bonds but is an intermolecular force. Hydrogen bonding occurs when a hydrogen atom is bonded to a very electronegative atom. The only atoms sufficiently electronegative are atoms of fluorine, oxygen, and nitrogen (though hydrogen bonding involving chlorine has been claimed in a few cases). The large electronegativity difference between atoms of these three elements and hydrogen is so great that the lone hydrogen electron is nearly pulled away to produce an ion. Since hydrogen has only one electron, pulling it away leaves an unshielded nucleus, which yields a δ+ greater than the electronegativity difference would indicate. This greater partial charge yields a greater intermolecular attraction than for a normal dipole-dipole force. Later we will see that the presence of hydrogen bonding leads to many unusual properties for water.

> For hydrogen bonding to occur, a hydrogen atom must be directly attached to one of the three very electronegative atoms (N, O, or F) and not just being present in the same molecule. For example, there is no hydrogen bonding between molecules of CH_3F.

EXERCISE

14·1

Types of intermolecular forces

1. All of the following molecules are nonpolar, so the only intermolecular force present is the London dispersion forces between the molecules. Which of these molecules has the strongest London dispersion forces?

   CH_4, C_8H_{18}, C_5H_{12}, C_2H_6

2. All of the following molecules are polar, so the main intermolecular force present is the dipole-dipole between the molecules. Which of these molecules has the strongest dipole-dipole forces?

   HI, ClF, HCl, NO

3. Which of the following do not have hydrogen bonding?

   NH_3, H_2S, HCN, $C_2H_2F_2$, HNO_3, H_2O_2

4. Which of the following is expected to have the strongest intermolecular force?

   O_2, HF, Xe, PH_3, NO_2

Consequences of intermolecular forces

The strength of the intermolecular forces is important to many of the properties of chemical substances. For example, the stronger the intermolecular forces are, the higher the melting and boiling points of a substance. In general, intermolecular forces are weaker than ionic, covalent, or metallic bonding, which is why substances such as sodium chloride, diamond, and platinum have much higher melting and boiling points than helium, hydrogen chloride, and water. Less obvious is hardness, which is influenced by other factors. We will see in the chapter on solutions that intermolecular forces are a factor in the formation of a solution; however, for solutions it is important to consider the intermolecular forces present in both the solvent and solute plus the intermolecular forces between the solvent and solute molecules. In addition, many solutions involve intermolecular forces not seen yet, such as the intermolecular forces between water and sodium chloride.

Consequences of intermolecular forces

1. Which of the following is expected to have the lowest melting point?
   CH_4, C_8H_{18}, C_5H_{12}, C_2H_6

2. Which of the following is expected to have the lowest melting point?
   HI, HCl, NO

3. Which of the following is expected to have the lowest melting point?
   O_2, HF, Xe, PH_3, NO_2

4. Which of the following is expected to have the highest boiling point?
   CH_4, C_8H_{18}, C_5H_{12}, C_2H_6

5. Which of the following is expected to have the highest boiling point?
   HBr, HCl, NO

6. Which of the following is expected to have the highest boiling point?
   O_2, HF, Xe, PH_3, NO_2

7. Which of the following is expected to have the highest boiling point?
   N_2, HF, HCl, SiH_4, NaCl

8. Which of the following is expected to be harder in the solid state?
   $NaNO_2$, HF, H_2O, SiH_4, BrF

9. In addition to the polarity of the bonds, what additional information is needed to determine if molecules of PF_2Cl_3 are polar or not?

10. If you knew that a molecule of KrF_2 was nonpolar (0 dipole moment), what would you know about the shape of the molecule?

11. If you knew that a molecule of NO_2 was nonpolar (nonzero dipole moment), what would you know about the shape of the molecule?

Molecular structure (VSEPR)

The structure (shape) of molecules is important to the prediction of many of their properties. One method of predicting the shape of a molecule is to use the VSEPR method. This acronym is a contraction of *valence shell electron pair repulsion* theory. This name indicates that only electrons in the valence shell are considered (actually, only the electrons in the valence shell of the "central atom"). In the chapter on Lewis structures, we saw that the electrons in compounds usually occur as electron pairs (with there usually being an octet of electrons). Since all electrons have a negative charge, the pairs experience repulsion. The idea behind this theory is that the electron pairs rearrange themselves to minimize repulsion between the pairs. Unlike valence bond theory (VBT) discussed in the next chapter, there is no theoretical basis for how VSEPR works, and some chemists consider the results being a coincidence. No matter what, the theory does repeatedly make correct predictions.

Electron groups

To utilize VSEPR, one *must* begin with a correct Lewis structure. Consider the following Lewis structures:

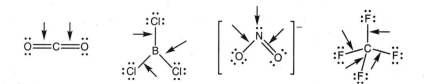

The Lewis structure for the nitrite ion is only one of two resonance forms. In general, all resonance forms give the same results. The arrows point at the electron "pairs" that must be counted when applying VSEPR. All electron pairs not on the central atom are ignored. In addition, multiple bonds, while containing more than two electrons, are counted only as one pair, which is why some sources refer to "electron groups" instead of the possibly misleading electron pairs. All the arrows, except the one at the top of the NO_2^- structure, are bonding pairs (groups). The one at the top of the nitrite ion structure is a lone pair (group).

Once the Lewis structure is drawn, the next step is to count the electron groups around the central atom. In these four examples, from left to right, these are 2, 3, 3, and 4. Next, determine how many lone groups are present around the central atom, which in this case, from left to right, are 0, 0, 1, 0. Once you know the number of electron groups and the number of lone groups, it is possible to apply VSEPR using these numbers.

EXAMPLE 1

Before going on to the application of VSEPR, let's do another set of examples. Consider the following set of correct Lewis structures:

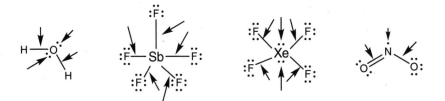

Again the arrows point at the electron groups. The total groups, from left to right, are 4, 5, 6, 3, and the number of lone groups, from left to right, are 2, 0, 2, and 1. Note that the "lone group" on the nitrogen dioxide consists of only one electron. It is important to realize that the shape of a molecule shown in the Lewis structure is not necessarily related to the actual shape of the molecule. In addition, if the Lewis structure of a molecule is drawn differently, the same result occurs. For example, if water were drawn as:

The result is the same (4 total groups and 2 lone groups).

To summarize the results before moving on, we will summarize the number of electron groups and lone groups for each of the previous examples:

CO_2	2	0	BCl_3	3	0	NO_2^-	3	1	CF_4	4	0
H_2O	4	2	SbF_5	5	0	XeF_4	6	2	NO_2	3	1

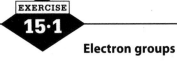

EXERCISE

15·1

Electron groups

1. Determine the number of total groups and the number of lone groups around the central atom in CH_4.

2. Determine the number of total groups and the number of lone groups around the central atom in NH_3.

3. Determine the number of total groups and the number of lone groups around the central atom in ClO_2^-.

4. Determine the number of total groups and the number of lone groups around the central atom in SF_4.

5. Determine the number of total groups and the number of lone groups around the central atom in KrF_2.

Molecular geometry and polarity

The total electron groups around the central atom gives the "parent" or orbital geometry around the central atom. The presence of lone groups may cause slight variations in these "ideal" geometries.

The results of the examples in the previous section include two, three, four, five, and six electron groups. (One group is possible as in H_2; however, one group means only two atoms, and no matter what, two atoms are linear.)

If there are two groups, their mutual repulsion leads to them being on opposite sides of the central atom, which makes the angle between them 180°. This geometry is described as linear. The mutual repulsion between three groups leads to three 120° angles, with the group at the corners of an equilateral triangle. This geometry is described as trigonal planar. More than three pairs requires a three-dimensional structure. Four groups will arrange themselves at the corners of a tetrahedron, which means the angles are 109.5°. This geometry is described as tetrahedral. Five pairs leads to the only situation where all angles are not identical. Three of the five groups arrange themselves with a trigonal planar arrangement (120°) and the remaining two groups are located directly above and directly below the central atom. The bond angle between these two axial groups and the three equatorial groups is 90°. The overall structure is a trigonal bipyramid, or trigonal bipyramidal. Finally, six groups arrange themselves at the corners of an octahedron, which means that the angles between the groups is 90°. The orbital geometry for six pairs is octahedral. There are structures for more groups; however, these rarely occur for the main group elements and, for this reason, will not be covered here. The following diagrams illustrate this discussion:

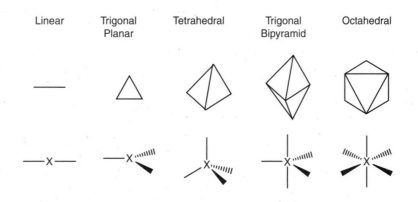

Linear   Trigonal Planar   Tetrahedral   Trigonal Bipyramid   Octahedral

A solid wedge indicates a group that is in front of this page, while a dashed wedge indicates a group behind this page. In most cases these are simplified to:

The other atoms attach (bond) to these groups. The groups do not rearrange to accommodate the atoms unless one of the atoms being attached is very large. Lone groups must also occupy the positions indicated by these arrangement. For now, we will only consider examples with no lone groups. If the central atom is very small and the atoms attached to it are very large, there may be a slight distortion from the ideal structure.

EXAMPLE 2

Consider the CO_2 example used previously. The two groups around the C are in a linear orientation; thus CO_2 is a linear molecule. The bonds between the carbon and the oxygens are polar covalent; however, their polarities are equal and opposed to each other to give a nonpolar molecule as indicated here:

If one of the oxygen atoms were to be replaced by an atom of another element, the two bond polarities would no longer be equal, which would make the molecule polar.

The geometry of the other structures leads to a more involved method to describe the polarity of the species. Suffice it to say that if the position of all the groups present are occupied by identical atoms, the molecule is nonpolar. If some of the peripheral atoms are replaced by atoms of different elements, the molecule may be polar or nonpolar depending upon which atoms are replaced. As noted previously, the presence of lone groups may modify your predictions.

EXERCISE
15·2

Molecular geometry and polarity

1. Determine the molecular geometry and polarity of CBr_4.

2. Determine the molecular geometry and polarity of COS.

3. Determine the molecular geometry and polarity of SbF_5.

4. Determine the molecular geometry and polarity of SF_6.

5. Determine the molecular geometry and polarity of CH_2Cl_2.

Lone pairs

Up to now, we have treated all electron groups as being equal. The presence of two types of groups (bonding and nonbonding) leads to subtle variations in the previous predictions.

Lone pairs of electrons always reside around the central atom, which means that they always repel other groups. Bonding pairs are being shared by the two atoms involved in the bond; therefore, their repulsion is most important when this sharing brings the pair closer to the central atom. The presence of a single lone group tends to increase the angle between it and the bonding group, which leads to a corresponding decrease in the angles between the bonding groups. With two or more lone groups, it is very important to minimize the repulsion between these lone groups. Of secondary importance is minimizing the repulsion between the lone and bonding groups. The repulsion between the bonding groups is the least important.

EXAMPLE 3

Let's consider the Lewis structure of XeF$_4$ drawn previously:

There are two lone groups around the central Xe atom. These two groups will pick two octahedral positions that are as far as possible from each other:

This arrangement places the lone groups 180° apart. This works for any two groups on the opposite sides of the central atom. If the two groups were not on the opposite side of the central atom, the angle between them would be 90°, which means greater repulsion than 180°. The lone group at the top of the diagram "pushes" the four bonding groups down; however, the lone group on the opposite side "pushes" the bonding groups up. These two potential distortions cancel each other to leave the molecule undistorted. Since the four fluorine atoms are on opposite sides (two pairs of fluorine atoms 180° apart), their bond polarities cancel to leave the molecule nonpolar. The molecular geometry (arrangement of the atoms) of XeF$_4$ is described as square planar, because the four fluorine atoms are in a plane at the corners of a square.

EXAMPLE 4

Another involved example is ClF$_3$. The Lewis structure for the compound is:

As with all Lewis structures, this may be drawn in more ways. The key is that all correct Lewis structures have five groups, two of which are lone groups, around the central chlorine. Using VSEPR, there are three possible ways of arranging the lone and bonding groups:

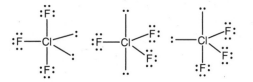

From left to right, the lone pairs are both equatorial, both axial, and one axial and one equatorial. Note that the central structure would be nonpolar, while the other two would be polar. Using VSEPR, it is possible to choose the most likely of the three arrangements. Recall that for five groups, the possible angles are 90° and 120°. The smaller angle (90°) is the key as it leads to more repulsion than the larger angle. (Throughout the remainder of this discussion, only the 90° angles need to be considered.) The structure farthest to the right has two lone pairs at 90°; therefore, this is the least likely arrangement and should be eliminated from the choices, as neither of the others has lone pairs at 90°. Next, we will consider the repulsion between the lone and bonding groups (also only at 90°). If we now move to the leftmost structure, the bonding group to the upper atom has two lone groups at 90° plus two more between the bonding group to the F at the bottom. This makes four total repulsions to be counted. Moving on to the central structure, there are six 90° repulsions between the lone and bonding groups. Since the left structure has four repulsions to be considered and the central has six repulsions, the left structure with fewer repulsions is the preferred structure. From this structure, the bond polarities to the two axial fluorine atoms cancel, leaving the bond polarity of the bond to the remaining fluorine uncanceled, which makes the molecule polar. The molecular geometry of ClF_3 is described as T-shaped because, as drawn, the atoms form a T on its side.

The procedure in the preceding two examples may be applied to other possible structures that are possible whenever lone pairs are present.

The new structural descriptions when lone pairs are present are summarized in the following table:

Total	Lone groups	Molecular groups	Polarity geometry
3	1	Bent (or angular)	Polar
4	1	Trigonal pyramid	Polar
	2	Bent (or angular)	Polar
5	1	Irregular tetrahedral (or see-saw)	Polar
	2	T-shaped	Polar
	3	Linear	Nonpolar
6	1	Square pyramid	Polar
	2	Square planar	Nonpolar

Other combinations lead to diatomic molecules (linear, nonpolar if both atoms are the same, and polar otherwise) or are nonexistent. The nonpolar examples in the proceeding table will be polar if the atoms attached to the central atom are not identical.

Again, when there are no lone pairs, the molecular geometry is identical to the orbital geometry. If all atoms attached to the central atom are identical, the molecule is nonpolar.

EXAMPLE 5

Another variation not seen earlier is for molecules such as HNO_3. The correct Lewis structure of one of the resonance forms is:

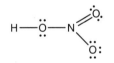

In this structure both the O attached to the H and the N may be treated as central atoms. They should be done so individually. The structure around the O is bent, while the structure around the N is trigonal planar. Normally, polarities are not predicted for these complicated structures and are not asked for, but you can use the given methods to make a prediction.

Note, the other resonance form of HNO_3 gives the same result as do the different resonance forms of all molecules with resonance.

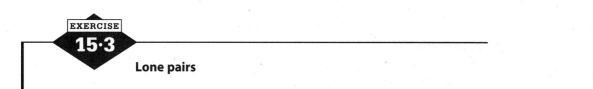

Lone pairs

1. Determine the molecular geometry and polarity of BeI_2. (Note, due to the small electronegativity difference, this is one of the few metal-nonmetal compounds that may be considered covalent instead of ionic.)

2. Determine the molecular geometry and polarity of NOCl. (As expected, the N is the central atom.)

3. Determine the molecular geometry and polarity of HNO_2. (Only consider the groups around the N.)

4. Determine the molecular geometry and polarity of IF_5.

5. Determine the molecular geometry and polarity of $COCl_2$. (As expected, the C is central.)

6. Determine the molecular geometry and polarity of H_2O_2. (Treat each oxygen separately.)

7. Determine the molecular geometry and polarity of CF_2Cl_2.

8. Determine the molecular geometry and polarity of ClF_4^-.

9. Determine the molecular geometry and polarity of BrF_3.

10. Determine the molecular geometry and polarity of XeF_2.

Molecular structure (VBT) ◆·16·◆

An alternative to VSEPR for the determination of molecular geometry is the valence bond theory (VBT). Unlike VSEPR, VBT has a firm theoretical basis in the advanced mathematics of quantum mechanics. VBT can go far beyond predicting molecular geometry and is important in many advanced chemistry courses. However, these advanced applications are far beyond what you need to know for a general chemistry course. The coverage here will not utilize any of the math required for more advanced applications.

Since both VSEPR and VBT can predict molecular geometry, one may be used to determine the answers predicted by the other. Another difference is that while VSEPR requires a correct Lewis structure, VBT is based upon the correct electron configuration of the central atom. A correct Lewis structure may help you understand why VBT works.

Using VBT requires the atomic orbitals in the valence shell to be modified to produce hybrid orbitals. This process alters the shape of the orbitals but does not alter the number of orbitals. For example, the valence shell of carbon contains one $2s$ and three $2p$ orbitals (four total). After hybridization, the carbon atom will still have four orbitals; however, some or all of them will be modified from the original atomic orbitals. Hybrid orbitals have a different shape and orientation than the original atomic orbitals.

Hybridization

EXAMPLE 1

Our first example showing hybridization will be BeI_2. Unlike most metal-nonmetal combinations, the electronegativity difference between Be and I (1.09) is small enough for the bond to be considered covalent. The correct Lewis structure for BeI_2 is:

$$\ddot{\underset{\cdot\cdot}{:}}\overset{\cdot\cdot}{\underset{\cdot\cdot}{I}} \overset{\cdot}{\underset{\times}{:}} Be \overset{\times}{\underset{\cdot}{:}} \overset{\cdot\cdot}{\underset{\cdot\cdot}{I}} :$$

This Lewis structure does not use lines but shows the individual electrons making the bonds. The electrons symbolized by an × were originally Be electrons, while the dots were I electrons. We will refer back to this structure later.

To begin the determination of the hybridization of Be in BeI_2, it is necessary to determine the electron configuration of beryllium (the central atom). The electron configuration of Be is $1s^2 2s^2$. We will ignore the core electrons and concern ourselves only with the complete valence shell. The complete valence shell for Be is $2s^2 2p^0$. This complete configuration may be represented by the following orbital diagram:

This orbital diagram shows the two Be valence electrons as a pair; however, the Lewis structure of BeI_2 shows the two Be electrons separated. Hybridization will rectify this difference. The addition of sufficient energy to one of the Be valence electrons will excite this electron to an excited state as shown here:

$$\underset{2s}{\uparrow}\quad\underset{2p}{\downarrow\,\text{—}\,\text{—}}$$

Each of these unpaired electrons can pair with an iodine electron to form a covalent bond as:

$$\text{⇅}\quad\text{⇅}\quad\text{—}\quad\text{—}$$

Through the excitation of one Be electron, it is possible to form two covalent bonds. The bond energies from the formation of these two bonds are more than sufficient to excite one of the Be electrons. In this case, a $2s$ orbital and a $2p$ orbital from Be are used, and the hybridization is called sp hybridization.

The next question is, what happens to the atomic orbitals during hybridization? To begin, let's recall the shapes of the atomic orbitals used:

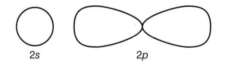

Hybridization of these two orbitals requires a blending of these two shapes to produce two sp hybrid orbitals:

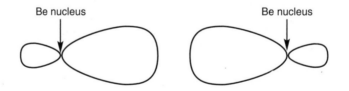

Note that we began with two orbitals (one $2s$ and one $2p$) and ended with two sp hybrid orbitals. This diagram shows the two sp orbitals separately, but as the Be nucleus labels show, they should be overlain. Based upon the fact that the original $2p$ orbital was aligned along the x-, y-, or z-axis (linear) the two hybrid orbitals will form a linear combination. For this reason, all molecules utilizing sp hybridization are linear (180° bond angle).

There are other types of hybridization, all of which blend the shapes of the orbitals used to form the hybrid orbitals. However, for other hybridizations the blending of the different shapes is more difficult to visualize. In the following examples, we will not worry about the blending of shapes.

Before moving on, we need to examine how orbitals (atomic or hybridized) form bonds. Covalent bonds form when two atoms share electrons, this sharing is accomplished through the orbitals overlapping as:

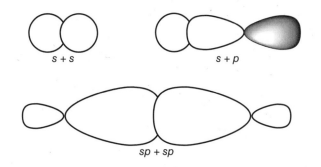

All of these, plus others have the overlap being along the line connecting the nuclei of the atoms involved. These are labeled sigma (σ) bonds. All single bonds between two atoms are σ bonds.

Double (and triple bonds) are different. One of the bonds in a double (or triple) bond is a σ bond. This bond consists of a concentration of electrons directly between the two nuclei involved, which means that the second (and third) bond cannot utilize this space. Therefore, there must be another type of bond. This bond is formed by the side-by-side overlap of two orbitals as:

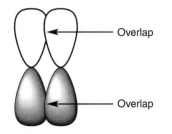

This type of bond is a pi (π) bond. A second π bond may form at a right angle to this one. It is also possible to form π bonds in other ways:

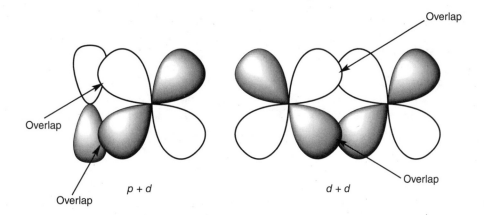

In all cases, a π bond utilizes *unhybridized* atomic orbitals.

EXAMPLE 2

The next example will involve HNO_2. The Lewis structure of nitrous acid is:

$$H—\ddot{O}—\ddot{N}=\ddot{O}:$$

Both the nitrogen and the oxygen attached to the hydrogen atom are central atoms. We will begin with the nitrogen ($1s^2 2s^2 2p^3$), which has the following orbital diagram for its valence shell:

↿⇂ ↿ ↿ ↿
2s   2p

This diagram appears not to need hybridization; however, the general rule for the second period elements is to hybridize the orbitals even though it seems unnecessary. This is due to the small size of the second period elements. If an element like nitrogen did not hybridize, the bond angles would be the same as for the p-orbitals, which are 90° apart. For small central atoms, a 90° bond angle brings the connecting atoms too close together to be stable. Hybridization increase the bond angle to stabilize the structure.

There is a complication in the hybridization of nitrogen. The Lewis structure of HNO_2 shows a double bond between the nitrogen and the oxygen on the right. A double bond consists of a σ plus a π bond. The σ bond may use hybridized orbitals; however, the π bond requires unhybridized orbitals. So when the orbitals on the nitrogen are hybridized, one of the p-orbitals must be "saved" back. The nitrogen will hybridize a 2s and two of the 2p orbitals and leave one of the 2p orbitals unhybridized. The combination of an s and two p orbitals is sp^2 hybridization. (Unlike electron configurations the superscripts stand for the number of orbitals being used and not on the number of electrons.) An sp^2 hybridization always gives a trigonal planer arrangement of the hybrid orbitals (120° bond angle).

Returning to the orbital diagram for N:

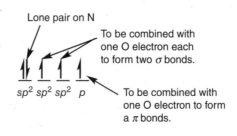

Moving on to the oxygen atom, which has the following orbital diagram:

↿⇂ ↿⇂ ↿ ↿
2s   2p

As with N, O is a second period elements requiring hybridization even though the appearance is the same. This O forms only two single bonds and no double bonds, so it may use all four of the orbitals to form sp^3 hybrid orbitals. The arrangement of the four sp^3 orbitals is always tetrahedral (109.5° bond angle).

For the second period elements, the valence shell contains one 2s and three 2p orbitals which can accommodate a maximum of 8 electrons (an octet). However, starting with the third period, the valence shell expands to include one 3s, three 3p, and five 3d orbitals, which can accommodate a maximum of 18 electrons. While there are some situations where a transition element has 18 electrons around one atom, such a situation is not seen for the representative elements. Nevertheless, representative elements in the third period and below can utilize some of the d-orbitals to exceed an octet.

EXAMPLE 3

Let us consider the compounds PF_5 and SF_6. The Lewis structures of these two compounds are:

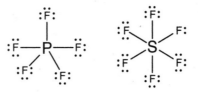

The orbital diagrams for the central atoms are:

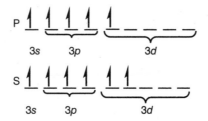

For the P to form 5 bonds and the S to form 6 bonds, hybridization must occur with the orbital diagrams changing to:

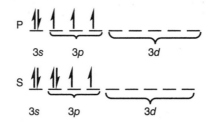

Phosphorus must hybridize one 3s, three 3p, and one 3d to produce five sp^3d hybrid orbitals. These five orbitals have a trigonal planar orientation (120° and 90° bond angles).

Sulfur must hybridize one 3s, three 3p, and two 3d to produce six sp^3d^2 hybrid orbitals. These six orbitals have an octahedral orientation (90° bond angles).

In all cases, the hybrid orbitals may either form a σ bond or hold a lone pair of electrons. Minor adjustments in the structures may occur for the same reasons they occur in VSEPR theory. The descriptive names of the structures are the same in both VSEPR and VBT. Resonance does not alter the hybridization; however, hybridization helps explain why resonance makes the species more stable.

Hybridization

1. Determine the hybridization of O in H_2O.

2. Determine the hybridization of N in NH_3.

3. Determine the hybridization of C in CH_4.

4. Determine the hybridization of N in NO_2.

5. Determine the hybridization of N in HNO_3.

6. Determine the hybridization of C in CO_2.

7. Determine the hybridization of Cl in ClF_3.

8. Determine the hybridization of Xe in XeF_4.

9. Determine the hybridization of S in H_2SO_4.

10. Determine the hybridization of S in SO_3.

11. Determine the hybridization of S in SF_4.

12. Determine the hybridization of Si in SiF_4.

13. Determine the hybridization of Si in SiF_6^{2-}.

14. Determine the hybridization of Kr in KrF_2.

15. SiF_4 may react with F^- to produce SiF_6^{2-}; however, CF_4 will not react with F^- to produce CF_6^{2-}. All other elements in the same family (Ge, Sn, and Pb) behave like Si. Explain the difference in terms of hybridization.

16. What are the hybridizations of each of the two carbon atoms in acetic acid?

17. Explain in terms of hybridization why the bond angle in H_2O is about 105° and the bond angle in H_2S is about 94°.

Mole relationships

One of the key concepts in chemistry is moles. A **mole** (mol) is the SI unit for the amount of a substance. The substance may be anything from atoms to molecules to grains of sand. A mole is analogous to a dozen or a gross. A mole of a substance contains the same number of particles as there are atoms in exactly 12.0000. . . grams of pure carbon-12. This number of atoms is known as **Avogadro's number**. To go with our analogy, if a mole were a dozen, then Avogadro's number would be 12. In chemistry Avogadro's number is 6.0221429×10^{23} mol^{-1}. (Unlike the exact number of 12 dozen^{-1}, Avogadro's number is not an exact number.) You will rarely, if ever, need to use all the significant figures; indeed, you will not need to use Avogadro's number itself very often.

Many chemistry problems involving chemicals will utilize moles in at least one step, and it is common to begin a problem by converting something to moles. If you have a balanced chemical equation and know the moles of any substance in the reaction, then you know the moles of everything else in the reaction.

Often the last step in a chemistry problem will involve changing the moles to the quantity being sought.

Finding moles

Here we will examine some of the methods that are used to determine the number of moles. In other chapters, you will learn additional methods for determining the number of moles present. Add these other methods to your list of how to find moles.

As stated previously Avogadro's number is not used too often. This is especially true if you are dealing with a chemical reaction. Using Avogadro's number to convert to or from moles means that you either have or want the number of the particles under consideration. However, to know the number of particles, you will need to count them, but even if counted one per second for every minute, hour, day, and year since the formation of Earth, you would have only a fraction of Avogadro's number. Even though uncommon, here are two examples where Avogadro's number is used.

EXAMPLE 1

How many silver atoms are in 0.0050 mol of silver atoms?
This is a one-step problem solved as:

$$(0.0050 \text{ mol Ag atoms}) \left(\frac{6.022 \times 10^{23} \text{ Ag atoms}}{1 \text{ mol Ag atoms}} \right) = 3.0 \times 10^{21} \text{ Ag atoms}$$

We started with the given (0.0050 mol Ag atoms) and ended with what we sought (the number of Ag atoms). Notice that we made sure that all unwanted units canceled (this cannot be done if you do not include them), and that the final answer contains only the correct number of significant figures (2). Failing to make sure that only the correct units remain and that the significant figures are correct are common reasons why students lose credit on exams.

In this case, this was a one-step problem. This one step may be the only step in a problem, or it may be one of many steps in a longer problem.

EXAMPLE 2

How many mol of iron are in a sample containing 4.9 billion (4.9×10^9) iron atoms?
This is also a one-step problem, which may be solved as:

$$(4.9 \times 10^9 \text{ Fe atoms}) \left(\frac{1 \text{ mol Fe}}{6.022 \times 10^{23} \text{ Fe atoms}} \right) = 8.1 \times 10^{-15} \text{ mol Fe}$$

(correct units and significant figures)

Notice that in the second example Avogadro's number is the inverse of Avogadro's number in the first example. This inversion was required to arrive at the appropriate units. You might miss this required inversion if you did not include the units; this would be especially likely if this were one step in a longer problem.

Now that we have seen some applications of using Avogadro's number, let us move on to a much more common method, the use of molar masses. The **molar mass** of a substance is the molecular mass expressed in terms of grams per mole. Molar masses utilize the atomic masses of the elements given on most periodic tables. For example, the atomic mass of hydrogen is given as 1.00794, which means 1.00794 amu atom^{-1}, or 1.00794 g mol^{-1}. (This implies [correctly] that Avogadro's number is the conversion factor relating amu to g.) For an H_2 molecule (or mol), the value would be double the amu (g).

For any substance, the molar mass may be determined by locating the atomic masses of each element present times the number of each element in the compound from the periodic table, or some other table. For example, let us determine the molar mass of K_2SO_4 to two decimal places.

$$\text{Molar mass } K_2SO_4 = 2 \text{ (mass K)} + \text{(mass S)} + 4 \text{ (mass O)} = 174.26 \text{ g mol}^{-1}$$

(Since the molar mass is required to have two decimal places, each of the atomic masses used in its determination must also have at least two decimal places.)

The molar mass of a substance allows the direct conversion of the mass of a substance to moles or the direct conversion of the moles of a substance to grams. Here are two examples using the molar mass of a substance.

EXAMPLE 3

The molar mass of H_2O is 18.015 g mol^{-1}. How many moles of water are present in a 50.000 g sample of water?
This is a one-step problem, which may be solved as:

$$(50.000 \text{ g } H_2O) \left(\frac{1 \text{ mol } H_2O}{18.015 \text{ g } H_2O} \right) = 2.7755 \text{ mol } H_2O$$

If the molar mass of water were not given, there would be an additional step to determine the molar mass of water. Note that the units have been canceled appropriately to yield only the desired units. Since both numbers used in the calculation have five significant figures, so does the answer.

EXAMPLE 4

The molar mass of glucose, $C_6H_{12}O_6$, is 180.1449 g mol^{-1}. A teaspoon of this sugar contains about 0.0224 mol of glucose. How many grams of sugar are in a teaspoon of glucose?

$$(0.0224 \text{ mol } C_6H_{12}O_6)\left(\frac{180.1449 \text{ g } C_6H_{12}O_6}{1 \text{ mol } C_6H_{12}O_6}\right) = 4.04 \text{ g } C_6H_{12}O_6$$

(Note that this answer was calculated using the numbers given before rounding to the correct significant figures.) Again all units except the desired units are canceled. If this did not happen, a mistake was made. As with the preceding examples, this single step may be an entire problem or one step in a longer problem.

The final method for determine moles here will be to use a mole ratio. A mole ratio may be used to relate the moles of any substance in a compound to the moles of any other substance in the same compound, or a mole ratio may be used to relate the moles of any substance in a reaction to the moles of any other substance in the same reaction. However, mole ratios from a reaction *require* a balanced chemical equation.

If we consider the compound $C_6H_{12}O_6$, 1 mol of this compound contains 6 mol of C, 12 mol of H, and 6 mol of O. These relationships lead to a variety of mole ratios such as:

$$\frac{1 \text{ mol } C_6H_{12}O_6}{6 \text{ mol } C} \qquad \frac{12 \text{ mol } H}{1 \text{ mol } C_6H_{12}O_6} \qquad \frac{6 \text{ mol } C \, 12 \text{ mol } H}{12 \text{ mol } H \, 6 \text{ mol } O}$$

There are other possibilities. These single steps may serve to solve the problem in one step or be one of a series of steps. Note that all numbers from formulas are exact numbers not measured numbers.

Consider the following balanced chemical equation: 2 $KClO_3(s) \rightarrow$ 2 KCl(s) + 3 $O_2(g)$. This equation says, in molar terms, that 2 mol $KClO_3$ produces 2 mol KCl and 3 mol O_2, which leads to a series of mole ratios including:

$$\frac{2 \text{ mol } KClO_3}{2 \text{ mol } KCl} \qquad \frac{3 \text{ mol } O_2}{2 \text{ mol } KClO_3} \qquad \frac{3 \text{ mol } O_2}{2 \text{ mol } KCl}$$

If the equation given were not balanced, balancing would be required before writing the mole ratios. Notice that the coefficients in each conversion are identical to the coefficients in the balanced chemical equation. Note that all numbers from balanced chemical equations are exact numbers, not measured numbers.

Here are two more examples using mole ratios.

EXAMPLE 5

It is possible to isolate copper from Cu_2O. How many moles of copper may be isolated from 10.5 mol of Cu_2O?

It is possible to solve this problem in one step as:

$$(10.5 \ \cancel{\text{mol Cu}_2\text{O}})\left(\frac{2 \text{ mol Cu}}{1 \ \cancel{\text{mol Cu}_2\text{O}}}\right) = 21.0 \text{ mol Cu}$$

As required, the units and significant figures are correct.

EXAMPLE 6

One way to isolate silver from Ag_2O is through the following decomposition reaction:

$$2 \text{ Ag}_2\text{O(s)} \rightarrow 4 \text{ Ag(s)} + \text{O}_2\text{(g)}$$

How many moles of silver may be produced from 7.75 mol of Ag_2O?

$$(7.75 \ \cancel{\text{mol Ag}_2\text{O}})\left(\frac{4 \text{ mol Ag}}{2 \ \cancel{\text{mol Ag}_2\text{O}}}\right) = 15.5 \text{ mol Ag}$$

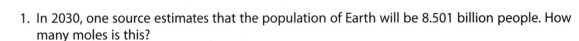

EXERCISE

17·1

Finding moles

1. In 2030, one source estimates that the population of Earth will be 8.501 billion people. How many moles is this?

2. Data from another source estimates that there are 1.2×10^{-5} mol of sand grains on Earth. How many grains of sand is this?

3. The molar mass of $KClO_3$ is 122.550 g mol^{-1}. How many moles of $KClO_3$ are in 15.45 g of $KClO_3$?

4. A certain reaction produces 15.25 mol of O_2. If the molar mass of O_2 is 32.00 g mol^{-1}, how many grams of O_2 were produced?

5. How many moles of carbon are in 15.2 mol of table sugar (sucrose = $C_{12}H_{22}O_{11}$)?

6. Respiration involves the oxidation of glucose. The balanced chemical equation for this reaction is:
   $$C_6H_{12}O_6\text{(aq)} + 6 \text{ O}_2\text{(g)} \rightarrow 6 \text{ CO}_2\text{(g)} + 6 \text{ H}_2\text{O(l)}$$

How many moles of CO_2 are produced when 1.53 mol of glucose is oxidized through the preceding reaction?

Mole conversions

The conversions in this section utilize two or more of the steps covered in the preceding sections. As you learn additional methods of determining/using moles, you will be able to add one or more of the steps in the preceding section to the new methods.
   Here are a few examples:

EXAMPLE 7

How many xenon atoms are present in a 27.1 g sample of Xe?

$$(27.1 \ \cancel{g \ Xe})\left(\frac{1 \ \cancel{mol \ Xe}}{131.293 \ \cancel{g \ Xe}}\right)\left(\frac{6.022 \times 10^{23} \ Xe \ atoms}{1 \ \cancel{mol \ Xe}}\right) = 1.24 \times 10^{23} \ Xe \ atoms$$

The first step uses Example 3 and the second step uses Example 1. The units cancel as shown to yield the correct unit. The significant figures are correct. In other problems, the order may differ to fit the problem.

EXAMPLE 8

The molar mass of MnO_2 is 86.9368 g mol^{-1}. How many moles of oxygen atoms are in a 15.25 g sample of MnO_2?

$$(15.25 \ \cancel{g \ MnO_2})\left(\frac{1 \ \cancel{mol \ MnO_2}}{86.9368 \ \cancel{g \ MnO_2}}\right)\left(\frac{2 \ mol \ O}{1 \ \cancel{mol \ MnO_2}}\right) = 0.3508 \ mol \ O$$

The first step is from Example 3 and the second step is from Example 5. Canceling gives the correct units, rounding of the final answer gives the correct significant figures.

EXAMPLE 9

The molar mass of glucose, $C_6H_{12}O_6$, is 180.1449 g mol^{-1}. Glucose burns in air according to the following equation:

$$C_6H_{12}O_6(s) + 6 \ O_2(g) \rightarrow 6 \ CO_2(g) + 6 \ H_2O(l)$$

How many CO_2 molecules are produced by burning 3.27 grams of glucose?

$$(3.27 \ \cancel{g \ C_6H_{12}O_6})\left(\frac{1 \ \cancel{mol \ C_6H_{12}O_6}}{180.1449 \ \cancel{g \ C_6H_{12}O_6}}\right)\left(\frac{6 \ \cancel{mol \ CO_2}}{1 \ \cancel{mol \ C_6H_{12}O_6}}\right)\left(\frac{6.022 \times 10^{23} \ molecules \ CO_2}{1 \ \cancel{mol \ CO_2}}\right)$$

$$= 6.56 \times 10^{22} \ CO_2 \ molecules$$

Step 1 = Example 3, Step 2 = 17.5, and Step 3 = Example 1. Canceling gives the correct units and rounding the final answer gives the correct significant figures.

EXERCISE

17·2

Mole conversions

Use the following molar masses in the next three questions: Fe_2O_3 = 159.688 g mol^{-1}, SO_2 = 64.064 g mol^{-1}, As_2S_3 = 238.038 g mol^{-1}, O_2 = 32.00 g mol^{-1}, and CO_2 = 44.0095 g mol^{-1}.

1. One of the reactions involved in the smelting of iron ore is:

   $$Fe_2O_3(s) + 3 \ CO(g) \rightarrow 2 \ Fe(l) + 3 \ CO_2(g)$$

   How many grams of CO_2 form when 1,550 g of Fe_2O_3 react according to the given equation?

2. When As_2S_3 is heated in air, the following reaction occurs:

$$2\ As_2S_3(s) + 9\ O_2(g) \rightarrow 2\ As_2O_3(s) + 6\ SO_2(g)$$

How many SO_2 molecules may form from the oxidation of 127 g of As_2S_3?

3. Processing lead ore containing PbS involves the following reaction:

$$2\ PbS(s) + 3\ O_2(g) \rightarrow 2\ PbO(s) + 2\ SO_2(g)$$

How many grams of oxygen reacted if 266 g of SO_2 were formed?

Use the following chemical equation for the next three questions:

$$2\ KMnO_4(aq) + 5\ H_2C_2O_4(aq) + 3\ H_2SO_4(aq) \rightarrow 2\ MnSO_4(aq) + K_2SO_4(aq) + 10\ CO_2 + 8\ H_2O(l)$$

| 158.0339 | 112.4199 | 98.078 | 151.001 | 174.258 | 44.0095 | 18.015 |

The molar masses are beneath the respective formulas in the equation. All molar masses are in terms of g mol^{-1}.

4. How many grams of $H_2C_2O_4$ will react with 1.000 g of $KMnO_4$?

5. How many molecules of H_2O will form from the reaction of 1.000 g of $KMnO_4$?

6. If the reaction produces 2.00 g of K_2SO_4, how many grams of $MnSO_4$ were produced?

Use the following chemical equation for the next three questions:

$$3\ CaCO_3(s) + 2\ H_3PO_4(aq) \rightarrow Ca_3(PO_4)_2(s) + 3\ H_2O(l) + 3\ CO_2(g)$$

| 100.087 | 97.9952 | 312.283 | 18.015 | 44.0095 |

The molar masses are beneath the respective formulas in the equation. All molar masses are in terms of g mol^{-1}.

7. How many CO_2 molecules form at the same time that 100.00 g of $Ca_3(PO_4)_2(s)$ form?

8. How many grams of H_2O form from the complete reaction of 2.5000 g of $CaCO_3$?

9. How many grams of H_3PO_4 will react with 50.00 g of $CaCO_3$?

10. How many grams of CO will form when 7.500 g of O_2 reacts according to the following equation?

$$2\ C(s) + O_2(g) \rightarrow 2\ CO(g)$$

11. $3\ Ca(OH)_2(aq) + 2\ H_3PO_4(aq) \rightarrow Ca_3(PO_4)_2(s) + 6\ H_2O(l)$

When 6.25 g of $Ca(OH)_2$ react according to the given reaction, how many grams of H_2O will form?

12. Butane, C_4H_{10}, reacts with O_2 to produce CO_2 and H_2O as:

$$C_4H_{10}(g) + O_2(g) \rightarrow CO_2(g) + H_2O(g)$$

How many grams of CO_2 form when 2.00 g of C_4H_{10} react?

Stoichiometry

Stoichiometry examines the mole relationships between substances. In many cases, these relationships involve chemical equations. If a chemical equation is involved, it must be balanced. If there is no chemical equation, then there must be some other connection between the moles that are being related. For example, the moles may be related by the fact that they are components of the same compound.

Stoichiometry relates to moles. As we have seen, there are numerous ways to determine the moles of a substance. There are other methods for determining moles, which you will see in other chapters. Once moles are determined, by whatever means, they may be used in a stoichiometry problem. In addition, stoichiometry will allow moles of one substance to be converted into moles of another substance for use in another problem.

Stoichiometry

EXAMPLE 1

How many moles of H are in 5.0 mol of glucose, $C_6H_{12}O_6$?
This problem requires a mole ratio for the conversion:

$$(5.0 \ \cancel{\text{mol } C_6H_{12}O_6})\left(\frac{12 \ \text{mol H}}{1 \ \cancel{\text{mol } C_6H_{12}O_6}}\right) = 60. \ \text{mol H}$$

This mole ratio comes from the chemical formula, which shows that each $C_6H_{12}O_6$ contains 12 H atoms.

The following example is a common basic type of stoichiometry problem.

EXAMPLE 2

How many moles of CO will form when 5.00 mol of O_2 gas react with an excess of solid carbon? The balanced chemical equation is:

$$2 \ C(s) + O_2(g) \rightarrow 2 \ CO(g)$$

To help with solving this problem, it will help to add the additional information from the problem:

2 C(s)	+	$O_2(g) \rightarrow$	2 CO(g)
		5.00 mol O_2	? mol CO

The quantity of O_2 is transferred from the question to below the O_2 in the chemical equation, and the actual question (how many moles of CO) is represented by a question mark below the CO. To solve the problem, it is necessary to begin with the 5.00 mol O_2 and then proceed to the question mark. This may be done as:

$$(5.00 \; \cancel{\text{mol } O_2})\left(\frac{2 \text{ mol CO}}{1 \; \cancel{\text{mol } O_2}}\right) = 10.0 \text{ mol CO}$$

The second term in this calculation, $\left(\dfrac{2 \text{ mol CO}}{1 \text{ mol } O_2}\right)$, is a mole ratio derived directly from the balanced chemical equation, and arranged so that the undesired unit, mol O_2, cancels to leave only the desired units, mol CO.

In this problem, you were given the number of moles; in other problems it may be necessary to calculate the number of moles before using the mole ratio, or it may be necessary to convert the moles found to something else. Let's see an example of this type of problem.

EXAMPLE 3

How many grams of CO will form when 160 g of O_2 gas react with an excess of solid carbon? The balanced chemical equation is:

$$2 \text{ C(s)} + O_2(g) \rightarrow 2 \text{ CO(g)}$$

As with the preceding example, we will begin by transferring the information from the problem to the balanced chemical equation:

$$2 \text{ C(s)} + \quad O_2(g) \rightarrow \quad 2 \text{ CO(g)}$$
$$160 \text{ g } O_2 \qquad ? \text{ g CO}$$

Again we will proceed from the value given (160 g O_2) to the question mark. Since we do not have moles, we cannot use the direct approach used in the preceding problem. In this case, we must seek an intermediate goal first. The intermediate goal will be to find the moles of O_2 as shown:

$$(160 \; \cancel{\text{g } O_2})\left(\frac{1 \text{ mol } O_2}{32 \; \cancel{\text{g } O_2}}\right) =$$

The conversion, $\left(\dfrac{1 \text{ mol } O_2}{32 \text{ g } O_2}\right)$, is the molar mass of O_2. It is not necessary to calculate the actual moles of O_2 at this point, as we can save time by continuing with the problem. Since we now have moles of O_2, we can use the same approach as we did in the preceding problem:

$$(160 \; \cancel{\text{g } O_2})\left(\frac{1 \; \cancel{\text{mol } O_2}}{32 \; \cancel{\text{g } O_2}}\right)\left(\frac{2 \text{ mol CO}}{1 \; \cancel{\text{mol } O_2}}\right) =$$

Again, we will postpone calculating a value since we do not have the correct units yet. We need to use the molar mass of CO to change from moles of CO to grams of CO:

$$(160 \; \cancel{\text{g } O_2})\left(\frac{1 \; \cancel{\text{mol } O_2}}{32 \; \cancel{\text{g } O_2}}\right)\left(\frac{2 \; \cancel{\text{mol CO}}}{1 \; \cancel{\text{mol } O_2}}\right)\left(\frac{28 \text{ g CO}}{1 \; \cancel{\text{mol CO}}}\right) = 280 \text{ g CO}$$

Once we have canceled all the undesired units and have only the desired units, we can finish the problem by doing the calculation.

EXAMPLE 4

How many grams of HCl(g) will form from 40.0 g of Cl_2(g) and excess H_2(g)? The reaction is:

$$Cl_2(g) + H_2(g) \rightarrow HCl(g)$$

In this case, we need to begin by balancing the equation to:

$$Cl_2(g) + H_2(g) \rightarrow 2\,HCl(g)$$
$$40.0\ g\ Cl_2 \qquad\qquad ?\ g\ HCl$$

Here, we have also added the information from the problem. Finishing the problem as we saw in the previous example gives:

$$(40.0\ \cancel{g\ Cl_2})\left(\frac{1\ \cancel{mol\ Cl_2}}{70.9\ \cancel{g\ Cl_2}}\right)\left(\frac{2\ \cancel{mol\ HCl}}{1\ \cancel{mol\ Cl_2}}\right)\left(\frac{36.5\ g\ HCl}{1\ \cancel{mol\ HCl}}\right) = 41.2\ g\ HCl$$

If you get an answer less than 40.0 g, you made an error in the calculation.

EXAMPLE 5

How many grams of Cl_2(g) are required to produce 50.0 g of HCl(g)? Assume the H_2(g) is in excess. The reaction is:

$$Cl_2(g) + H_2(g) \rightarrow HCl(g)$$

Adding the information to the problem gives us:

$$Cl_2(g) \quad + \quad H_2(g) \rightarrow 2\,HCl(g)$$
$$?\ g\ Cl_2 \qquad\qquad\qquad 50.0\ g\ HCl$$

Even though the question mark is on the opposite side of the reaction arrow, it does not alter how the problem is solved.

$$(50.0\ \cancel{g\ HCl})\left(\frac{1\ \cancel{mol\ HCl}}{36.5\ \cancel{g\ HCl}}\right)\left(\frac{1\ \cancel{mol\ Cl_2}}{2\ \cancel{mol\ HCl}}\right)\left(\frac{70.9\ g\ Cl_2}{1\ \cancel{mol\ Cl_2}}\right) = 48.6\ g\ Cl_2$$

Note, if the grams of Cl_2 you got exceeded 50.0 g, you made an error.

When you do problems of these types on your own, you will need to make sure you keep track of the units and round the final answers to the appropriate number of significant figures.

EXERCISE
18·1

Stoichiometry

1. Joseph Priestly first discovered oxygen gas through the following reaction:

$$2\,HgO(s) \rightarrow 2\,Hg(l) + O_2(g)$$

How many grams of O_2(g) may be produced from the decomposition of 50.0 g of HgO(s)?

2. Wilhelm Scheele discovered chlorine gas by the following reaction:

$$MnO_2(s) + 4\,HCl(aq) \rightarrow MnCl_2(aq) + 2\,H_2O(l) + Cl_2(g)$$

How many grams of Cl_2(g) could be prepared by the reaction of 100.0 g HCl(aq) with excess MnO_2(s)?

3. How many grams of HCl(aq) are needed to produce 75.0 g of Cl_2? Assume the reaction in problem 2 is used and that there is excess $MnO_2(s)$.

4. How many grams of $H_2SO_4(l)$ are necessary to prepare 100.0 g of $MnSO_4(s)$? Assume that both the $H_2C_2O_4(aq)$ and the $H_2SO_4(aq)$ are in excess. The reaction is:

$$2\,KMnO_4(aq) + 5\,H_2C_2O_4(aq) + 3\,H_2SO_4(l) \rightarrow 2\,MnSO_4(s) + 3\,H_2O(aq) + K_2SO_4(aq) + 10\,CO_2(g)$$

5. Using the reaction in the previous problem, how many grams of $H_2C_2O_4$ are necessary to react with 25.00 g of $KMnO_4$? Assume that the $H_2SO_4(l)$ is in excess.

Limiting reactants

Did you notice that in the previous examples that used more than one reactant there was also a comment stating that any other reactants were in excess? The reason for this was that if the other reactants were not in excess, there might not be sufficient reactant to produce the amount of product expected from the given reactant. The problems were set up so that the given material limited (controlled) how far a reaction will proceed. This makes the given reactant the limiting reactant or limiting reagent. The **limiting reagent** controls a chemical reaction.

What do you do if you do not know which reactant is limiting? This requires an additional step in the problem to determine which reactant is limiting. The next example will show how to do this.

EXAMPLE 6

The following reaction is to be used to synthesize chlorine gas:

$$MnO_2(s) + 4\,HCl(aq) \rightarrow MnCl_2(aq) + 2\,H_2O(l) + Cl_2(g)$$

A chemist weighs 25.25 g of $MnO_2(s)$ and adds it to a solution containing 38.50 g of HCl. How many grams of $Cl_2(g)$ will form?

There are two basic methods for solving this problem. We will begin with the longer of the two.

$$MnO_2(s) + 4\,HCl(aq) \rightarrow MnCl_2(aq) + 2\,H_2O(l) + Cl_2(g)$$

$$25.25\,g \qquad 38.50\,g \qquad\qquad\qquad\qquad\qquad ?\,g$$

Two reactants are given; therefore, we need two calculations:

$$(25.25\ \text{g MnO}_2)\left(\frac{1\ \text{mol MnO}_2}{86.9368\ \text{g MnO}_2}\right)\left(\frac{1\ \text{mol Cl}_2}{1\ \text{mol MnO}_2}\right)\left(\frac{70.906\ \text{g Cl}_2}{1\ \text{mol Cl}_2}\right) = 20.59\ \text{g Cl}_2\,(g)$$

$$(38.50\ \text{g HCl})\left(\frac{1\ \text{mol HCl}}{36.461\ \text{g HCl}}\right)\left(\frac{1\ \text{mol Cl}_2}{4\ \text{mol HCl}}\right)\left(\frac{70.906\ \text{g Cl}_2}{1\ \text{mol Cl}_2}\right) = 18.72\ \text{g Cl}_2\,(g)$$

The *smaller* of these two answers is the answer to the problem, as it indicates that HCl is the limiting reagent. The other reagent is in excess.

A shorter method begins by determining the moles of each of the reactants:

$$(25.25\ \text{g MnO}_2)\left(\frac{1\ \text{mol MnO}_2}{86.9368\ \text{g MnO}_2}\right) = 0.2904\ \text{mol MnO}_2$$

$$(38.50\ \text{g HCl})\left(\frac{1\ \text{mol HCl}}{36.461\ \text{g HCl}}\right) = 1.056\ \text{mol HCl}$$

Next, divide each of the moles by the coefficient of that substance in the balanced chemical equation:

$$(25.25 \text{ g MnO}_2)\left(\frac{1 \text{ mol MnO}_2}{86.9368 \text{ g MnO}_2}\right)\left(\frac{1}{1}\right) = 0.2904 \text{ mol MnO}_2$$

$$(38.50 \text{ g HCl})\left(\frac{1 \text{ mol HCl}}{36.461 \text{ g HCl}}\right)\left(\frac{1}{4}\right) = 0.2640 \text{ mol HCl}$$

The smaller value after this division is the limiting reactant, and it is only necessary to finish the problem with that one substance as:

$$(1.056 \text{ mol HCl})\left(\frac{1 \text{ mol Cl}_2}{4 \text{ mol HCl}}\right)\left(\frac{70.906 \text{ g Cl}_2}{1 \text{ mol Cl}_2}\right) = 18.72 \text{ g Cl}_2 (g)$$

EXAMPLE 7

The following reaction is to be used to synthesize chlorine gas:

$$MnO_2(s) + 4\,HCl(aq) \rightarrow MnCl_2(aq) + 2\,H_2O(l) + Cl_2(g)$$

A chemist weighs 30.25 g of $MnO_2(s)$ and adds it to a solution containing 45.00 g of HCl. How many grams of the excess reactant will remain after the reaction?

We will begin by identifying the limiting reactant:

$$(30.25 \text{ g MnO}_2)\left(\frac{1 \text{ mol MnO}_2}{86.9368 \text{ g MnO}_2}\right)\left(\frac{1}{1}\right) = 0.3480 \text{ mol MnO}_2$$

$$(45.00 \text{ g HCl})\left(\frac{1 \text{ mol HCl}}{36.461 \text{ g HCl}}\right)\left(\frac{1}{4}\right) = 0.3085 \text{ mol HCl}$$

From these calculations, we know that the HCl is limiting; therefore, MnO_2 is in excess. We now need to determine how many grams of $MnO_2(s)$ will remain after the reaction. The next step is to determine how many grams of $MnO_2(s)$ will react with the limiting reactant:

$$(45.00 \text{ g HCl})\left(\frac{1 \text{ mol HCl}}{36.461 \text{ g HCl}}\right)\left(\frac{1 \text{ mol MnO}_2}{4 \text{ mol HCl}}\right)\left(\frac{86.9368 \text{ g MnO}_2}{1 \text{ mol MnO}_2}\right) = 26.82 \text{ g MnO}_2 (s)$$

The original $MnO_2(s)$ minus the amount of $MnO_2(s)$ that reacted tells you how much is left:

$$30.25 \text{ g MnO}_2(s) - 26.82 \text{ g MnO}_2(s) = 3.43 \text{ g MnO}_2(s)$$

When you are given the amount of more than one reactant in a problem, it is very likely that you will need to determine which reactant is limiting to finish the problem. Once you know which reactant is limiting, any further calculations in the problem will be based upon the limiting reactant.

Limiting reactants

1. How many grams of HCl(g) form from the reaction of 40.00 g Cl_2(g) and 5.000 grams of H_2(g)? The reaction is:

   Cl_2(g) + H_2(g) → 2 HCl(g)

2. How many grams of CO will form when 160 g of O_2 gas react with 150 g of solid carbon? The balanced chemical equation is:

   2 C(s) + O_2(g) → 2 CO(g)

3. The following reaction may be used to synthesize $MnSO_4$(s):

   2 $KMnO_4$(aq) + 5 $H_2C_2O_4$(aq) + 3 H_2SO_4(l) → 2 $MnSO_4$(s) + 3 H_2O(aq) + K_2SO_4(aq) + 10 CO_2(g)

   The chemist begins with a solution containing 25.00 g of $KMnO_4$(aq), a solution containing 35.00 g of $H_2C_2O_4$(aq), and excess H_2SO_4(l). How many grams of $MnSO_4$(s) may be formed?

4. A chemist wishes to use the following reaction to synthesize $MnSO_4$(s):

   2 $KMnO_4$(aq) + 5 $H_2C_2O_4$(aq) + 3 H_2SO_4(l) → 2 $MnSO_4$(s) + 3 H_2O(aq) + K_2SO_4(aq) + 10 CO_2(g)

   The chemist begins with a solution containing 15.00 g of $KMnO_4$(aq), a solution containing 15.00 g of $H_2C_2O_4$(aq), and 15.00 g of H_2SO_4(l). How many grams of $MnSO_4$(s) may be formed?

5. A chemist wishes to use the following reaction to synthesize $MnSO_4$(s):

   2 $KMnO_4$(aq) + 5 $H_2C_2O_4$(aq) + 3 H_2SO_4(l) → 2 $MnSO_4$(s) + 3 H_2O(aq) + K_2SO_4(aq) + 10 CO_2(g)

   The chemist begins with a solution containing 18.00 g of $KMnO_4$(aq), a solution containing 19.00 g of $H_2C_2O_4$(aq), and some H_2SO_4(l). How many grams of H_2SO_4(l) are needed?

Actual and percent yield

In many of the previous examples, we calculated the amount of material that would be produced in a reaction. This amount is the theoretical yield. The **theoretical yield** is the amount calculated to form in a reaction. The amount of product that will actually form equals the theoretical yield if everything goes perfectly. However, reactions often do not go perfectly, and the amount actually formed is less than the theoretical yield. The **actual yield** is the amount actually formed in a reaction. To determine how close the reaction was to being "perfect," it is necessary to calculate the percent yield using the relationship:

$$\text{Percent yield} = \frac{\text{Actual yield}}{\text{Theoretical yield}} \times 100\%$$

If the reaction goes perfectly, the percent yield is 100%. If you calculate a percent yield over 100%, you made an error.

EXAMPLE 8

What is the percent yield of HCl(g), if 75.0 g of Cl_2(g) and excess H_2(g) react to produce 62.5 g of HCl? The balanced reaction is:

$$Cl_2(g) \quad + \quad H_2(g) \quad \rightarrow \quad 2\,HCl(g)$$

$$\text{75.0 g } Cl_2 \qquad\qquad\qquad \text{62.5 g HCl} \qquad \text{? percent yield}$$

Here, we have also added the information from the problem. Notice that there are quantities beneath two substances in the reaction, which is similar to a limiting reactant problem; however, having one of the numbers on the product side differentiates this problem from a limiting reagent problem, which would have the two amounts on the reactant side.

Since we have the actual yield (62.5 g HCl), we still need to calculate the theoretical yield before moving on:

$$(75.0\ \text{g}\,\cancel{Cl_2})\left(\frac{1\ \text{mol}\,\cancel{Cl_2}}{70.9\ \text{g}\,\cancel{Cl_2}}\right)\left(\frac{2\ \text{mol}\,\cancel{HCl}}{1\ \text{mol}\,\cancel{Cl_2}}\right)\left(\frac{36.5\ \text{g HCl}}{1\ \text{mol}\,\cancel{HCl}}\right) = 77.2\ \text{g HCl}$$

We can now enter the values into the definition of percent yield:

$$\text{Percent yield} = \frac{\text{Actual yield}}{\text{Theoretical yield}} \times 100\% = \frac{62.5\ \text{g}\,\cancel{HCl}}{77.2\ \text{g}\,\cancel{HCl}} \times 100\% = 81.0\%$$

Let's modify the preceding example slightly in the next example.

EXAMPLE 9

What is the percent yield of HCl(g), if 85.0 g of Cl_2(g) and 15.0 g of H_2(g) react to produce 79.0 g of HCl? The reaction is:

$$Cl_2(g) \quad + \quad H_2(g) \quad \rightarrow \quad 2\,HCl(g)$$

$$\text{85.0 g } Cl_2 \quad \text{15.0 g } H_2 \quad \text{79.0 g HCl} \qquad \text{? percent yield}$$

Here, we have also added the information from the problem. Notice that there are quantities beneath two substances in the reaction, which is similar to a limiting reactant problem; however, having one of the numbers on the product side differentiates this problem from a limiting reagent problem, which would have the two amounts on the reactant side.

We have the actual yield (79.0 g HCl), so we still need to calculate the theoretical yield. However, we will need to do a limiting reactant problem to calculate the theoretical yield:

$$(85.0\ \text{g}\,\cancel{Cl_2})\left(\frac{1\ \text{mol } Cl_2}{70.906\ \text{g}\,\cancel{Cl_2}}\right)\left(\frac{1}{1}\right) = 1.20\ \text{mol } Cl_2$$

$$(15.00\ \text{g}\,\cancel{H_2})\left(\frac{1\ \text{mol } H_2}{2.016\ \text{g}\,\cancel{H_2}}\right)\left(\frac{1}{1}\right) = 7.44\ \text{mol } H_2$$

Thus the limiting reactant is Cl_2, and we will continue the problem with it:

$$(1.20\ \text{mol}\,\cancel{Cl_2})\left(\frac{2\ \text{mol}\,\cancel{HCl}}{1\ \text{mol}\,\cancel{Cl_2}}\right)\left(\frac{36.5\ \text{g HCl}}{1\ \text{mol}\,\cancel{HCl}}\right) = 87.6\ \text{g HCl}$$

We can now enter the values into the definition of percent yield:

$$\text{Percent yield} = \frac{\text{Actual yield}}{\text{Theoretical yield}} \times 100\% = \frac{79.0\ \text{g}\,\cancel{HCl}}{87.6\ \text{g}\,\cancel{HCl}} \times 100\% = 90.2\%$$

Actual and percent yield

1. The following reaction may be used to produce $MnCl_2(s)$ when $MnO_2(s)$ and $HCl(g)$ are mixed:

   $$MnO_2(s) + 4\,HCl(g) \rightarrow MnCl_2(s) + 2\,H_2O(g) + Cl_2(g)$$

   What is the percent yield if 40.0 g of $MnCl_2(s)$ form when 50.0 g of $HCl(g)$ are mixed with excess $MnO_2(s)$ and heated?

2. If excess carbon is burned with 175 g $O_2(g)$, the following reaction occurs:

   $$2\,C(s) + O_2(g) \rightarrow 2\,CO(g)$$

   If 295 g of CO form, what is the percent yield?

3. The first binary xenon compound was prepared by the following reaction:

   $$Xe(g) + 2\,F_2(g) \rightarrow XeF_4(s)$$

   What is the percent yield of $XeF_4(s)$ if 14.3 g of this compound formed from the reaction of 7.40 g of $F_2(g)$ react with 13.1 g of $Xe(g)$?

4. It is possible to produce $Ca_3(PO_4)_2(s)$ by the following reaction:

   $$3\,Ca(OH)_2(aq) + 2\,H_3PO_4(aq) \rightarrow Ca_3(PO_4)_2(s) + 6\,H_2O(l)$$

   What is the percent yield of $Ca_3(PO_4)_2(s)$ if 45.7 g of this compound formed when a solution containing 50.0 g of $Ca(OH)_2$ react with a solution containing 70.0 g of H_3PO_4?

5. The following reaction may be used to analyze samples of $H_2C_2O_4(s)$:

   $$2\,KMnO_4(aq) + 5\,H_2C_2O_4(aq) + 3\,H_2SO_4(l) \rightarrow 2\,MnSO_4(s) + 3\,H_2O(aq) + K_2SO_4(aq) + 10\,CO_2(g)$$

   What is the percent yield of $MnSO_4(s)$ if 4.50 g of this compound formed when mixing the following three solutions: a solution containing 10.0 g of $KMnO_4$, a solution containing 7.00 g of $H_2C_2O_4$, and a solution containing 15.0 g H_2SO_4?

Empirical formulas

Another of the many applications of stoichiometry is to determine the chemical formula of a substance. As with all other stoichiometry calculations, the key is moles. While there are more advanced methods available at present, the original determination of chemical formulas like CO_2 was done by methods similar to what we will examine here. Newer methods still do this; however, with appropriate advanced instruments, some of the work is done for you.

Using moles, these methods generate the empirical formula of the substance. The **empirical formula** is the simplest formula of a compound. With additional information, it is possible to convert the empirical formula to the actual formula of the compound.

EXAMPLE 10

A sample of gaseous substance was analyzed and found to contain 0.1052 mol of nitrogen and 0.2100 mol of oxygen. What is the empirical formula of this compound?

We begin by comparing the moles of each of the elements present:

$$0.1052 \text{ mol N}$$

$$0.2100 \text{ mol O}$$

Each of the moles is then divided by the smallest number of moles:

$$\frac{0.1052 \text{ mol N}}{0.1052} = 1.000 \text{ N}$$

$$\frac{0.2100 \text{ mol O}}{0.1052} = 1.996 \approx 2.000 \text{ O}$$

The moles of each element need to be converted to a whole number; usually, as is the case here, rounding is acceptable. However, if the value is not very close to a whole number (more than ± 0.1), you should not round. The next example will show how to deal with situations where rounding will not work.

Once each element is expressed as a whole number, those whole numbers become the subscripts in the empirical formula. In this case, the empirical formula is N_1O_2 or simply NO_2.

This may or may not be the actual formula of the compound. The actual formula must be a multiple of the empirical formula. For example, multiplied by 1 gives NO_2, multiplied by 2 gives N_2O_4 multiplied by 3 gives N_3O_6, and so on. To know which one is the actual formula requires the empirical formula plus some additional information.

EXAMPLE 11

A sample of gaseous substance was analyzed and found to contain 0.1052 mol of nitrogen and 0.1578 mol of oxygen. What is the empirical formula of this compound?

We begin by comparing the moles of each of the elements present:

$$0.1052 \text{ mol N}$$

$$0.1578 \text{ mol O}$$

Each of the moles is then divided by the smallest number of moles:

$$\frac{0.1052 \text{ mol N}}{0.1052} = 1.000 \text{ N}$$

$$\frac{0.1578 \text{ mol O}}{0.1052} = 1.500 \text{ O}$$

Again the moles of each element need to be converted to a whole number; however, a value of 1.5 is not close enough to a whole number to round. In situations such as this, it is necessary to multiply each of the values by the smallest integer possible to get all values to be equal to a whole number (at least close enough to round). In this case, the smallest multiple is 2, which gives:

$$\frac{0.1052 \text{ mol N}}{0.1052} = 1.000 \text{ N} \times 2 = 2.000$$

$$\frac{0.1578 \text{ mol O}}{0.1052} = 1.500 \text{ O} \times 2 = 3.000$$

Once each element is expressed as a whole number, those whole numbers become the subscripts in the empirical formula. In this case, the empirical formula is N_2O_3.

EXAMPLE 12

A sample of white solid substance was analyzed and found to contain 0.100 g of hydrogen, 1.20 g of carbon, and 3.15 g of oxygen. The molar mass of the substance is about 90 g mol^{-1}. What are the empirical and actual formulas of this compound?

Again, we need to compare the moles of each element; however, we need to convert the grams to moles before we can do this:

$$(0.100 \text{ gH})\left(\frac{1 \text{ mol H}}{1.008 \text{ gH}}\right) = 0.09921 \text{ mol H}$$

$$(1.20 \text{ gC})\left(\frac{1 \text{ mol C}}{12.01 \text{ gC}}\right) = 0.09992 \text{ mol C}$$

$$(3.15 \text{ gO})\left(\frac{1 \text{ mol O}}{15.999 \text{ gO}}\right) = 0.1969 \text{ mol O}$$

Each of these has one extra digit instead of being rounded to the correct number of significant figures. This is done to minimize rounding errors.

Next, we need to divide all three values by the smallest value:

$$\frac{0.1052 \text{ mol H}}{0.09921} = 1.000 \text{ H}$$

$$\frac{0.09992 \text{ mol C}}{0.09921} = 1.007 \approx 1.000 \text{ C}$$

$$\frac{0.1969 \text{ mol O}}{0.09921} = 1.998 \approx 2.000 \text{ O}$$

These numbers make the empirical formula HCO_2. The approximate molar mass of this formula is 45 g mol^{-1}. Comparing this molar mass to the given molar mass gives us the multiplier. Notice, we are using only approximate values, since the problem states that the molar mass of the substance is *about* 90 g mol^{-1}. This ratio is:

$$\frac{\text{actual molar mass}}{\text{empirical molar mass}} = \frac{90 \text{ g mol}^{-1}}{45 \text{ g mol}^{-1}} = 2$$

The "2" is the multiplier. (Note, the multiplier must be an integer $\geq$ 1.) Since in this case the multiplier is 2, the actual formula must be two times the empirical formula, or 2 × HCO_2 = $H_2C_2O_4$.

EXAMPLE 13

A 1.778 g sample of an unknown solid was burned in oxygen, and 2.842 g of CO_2, 0.2609 g of H_2O, and 0.4056 g of N_2 were produced. From these values, it was possible to determine that the mass of O present in the original sample was 0.5676 g O (since the sample was burned in O_2, the original O had to be determined separately). The compound contained C, H, N, and O. Determine the empirical formula of this compound.

> Do not be intimidated by a long problem like this. Just focus on the simple steps one at a time.

The compound contains four elements (C, H, N, and O), which means we need to determine the moles of each of these four elements:

$$(2.842 \text{ g } CO_2)\left(\frac{1 \text{ mol } CO_2}{44.009 \text{ g } CO_2}\right)\left(\frac{1 \text{ mol } C}{1 \text{ mol } CO_2}\right) = 0.06458 \text{ mol C}$$

$$(0.2609 \text{ g } H_2O)\left(\frac{1 \text{ mol } H_2O}{18.015 \text{ g } H_2O}\right)\left(\frac{2 \text{ mol } H}{1 \text{ mol } H_2O}\right) = 0.02896 \text{ mol H}$$

$$(0.4056 \text{ g } N_2)\left(\frac{1 \text{ mol } N_2}{28.028 \text{ g } N_2}\right)\left(\frac{2 \text{ mol } N}{1 \text{ mol } N_2}\right) = 0.02896 \text{ mol N}$$

$$(0.5676 \text{ g O})\left(\frac{1 \text{ mol O}}{15.999 \text{ g O}}\right) = 0.03548 \text{ mol O}$$

As always, we need to divide each of these by the smallest value:

$$\left(\frac{0.06458 \text{ mol C}}{0.02896 \text{ mol N}}\right) = 2.230 \text{ mol C/mol N}$$

$$\left(\frac{0.02896 \text{ mol H}}{0.02896 \text{ mol N}}\right) = 1.000 \text{ mol H/mol N}$$

$$\left(\frac{0.02896 \text{ mol N}}{0.02896 \text{ mol N}}\right) = 1.000$$

$$\left(\frac{0.03548 \text{ mol O}}{0.02896 \text{ mol N}}\right) = 1.225 \text{ mol O/mol N}$$

Next, as always, multiply each of these by the lowest integer to get everything close to a whole number:

$$2.230 \times 4 = 8.820 \text{ C} \approx 9$$
$$1.000 \times 4 = 4.000 \text{ H} = 4$$
$$1.000 \times 4 = 4.000 \text{ N} = 4$$
$$1.225 \times 4 = 4.900 \text{ O} \approx 5$$

This makes the empirical formula = $C_9H_4N_4O_5$.

If you did not maintain your significant figures, you might have a problem. For example, if you used only two digits in the last step, you would have been multiplying by 5 instead of 4, and this would give you a wrong answer of $C_{11}H_5N_5O_6$.

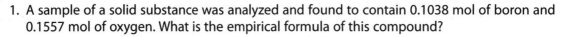

EXERCISE 18·4

Empirical formulas

1. A sample of a solid substance was analyzed and found to contain 0.1038 mol of boron and 0.1557 mol of oxygen. What is the empirical formula of this compound?

2. A sample of a solid substance was analyzed and found to contain 0.2035 mol of iron and 0.2713 mol of oxygen. What is the empirical formula of this compound?

3. A sample of a colorless liquid substance was analyzed and found to contain 0.200 g of hydrogen and 3.20 g of oxygen. The molar mass of the substance is about 34 g mol^{-1}. What are the empirical and actual formulas of this compound?

4. A sample of a white solid substance was analyzed and found to contain 3.603 g of carbon, 0.6042 g of hydrogen, and 4.790 g of oxygen. The molar mass of the substance is about 180 g mol^{-1}. What are the empirical and actual formulas of this compound?

5. A 3.487 g sample of an unknown solid was burned in oxygen, and 5.8312 g of CO_2, 0.5968 g of H_2O, and 0.5570 g of N_2 were produced. From these values, it was possible to determine that the mass of O present in the original sample was 1.2719 g of O (since the sample was burned in O_2, the original O had to be determined separately). The compound contained C, H, N, and O. Determine the empirical formula of this compound.

Molarity

In many situations chemists deal with solutions. A **solution** is a homogeneous mixture of one or more solutes dissolved in a solvent. A solution may be solid (steel), liquid (saltwater), or gas (air). In this chapter, we will be concerned only with liquid solutions. For the time being, we will limit ourselves to solutions where the solvent is water. These solutions are known as aqueous solutions.

Normally an aqueous solution may be saturated or unsaturated. A **saturated solution** contains the maximum amount of solute that will normally dissolve. An **unsaturated solution** is a solution containing less solute than a saturated solution. In some situations, it is possible to produce an unstable situation known as a supersaturated solution. A **supersaturated solution** contains more solute than a saturated solution. These terms are more useful than relative terms like *dilute* and *concentrated*. However, these terms have limitations; for example, the amount of solute necessary to produce a saturated solution will vary when factors such as temperature and/or pressure are changed.

While there are situations where relative terms are acceptable, there are many situations where a more quantitative method is necessary to express the concentration of the solute in a given amount of solution. A common way of expressing the concentration of a solution quantitatively is molarity. **Molarity** is the moles of solute per liter (dm^3) of solution. Molarity may be written as:

$$M = \frac{\text{Moles solute}}{\text{Liters of solution}}$$

This definition relates the molarity, moles, and volume of the solution. If any two of these is known, it is possible to determine the third. Moles may be determined by any of the methods (and others) discussed in the chapter on molar relationships. In addition, using the molarity and the volume to determine the moles gives an additional method for finding the moles, to be added to the methods discussed in the chapter on molar relationships.

Molarity

As just mentioned, if any two of the three terms—molarity (M), moles (n), and volume (V)—are known, the third may be determined. (Do not forget, if the volume is expressed in a unit other than liters, it will need to be converted to liters.) In general, unless told otherwise, you may assume that the volume of the solution is equal to the volume of the solvent.

Finding molarity from moles and volume

We shall begin with the definition of molarity:

$$M = \frac{\text{Moles solute}}{\text{Liters of solution}} = \frac{n}{V}$$

To determine the molarity, it is only necessary to enter the moles of the solute (n) and the volume of the solution (V) into the appropriate positions in the definition.

Finding moles from molarity and volume

In this situation, it will help if you replace the molarity (M) with its definition $\left(\dfrac{n}{V}\right)$. First, write down the volume and then to get the proper unit cancellation, write down the molarity using its definition:

$$\text{Moles} = (\cancel{V})\left(\frac{n}{\cancel{V}}\right) = n$$

Finding volume from molarity and moles

Again, it will help to write down the molarity in terms of its definition. First, write down the moles (n), and then to get the appropriate unit cancellation, write down the molarity inverted:

$$\text{Volume} = (\cancel{n})\left(\frac{V}{\cancel{n}}\right) = V$$

Now, we will revisit these three methods with three examples utilizing numbers.

EXAMPLE 1

Determine the molarity of a solution prepared by adding 0.375 mol of NaCl to 0.500 L of water:

$$\text{M NaCl} = \frac{n}{V} = \frac{0.375 \text{ mol NaCl}}{0.500 \text{ L}} = 0.750 \text{ M NaCl}$$

Be careful to express your answer to the correct number of significant figures as done here.

EXAMPLE 2

Determine the moles of KNO_3 in 0.600 L of a 0.125 M KNO_3 solution:

$$\text{M } KNO_3 = (0.600 \text{ \cancel{L}})\left(\frac{0.125 \text{ mol } KNO_3}{\cancel{L}}\right) = 0.0750 \text{ mol } KNO_3$$

In this case, the volumes (L) cancel, leaving the desired number of moles. As always, make sure you get the appropriate number of significant figures.

EXAMPLE 3

Determine the volume of a 0.500 M HCl solution that contains 0.300 mol of HCl:

$$\text{Volume} = (0.300 \ \text{mol HCl})\left(\frac{L}{0.500 \ \text{mol HCl}}\right) = 0.600 \ \text{L HCl solution}$$

In this case, moles of HCl cancel, leaving the desired unit (L HCl solution). Again, report the answer using the correct number of significant figures.

There are situations where it is desired to alter the molarity of a solution. This may be done in several ways, such as adding/removing solute or adding/removing solvent. The removal of either solvent or solute from a solution is inconvenient in most cases. The addition of more solute is not often done. However, the addition of more solvent is commonly done. Adding more solvent is a process known as dilution. Dilution changes both the molarity and the volume of a solution, leaving the moles of solute unchanged. As we saw in Example 2, multiplying the volume of a solution by the molarity yields the moles. Since the moles are constant, the product of MV for any dilution is a constant. It is possible to express this as:

$$(MV)_{\text{before dilution}} = n = (MV)_{\text{after dilution}}$$

However, this is a little cumbersome, so simplifications are used. For example:

$$M_1 V_1 = M_2 V_2$$

where the "1" subscripts refer to the values before dilution and the "2" subscripts refer to the values after dilution. It is possible to solve for any of the terms in this equation if the other three are known. Here are three examples.

EXAMPLE 4

Determine the molarity of a solution prepared by adding sufficient water to 250.0 mL of a 1.000 M HCl solution to produce a 350.0 mL solution.

A little organization will simplify this problem. This is done by listing the variables and their known values. Also, it will help to group the before and after values separately.

$$M_1 = 1.000 \ \text{M HCl} \qquad M_2 = ?$$
$$V_1 = 250.0 \ \text{mL} \qquad V_2 = 350.0 \ \text{mL}$$

The question mark (unknown value) indicates that the original equation ($M_1V_1 = M_2V_2$) should be rearranged to isolate M_2. This rearrangement gives:

$$M_2 = \frac{M_1 V_1}{V_2}$$

Once rearranged, it is necessary to enter the various terms into the appropriate positions as:

$$M_2 = \frac{(1.000 \ \text{M HCl})(250.0 \ \text{mL})}{(350.0 \ \text{mL})} = 0.7143 \ \text{M HCl}$$

Notice that the volume units (mL) cancel to leave the desired unit (M HCl). Note, if the volumes did not have matching units, it would be necessary to convert one to the same units as the other. As always, be careful to report the answer to the correct number of significant figures.

A dilution always results in a concentration lower than the original.

EXAMPLE 5

What was the original concentration of a solution if the addition of 100.0 mL of water to 500.0 mL of the solution produced a 0.2500 M KCl solution?

Beginning by sorting the information:

$$M_1 = ? \qquad\qquad M_2 = 0.2500 \text{ M KCl}$$
$$V_1 = 500.0 \text{ mL} \qquad\qquad V_2 = (500.0 + 100.0) = 600.0 \text{ mL}$$

Do not forget to add the two volumes.
   Rearranging to isolate M_1:

$$M_1 = \frac{M_2 V_2}{V_1}$$

Entering values:

$$M_2 = \frac{(0.2500 \text{ M KCl})(600.0 \text{ mL})}{(500.0 \text{ mL})} = 0.3000 \text{ M KCl}$$

Again, the units cancel to give the desired unit. The significant figures are also correct.

EXAMPLE 6

A concentrated hydrochloric acid solution is 12.0 M HCl. What volume of this solution is necessary to produce 1.000 L of 1.00 M HCl?
   Beginning as usual:

$$M_1 = 12.0 \text{ M HCl} \qquad\qquad M_2 = 1.00 \text{ M HCl}$$
$$V_1 = ? \qquad\qquad V_2 = 1.000 \text{ L}$$

Rearranging to isolate V_1:

$$V_1 = \frac{M_2 V_2}{M_1}$$

Entering values:

$$V_1 = \frac{(1.00 \text{ M HCl})(1.000 \text{ L})}{(12.0 \text{ M HCl})} = 0.0833 \text{ L HCl}$$

As always, if the units do not cancel to leave the appropriate unit, you have made an error. This is an excellent way to make sure you have set up a problem correctly. The significant units are correct.

EXAMPLE 7

Before moving on to some problems for you to solve on your own, we will explain what needs to be done when the two volume units do not match. We will illustrate how to deal with this "problem" by redoing Example 4 with $V_2 = 0.3500$ L instead of 350.0 mL.
   This volume change alters the table given in Example 4 to:

$$M_1 = 1.000 \text{ M HCl} \qquad M_2 = ?$$
$$V_1 = 250.0 \text{ mL} \qquad V_2 = 0.3500 \text{ L}$$

When we enter this information into the rearranged equation, we get:

$$M_2 = \frac{(1.000 \text{ M HCl})(250.0 \text{ mL})}{(0.3500 \text{ L})} =$$

At this point, no units cancel. What we need to do is to add one more step. This step will involve the relationship 1 L = 1,000 mL, this relationship, like all SI–SI relationships, involves exact numbers. Adding this relationship to the previous equation gives:

$$M_2 = \frac{(1.000 \text{ M HCl})(250.0 \text{ mL})}{(0.3500 \text{ L})}\left(\frac{1 \text{ L}}{1,000 \text{ mL}}\right) = 0.7143 \text{ M HCl}$$

Now the units cancel to give the same answer as in the original Example 17.4, and with the same number of significant figures.

EXERCISE
19·1

Molarity

1. A chemistry student prepares a KNO_3 solution by adding 0.150 mol of this compound to 0.5000 L of water. What is the molarity of this solution?

2. Another student has a 0.500 M $MgSO_4$ solution. What volume of this solution is necessary to provide 0.100 mol of $MgSO_4$?

3. A third student has a 3.000 M HCl solution. How many moles of HCl are in 150.0 mL of this solution?

4. A student needs to prepare 500.0 mL of a 1.250 M HNO_3 solution from a 5.000 M HNO_3 solution. How many milliliters of the concentrated solution are needed?

5. What is the concentration of a solution prepared by adding 500.0 mL of water to 0.7500 L of a 0.8000 M KCl solution?

More molarity calculations

When preparing a solution with a desired concentration from a solid solute, it is necessary to determine the moles needed. Unfortunately, chemical balances are calibrated to measure grams not moles. Therefore, it is necessary to weigh the solute (grams) and use the molar mass of the solute (g mol^{-1}) to convert the grams to moles. The following example, which is a modification of Example 1, will illustrate how to do this.

EXAMPLE 8

Determine the molarity of a solution prepared by adding 21.9 g of NaCl to 0.500 L of water.
We will begin the problem as:

$$M \text{ NaCl} = \frac{21.9 \text{ g NaCl}}{0.500 \text{ L}} =$$

Here the units do not match what we are seeking, so we need an additional step. This step uses the molar mass of NaCl, which is 58.443 g mol^{-1}. For the units to match, the molar mass needs to be inverted before incorporating it into the preceding partial solution:

$$\text{M NaCl} = \frac{21.9 \; \cancel{\text{g NaCl}}}{0.500 \; \text{L}} \left(\frac{\text{mol NaCl}}{58.443 \; \cancel{\text{g NaCl}}} \right) = 0.750 \; \text{M NaCl}$$

Adding this step allows for the cancellation of all but the desired units, which now leads us to the correct answer.

Another situation that may occur is to have a solute that is a liquid instead of a solid. It is still possible to weigh the solute as if it were a solid; however, it is often more convenient to measure the volume of the liquid instead of its mass. This approach requires the density of the liquid. This problem is similar to Example 8, plus an additional step (using the density).

EXAMPLE 9

Determine the molarity of a solution prepared by adding 125.0 mL of C_2H_5OH to 0.500 L of water. The density and molar mass of C_2H_5OH are 0.7890 g mL^{-1} and 46.2610 g mol^{-1}, respectively.

We will begin the problem as:

$$\text{M } C_2H_5OH = \left(\frac{125.0 \; \text{mL } C_2H_5OH}{0.500 \; \text{L}} \right) =$$

If we add a step utilizing the density we get:

$$\text{M } C_2H_5OH = \left(\frac{125.0 \; \cancel{\text{mL } C_2H_5OH}}{0.500 \; \text{L}} \right) \left(\frac{0.7890 \; \text{g } C_2H_5OH}{\cancel{\text{mL } C_2H_5OH}} \right) =$$

At this point the units are g L^{-1}. These are the same as where we started Example 8, so we will finish the problem as we did Example 8 (add a step utilizing the inverted molar mass):

$$\text{M } C_2H_5OH = \left(\frac{125.0 \; \cancel{\text{mL } C_2H_5OH}}{0.500 \; \text{L}} \right) \left(\frac{0.7890 \; \cancel{\text{g } C_2H_5OH}}{\cancel{\text{mL } C_2H_5OH}} \right) \left(\frac{\text{mol } C_2H_5OH}{46.2610 \; \cancel{\text{g } C_2H_5OH}} \right) = 4.26 \; \text{M } C_2H_5OH$$

As always the units that are no longer necessary must cancel, and the final answer must be rounded to the correct number of significant figures.

The same type of procedure (simply adding additional steps) works for any conversion problem as seen here.

Let's try a different example.

EXAMPLE 10

How many grams of NaCl (molar mass = 58.443 g mol^{-1}) are needed to prepare 0.1000 L of a 0.05000 M NaCl solution?

This problem will begin like Example 19.2:

$$\text{M NaCl} = (0.1000 \; \cancel{\text{L}}) \left(\frac{0.05000 \; \text{mol NaCl}}{\cancel{\text{L}}} \right) =$$

Now, we will add the molar mass step to make the final conversion:

$$\text{M NaCl} = (0.1000 \ \cancel{L})\left(\frac{0.05000 \ \cancel{\text{mol NaCl}}}{\cancel{L}}\right)\left(\frac{58.443 \text{ g NaCl}}{\cancel{\text{mol NaCl}}}\right) = 0.2922 \text{ g NaCl}$$

Finally, we have the correct units, so all you need to do is enter the numbers into a calculator and round to the correct number of significant figures.

> Remember the molar mass is not always given in a problem and must be calculated (as always, by summing the atomic masses of the elements present).

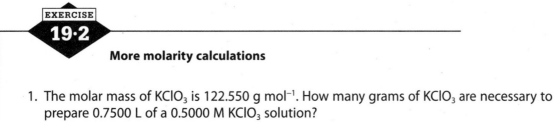

EXERCISE
19·2

More molarity calculations

1. The molar mass of $KClO_3$ is 122.550 g mol^{-1}. How many grams of $KClO_3$ are necessary to prepare 0.7500 L of a 0.5000 M $KClO_3$ solution?

2. What is the molarity of a solution prepared by adding 5.000 g of glucose, $C_6H_{12}O_6$, to 500.0 mL of water?

3. Acetic acid, $HC_2H_3O_2$, has a density of 1.05 g mL^{-1}. What is the concentration of an acetic acid solution prepared by adding 100.0 mL of acetic acid to sufficient water to make 1,000 mL of solution? The molar mass of acetic acid is 60.0520 g mol^{-1}.

4. What is the molarity of a solution prepared by adding 12.5 grams of $MgSO_4$ to 500.0 mL or water?

5. How many grams of acetic acid, $HC_2H_3O_2$, are in 500.0 mL of a 1.500 M acetic acid solution?

6. What volume of C_2H_5OH (0.7890 g mL^{-1} and 46.2610 g mol^{-1}) is present in 0.5000 L of 0.7500 M C_2H_5OH?

Titrations

A titration is a common laboratory experiment where two solutions are added together to complete a reaction. Any type of reaction may be involved. This section will concentrate on acid-base reactions.

A titration is a form of quantitative analysis. All forms of quantitative analysis involve the determination of how much of a substance is present in a sample. For acid-base titrations, the amount sought is commonly the concentration of an unknown solution; however, acid-base titrations are not limited to determining concentrations.

A **titration** is the addition of a known volume of one reactant to a set amount of another reactant until the reaction is complete. The reaction is complete at the equivalence point. At the equivalence point, both reactants are limiting reagents. An **equivalence point** is the theoretical end of a titration. In a titration, there needs to be a means of determining when the reaction is complete. One method of determining when a reaction is complete is to use an indicator. An **indicator** is a substance that indicates the end of a reaction, usually through a change in color.

The point in a titration when the indicator signals that the reaction is over is the endpoint. An **endpoint** is the experimental end of a titration. If the titration is done correctly, the endpoint and equivalence points should be the same.

A solution of one reactant is usually placed in a flask or a beaker and some indicator is added. A solution of the other reactant is usually placed in a burette, which allows a slow controlled release of the solution to the solution containing the indicator. (The concentration of one of the reactant solutions is known.) When the indicator indicates that the reaction is complete, the addition from the burette is stopped and the volume of solution that has been added is determined.

Let's work through two examples to clarify the preceding discussion.

EXAMPLE 11

A student wishes to determine the concentration of a solution of sulfuric acid, H_2SO_4. She carefully measures 25.00 mL of the unknown H_2SO_4 solution into a beaker and adds a little indicator. She then fills a burette with 0.1000 M NaOH solution. She then begins to slowly add the NaOH solution, stopping when the indicator changes color. She then reads the burette and determines that she has added 35.00 mL of the NaOH solution. What is the concentration of the H_2SO_4 solution?

The safest place to start is to make sure you have a balanced chemical equation. Once you have one, transfer the other information to below the appropriate formula in the balanced chemical equation. This gives:

$$H_2SO_4(aq) \quad + \quad 2\,NaOH(aq) \quad \rightarrow \quad Na_2SO_4(aq) \quad + \quad 2\,H_2O(l)$$
$$V = 25.00 \text{ mL} \qquad V = 35.00 \text{ mL}$$
$$M = ?\text{ M} \qquad\qquad M = 0.1000 \text{ M}$$

Next we move to the substance that we know the most about (NaOH). Since the key to chemical reactions is moles, our next step will be to find the moles of NaOH:

$$M\,H_2SO_4 = \left(\frac{0.1000 \text{ mol NaOH}}{\text{L}}\right)(35.00 \text{ mL})\left(\frac{1 \text{ L}}{1,000 \text{ mL}}\right) =$$

Now it is necessary to use a mole ratio to get from moles of NaOH to moles of H_2SO_4 (using the balanced chemical equation):

$$M\,H_2SO_4 = \left(\frac{0.1000 \text{ mol NaOH}}{\text{L}}\right)(35.00 \text{ mL})\left(\frac{1 \text{ L}}{1,000 \text{ mL}}\right)\left(\frac{1 \text{ mol } H_2SO_4}{2 \text{ mol NaOH}}\right) =$$

Now that we have the moles of H_2SO_4 (note, if we were looking for the moles, we could stop here, or if we wanted the grams, we could use the molar mass), we can determine the molarity by dividing by the volume of the H_2SO_4 solution:

$$M\,H_2SO_4 = \left(\frac{0.1000 \text{ mol NaOH}}{\text{L}}\right)(35.00 \text{ mL})\left(\frac{1 \text{ L}}{1,000 \text{ mL}}\right)\left(\frac{1 \text{ mol } H_2SO_4}{2 \text{ mol NaOH}}\right)\left(\frac{1}{25.00 \text{ mL}}\right) =$$

To finish the problem it is necessary to convert the milliliters H_2SO_4 to liters H_2SO_4:

$$M\,H_2SO_4 = \left(\frac{0.1000 \text{ mol NaOH}}{\text{L}}\right)(35.00 \text{ mL})\left(\frac{1 \text{ L}}{1,000 \text{ mL}}\right)\left(\frac{1 \text{ mol } H_2SO_4}{2 \text{ mol NaOH}}\right)\left(\frac{1}{25.00 \text{ mL}}\right)\left(\frac{1,000 \text{ mL}}{1 \text{ L}}\right)$$

$$= 0.07000 \text{ M } H_2SO_4$$

Notice, that in addition to canceling the units, we were also able to cancel a pair of 1,000. This indicates that since both volumes were in terms of milliliters, the two conversions were unnecessary:

$$M \: H_2SO_4 = \left(\frac{0.1000 \: \cancel{mol \: NaOH}}{\cancel{L}} \right) (35.00 \: \cancel{mL}) \left(\frac{1 \: mol \: H_2SO_4}{2 \: \cancel{mol \: NaOH}} \right) \left(\frac{1}{25.00 \: \cancel{mL}} \right)$$

$$= 0.07000 \: M \: H_2SO_4$$

A common error when being presented with this type of problem is that this must be a dilution problem since there are two volumes and two molarities. To see why this assumption is wrong, let's see what happens when we try to solve this as a dilution problem:

$$M_1 = \frac{M_2 V_2}{V_1} = \frac{(0.1000 \: M \: NaOH)(35.00 \: mL \: NaOH)}{(25.00 \: mL \: H_2SO_4)} = 0.1400 \: \frac{(M \: NaOH)(mL \: NaOH)}{(mL \: H_2SO_4)}$$

Not only does this method give the wrong numerical value, but it also gives useless units. You would have missed this error if you did not write down the units.

EXAMPLE 12

A student wishes to determine the concentration of a solution of phosphoric acid, H_3PO_4. He carefully measures 35.00 mL of the 0.1250 M KOH solution into a beaker and adds a little indicator. He then fills a burette with an H_3PO_4 solution with an unknown concentration. He then begins to slowly add the H_3PO_4 solution, stopping when the indicator changes color. He then reads the burette and determines that he has added 40.00 mL of the H_3PO_4 solution. What is the concentration of the KOH solution?

Again, we will begin with the balanced chemical equation:

$$H_3PO_4(aq) \quad + \quad 3 \: KOH(aq) \quad \rightarrow K_3PO_4(aq) + 3 \: H_2O(l)$$
$$V = 40.00 \: mL \qquad V = 35.00 \: mL$$
$$M = ? \: M \qquad \quad M = 0.1250 \: M$$

Next we move to the substance that we know the most about (KOH). Since the key to chemical reactions is moles, our next step will be to find the moles of KOH (except, as we saw in the last example, without the mL to L conversion):

$$M \: H_3PO_4 = \left(\frac{0.1250 \: mol \: KOH}{L} \right) (35.00 \: mL) =$$

It is now necessary to use a mole ratio to get from moles of KOH to moles of H_3PO_4 (using the balanced chemical equation):

$$M \: H_3PO_4 = \left(\frac{0.1250 \: \cancel{mol \: KOH}}{L} \right) (35.00 \: mL) \left(\frac{1 \: mol \: H_3PO_4}{3 \: \cancel{mol \: KOH}} \right) =$$

Now that we have the moles of H_3PO_4, we can determine the molarity by dividing by the volume of the H_3PO_4 solution:

$$M \: H_3PO_4 = \left(\frac{0.1250 \: \cancel{mol \: KOH}}{L} \right) (35.00 \: \cancel{mL}) \left(\frac{1 \: mol \: H_3PO_4}{3 \: \cancel{mol \: KOH}} \right) \left(\frac{1}{25.00 \: \cancel{mL}} \right) = 0.05833 \: M \: H_3PO_4$$

Titrations

1. Determine the molarity of a KOH solution if 30.00 mL of this solution reacts with 35.00 mL of a 0.1250 M H_2SO_4 solution. The reaction is:

   $2 KOH(aq) + H_2SO_4(aq) \rightarrow K_2SO_4(aq) + 2 H_2O(l)$

2. Determine the molarity of an HCl solution if 35.00 mL of this solution reacts with 42.50 mL of a 0.1000 M NaOH solution. The reaction is:

   $2 NaOH(aq) + HCl(aq) \rightarrow NaCl(aq) + H_2O(l)$

3. Determine the molarity of an H_3PO_4 solution, if 50.00 mL of this solution reacts with 45.50 mL of a 0.1000 M $Ba(OH)_2$ solution?

4. Determine the molarity of an $Ca(OH)_2$ solution, if 45.00 mL of this solution reacts with 40.50 mL of a 0.2000 M HCl solution?

5. How many moles of $Ca(OH)_2$ were in a solution, if the titration of this solution required 37.50 mL of 0.1000 M HCl to reach the endpoint?

6. How many grams of NaOH were in a solution, if the titration of this solution required 42.50 mL of 0.1250 M H_2SO_4 to reach the endpoint?

Gases

Under normal conditions, matter exists in three phases: solid, liquid, and gas. Under extreme conditions, there is a plasma state, the properties of which are beyond this book and will not be discussed further.

Properties of solids, liquids, and gases. (Particles may be atoms, molecules, or ions.)

Solid	Fixed shape	Fixed volume	Particles in contact	Slow diffusion
Liquid	Variable shape	Fixed volume	Particles in contact	Faster diffusion
Gas	Variable shape	Variable volume	Particles not in contact	Fastest diffusion

The variable shape of liquids and gases means that their shape conforms to the shape of the container. The variable volume of gases means that a gas fills whatever container it is in. Particles in contact means that there is very little free space between the components of a solid or liquid. Slow diffusion means that the particles in a solid do not move much. The diffusion in a gas may be measured in seconds, in a liquid, it may be measured in minutes or hours; while in a solid, diffusion may take years.

Whether a substance is a solid, liquid, or a gas depends upon two things. These two things are the strength of the intermolecular forces and the kinetic energy of the particles present. The intermolecular forces tend to hold the particles to form a solid, whereas the kinetic energy tends to move the particles apart to form a gas. The intermolecular forces depend upon the identity of the particles, whereas the kinetic energy depends upon the temperature, with higher temperatures resulting in higher kinetic energy. If we begin with a solid and begin heating it, sooner or later the kinetic energy will begin to overcome the intermolecular forces and the solid will melt to form a liquid. If heating is continued, sooner or later the kinetic energy will completely overcome the intermolecular forces and the liquid will boil and become a gas.

The properties of solids and liquids are discussed in a separate chapter.

In the study of gases, four variables are commonly used. These four variables are volume (V), pressure (P), temperature (T), and moles (n). Because the particles in a gas are not in contact, changes in pressure and temperature affect a gas significantly, while slight changes in either or both make a very small change in a solid or liquid.

Pressure

Pressure is a property not mentioned previously in this book. For this reason, we need to discuss pressure here.

Pressure is defined as the force per unit area or $P = \dfrac{F}{A}$. The SI unit of force is the newton (kg m s^{-2}), and the SI unit of area is m^2 (these are both derived SI units consisting of combinations of base units). Entering the two derived SI units into the definition of pressure gives $P = $ kg m^{-1} s^{-2}, which is a pascal (Pa).

Historically, pressure was first measured using a barometer. A barometer consists of a long narrow tube sealed at one end with the open end immersed in a liquid. The tube is evacuated (all air removed). Air pushing down on the surface of the liquid forces the liquid up into the tube. The height to which the liquid rises depends upon the atmospheric pressure and upon the density of the liquid. For example, a barometer using water as the liquid would need to be about 33 ft high, which is too high to be convenient. It was quickly determined that if you wanted a barometer with a reasonable height, it would be necessary to use the densest liquid possible at room temperature. This liquid is mercury, which is 13.6 times as dense as water. With mercury, the barometer needs only to be about 30 in. (760 mm high), which is much more convenient than 33 ft. Using a mercury-filled barometer, the pressure may be expressed as inches of mercury (watch many US local weather reports on TV) or in millimeters of mercury (mm Hg). The unit mm Hg is also known as a torr.

There is one other commonly used pressure unit. This is the atmosphere (atm), which is based on the standard atmosphere. A standard atmosphere is the average air pressure at sea level.

There are other pressure units, which will not be discussed here.

The following illustrates how the units discussed are related to each other:

$$1 \text{ atm} = 760 \text{ mm Hg} = 760 \text{ torr} = 101.325 \text{ kPa}$$

All but the last of these is an exact conversion.

EXAMPLE 1

A sample of gas in a balloon has a pressure of 1.75 atm. Convert this pressure to mm Hg.

$$P = (1.75 \ \text{atm})\left(\frac{760 \text{ mm Hg}}{1 \ \text{atm}}\right) = 1{,}330 \text{ mm Hg}$$

This conversion, like all simple unit conversions, simply involve knowing the relationship between the two units (in this case, 760 mm Hg = 1 atm).

In many problems concerning gases, these relationships will be unnecessary, as the units remain the same throughout the problem and no conversions are necessary.

EXERCISE
20·1

Pressure

1. A sample of a gas has a pressure of 0.755 atm. What is the pressure in kilopascals (kPA)?

2. A sample of a gas has a pressure of 1,077 torr. What is the pressure in kPa?

Temperature

For many problems, it does not matter what temperature units you use; however, there are limitations on the units used in examining the effect of temperature upon a gas.

The volume of a gas will decrease as the gas is cooled. However, there is a limit to this as gases condense to a liquid at some point. If we compare a plot of volume versus temperature for different gases we might see:

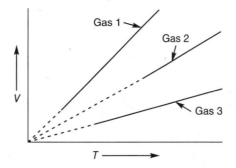

In this graph, when the gas condenses to a liquid, the solid line changes to a dotted line; the dotted lines are a continuation of the solid line. What this graph shows is that no matter what the substance is, the line will extrapolate to the same point. If we designate this temperature as 0, we have the beginnings of a temperature scale, which has no negative temperatures (every gas liquifies before reaching this temperature). The temperature scale is designated the Kelvin scale with 0 K = −273.15°C. (There is a similar scale, the Rankine scale, based on Fahrenheit temperatures with 0 K = 0°R = −459.67°F). This temperature is designated absolute zero. At absolute zero, the particles no longer have kinetic anergy to move, so all motion ceases. Absolute zero is the lowest possible temperature. Note, temperatures expressed with the Kelvin scale never have a degree,°.

The temperature of a gas may be measured using any temperature units. This is fine for reporting the temperature. However, any temperature calculations must use the Kelvin scale. If the problem involves the temperature of a gas, it is only possible to work the problem using an absolute temperature scale such as the Kelvin scale. Using any other temperature unit in the calculation is guaranteed to give you the wrong answer.

Conversions from °C to K, or vice versa, simply involve the relationship:

$$°C = K + 273.15$$

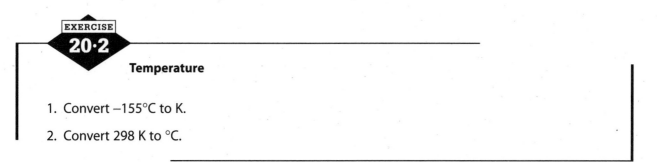

EXERCISE 20·2

Temperature

1. Convert −155°C to K.

2. Convert 298 K to °C.

Gas laws

The basic gas laws deal with the main four gas variables, P, V, T, and n. In general, the development dealt with using a pair of variables at a time, with the other two variables being held constant. Boyle's law (also called Mariotte's law) is concerned with the relationship between volume and pressure. Charles' law is concerned with the relationship between volume and temperature. Avogadro's law is concerned with the relationship between volume and moles. There are other gas laws of this type, which we will not discuss here.

Boyle's law says that for a gas (n and T constant), the volume is inversely proportional to the pressure, which means that $V \propto \dfrac{1}{P}$, or $PV = $ a constant for the sample (which we will call "a"). A common way of using Boyle's law is to do a before/after comparison and use the relationship:

$$P_1 V_1 = P_2 V_2$$

These two values must be equal since, as stated earlier, "$PV = $ a constant for the sample."

Charles' law says that for a gas (n and P constant) the volume is directly proportional to the temperature, which means that $V \propto T$, or $\dfrac{V}{T} = $ a constant for the sample (which we will call "b"). A common way of using Charles' law is to do a before/after comparison and use the relationship:

$$\frac{V_1}{T_1} = \frac{V_2}{T_2}$$

These two values must be equal since, as stated earlier, "$\dfrac{V}{T} = $ a constant for the sample."

Avogadro's law says that for a gas (P and T constant) the volume is directly proportional to the moles, which means that $V \propto n$, or $\dfrac{V}{n} = $ a constant for the sample (which we will call "c"). A common way of using Avogadro's law is to do a before/after comparison and use the relationship:

$$\frac{V_1}{n_1} = \frac{V_2}{n_2}$$

These two values must be equal since, as stated earlier, "$\dfrac{V}{T} = $ a constant for the sample."

What is to be done for the volume, pressure, moles, and temperature to all change? Let's see what can be done.

From the previous discussion, we can rewrite the three gas laws in terms of the constants (a, b, and c):

$$PV = \text{a} \qquad \frac{V}{T} = \text{b} \qquad \frac{V}{n} = \text{c}$$

Rearranging each of these for volume gives:

$$V = \frac{\text{a}}{P} \qquad V = \text{b}T \qquad V = \text{c}n$$

Combining these three equations:

$$V = \left(\frac{\text{a}}{P}\right)(\text{b}T)(\text{c}n) = \frac{(\text{abc})nT}{P} = \frac{RnT}{P}$$

R is a combination of the three constants (a, b, and c), and it is called the (ideal) gas constant. This equation is usually rewritten in the form $PV = nRT$; this form is commonly called the ideal gas equation or the ideal gas law. The value of R may be evaluated using:

$$R = \frac{PV}{nT}$$

The value and units for R will depend upon the units of the variables used to calculate it. Commonly, R is used as 0.08206 L atm mol^{-1} K^{-1}. Other units (and values) are possible, but it is usually simpler to convert the units given to the ones used in this value of R.

Another use of R is found by using the relationship:

$$\frac{P_1 V_1}{n_1 T_1} = R = \frac{P_2 V_2}{n_2 T_2}$$

This may be written as:

$$\frac{P_1 V_1}{n_1 T_1} = \frac{P_2 V_2}{n_2 T_2}$$

In this form, it is known as the combined gas law, since it came from combining Boyle's, Charles', and Avogadro's laws.

Let's try a few examples.

EXAMPLE 2

How many moles of oxygen gas are contained in a 10.50 L container, under a pressure of 2.500 atm, at a temperature of 300.0 K?

A simple way to begin solving the problem is by listing the variables given in the problem:

$n = ?$

$V = 10.50$ L

$P = 2.500$ atm

$T = 300.0$ K

Notice that there are no subscripts (a 1 or a 2) for any of the variables, which means that you need to use an equation that has no subscripts. The only choice from the above is $PV = nRT$, which needs to be rearranged to:

$$n = \frac{PV}{RT}$$

Now, all that is necessary to complete the problem is to plug the values into this equation:

$$n = \frac{(2.500 \ \cancel{atm})(10.50 \ \cancel{L})}{(0.08206 \ \cancel{L} \ \cancel{atm} \ mol^{-1} \cancel{K}^{-1})(300.0 \ \cancel{K})} = 1.066 \ mol \ O_2$$

Notice, that we used a value for R with four significant figures to match the other numbers in the problem.

EXAMPLE 3

A sample of a gas occupies 15.6 L at a pressure of 0.750 atm. What is the volume of the gas if the pressure is increased to 1.32 atm?

Again, we will start by listing the variables (assume any variable not listed is constant).

$V_1 = 15.6$ L $\qquad\qquad V_2 = ?$

$P_1 = 0.750$ atm $\qquad\quad P_2 = 1.32$ atm

(Make sure all the "1" subscripts are kept together and that all the "2" subscripts are kept together.) The presence of subscripts means that we need to use an equation with subscripts.

We can use either the following form of Boyle's law: $P_1V_1 = P_2V_2$ or the combined gas law (modified by deleting any variable that does not change): $\dfrac{P_1V_1}{n_1T_1} = \dfrac{P_2V_2}{n_2T_2}$

Notice that both equations lead to $P_1V_1 = P_2V_2$, which means that Boyle's law was "hiding" inside the combined gas law.

We can now rearrange the equation to solve for the unknown variable (V_2):

$$V_2 = \frac{P_1V_1}{P_2}$$

Now is just a matter of plugging the values into this equation:

$$V_2 = \frac{(0.750 \ \cancel{atm})(15.6 \ L)}{(1.32 \ \cancel{atm})} = 0.886 \ L$$

EXAMPLE 4

A sample of a gas has a volume of 5.52 L at 298 K. What must the temperature be if the volume decreases to 3.75 L?

The information in the problem gives:

$$V_1 = 5.52 \ L \qquad V_2 = 3.75 \ L$$
$$T_1 = 298 \ K \qquad T_2 = ?$$

Again, we need an equation with subscripts. We will begin with the combined gas law and remove the variables that do not change (are not listed $= n$ and P).

$$\frac{\cancel{P_1}V_1}{\cancel{n_1}T_1} = \frac{\cancel{P_2}V_2}{\cancel{n_2}T_2}$$

$$\frac{V_1}{T_1} = \frac{V_2}{T_2}$$

(Charles' law is "hidden" inside the combined gas law.)

Rearranging the equation to find the unknown (T_2):

$$T_2 = \frac{T_1V_2}{V_1}$$

Now we can enter the values from or lest:

$$T_2 = \frac{(298 \ K)(3.75 \ \cancel{L})}{(5.52 \ \cancel{L})} = 202 \ K$$

EXAMPLE 5

A sample of a gas occupies 8.50 L at a pressure of 755 torr and 295 K. What is the volume of the gas if the pressure is increased to 1.50 atm and the temperature is decreased to 0°C?

$$P_1 = 755 \ torr \qquad P_2 = 1.50 \ atm$$
$$V_1 = 8.50 \ L \qquad V_2 = ?$$
$$T_1 = 295 \ K \qquad T_2 = 0°C = 273 \ K$$

(Note that when doing a gas law problem, always make sure that the temperatures are in Kelvins; if not, convert them as soon as possible.)

Beginning with the combined gas law and canceling what does not change:

$$\frac{P_1 V_1}{n_1 T_1} = \frac{P_2 V_2}{n_2 T_2}$$

Rearranging to solve for V_2 gives:

$$V_2 = \frac{P_1 V_1 T_2}{T_1 P_2}$$

Entering the values from the table into this equation gives:

$$V_2 = \frac{(755 \text{ torr})(8.50 \text{ L})(273 \text{ K})}{(295 \text{ K})(1.50 \text{ atm})}$$

Unlike the preceding examples, all the unneeded units do not cancel; therefore, we will need to add a conversion step as:

$$V_2 = \frac{(755 \text{ torr})(8.50 \text{ L})(273 \text{ K})}{(295 \text{ K})(1.50 \text{ atm})}\left(\frac{1 \text{ atm}}{760 \text{ torr}}\right) = 5.21 \text{ L}$$

Notice that while you were given the equation forms of Boyle's, Charles', and Avogadro's laws, it is not necessary to learn them, as these three gas laws, and others, are "hidden" inside the combined gas law.

EXERCISE 20·3

Gas laws

1. What is the pressure exerted by a 5.00 mol of a gas contained in a 2.50 L container at 355 K?

2. A 10.0 L sample of a gas at 755 torr is compressed to a volume of 8.50 L. What is the new pressure?

3. A sample of a gas occupies 5.75 L at a certain temperature. Later the volume has increased to 7.25 L and the temperature is 25°C. What was the original temperature in °C?

4. A sample of a gas occupies 1275 mL at a temperature of 273 K at a pressure of 625 torr. Later the volume of the gas was found to be 1.10 L and the temperature was 298 K. What is the new pressure in atmospheres?

More gas laws

Two important additional gas laws are Dalton's law and Graham's law.

Dalton's law, or more commonly, Dalton's law of partial pressures, deals with mixtures of gases. Each gas in a mixture has its own pressure known as a partial pressure. Dalton's law says that the total pressure of a mixture equals the sum of the partial pressures of the gases in the mixture, or:

$$P_{\text{Total}} = P_A + P_B + P_C + \ldots.$$

In this equation, the letter subscripts (A, B, C,....) refer to the partial pressures of the various components.

EXAMPLE 6

A sample of hydrogen gas is collected over water at 25°C. If the total pressure of the mixture was 765 torr, what is the partial pressure of the hydrogen gas?

To work this problem, it is necessary to know the vapor pressure of water at 25°C. Such values are given in the problems or in tables. The vapor pressure of water at 25°C is 23.7695 torr. We will discuss vapor pressure more in another chapter. For now, a substance will partially vaporize at a given temperature; the amount vaporized is known as the vapor pressure. Vapor pressure is why the water in a glass will evaporate over time.

This problem involves two gases (H_2 and H_2O); therefore, Dalton's law becomes:

$$P_{Total} = P_{H_2} + P_{H_2O}$$

Rearranging this equation to solve for the pressure of hydrogen gives:

$$P_{H_2} = P_{Total} - P_{H_2O}$$

Entering the given values:

$$P_{H_2} = 765 \text{ torr} - 23.7695 \text{ torr} = 741 \text{ torr}$$

Graham's law, sometimes called Graham's law of diffusion or Graham's law of effusion. Diffusion deals with how fast two gases will mix, while effusion deals with how fast a gas will pass through a small hole. The time required for diffusion or effusion to occur depends upon how fast the gas particles are moving. Gas molecules move faster at higher temperatures. In addition, lighter gas molecules move faster than heavier gas particles. In an effusion or a diffusion experiment, both the pressure and temperatures are kept constant. Graham's law is commonly used in the following form:

$$\frac{Rate_1}{Rate_2} = \sqrt{\frac{MM_2}{MM_1}}$$

In this equation, the rates of diffusion or effusion are for gases 1 and 2. The MM are the molar masses of gases 1 and 2. The experiment is set up so that one of the two gases is known (meaning that its MM is known), and the other gas is an unknown gas (MM is unknown). It does not matter if the unknown gas is number 1 or number 2. The experiment to determine the rate is then run twice: once with the known gas and once with the unknown gas. In both cases, the experiment is run at the same temperature and with both gases at the same pressure. One way to determine the rate is to determine what volume of gas has effused or diffused over a certain amount of time.

EXAMPLE 7

In a Graham's law experiment, a sample of methane gas was tested against an unknown gas. When the methane was used, 75 mL of the gas effused in 15 s. For the unknown gas, 220 mL effused in 22 s. What is the molar mass of the unknown gas?

Collecting the information from the problem (and setting gas 1 to be methane, CH_4):

$$V_1 = 220 \text{ mL} \qquad V_2 = 75 \text{ mL}$$
$$T_1 = 22 \text{ s} \qquad T_2 = 15 \text{ s}$$
$$MM_1 = 16 \text{ g mol}^{-1} \qquad MM_2 = ?$$

Next, we need to determine Rate_1 and Rate_2:

$$\text{Rate}_1 = \frac{220 \text{ mL}}{22 \text{ s}} = 10. \text{ mL s}^{-1} \qquad \text{Rate}_2 = \frac{75 \text{ mL}}{15 \text{ s}} = 5.0 \text{ mL s}^{-1}$$

Since we are looking for MM_2, we need to rearrange the Graham's law equation:

$$\frac{\text{Rate}_1}{\text{Rate}_2} = \sqrt{\frac{MM_2}{MM_1}}$$

$$\left(\frac{\text{Rate}_1}{\text{Rate}_2}\right)^2 = \frac{MM_2}{MM_1}$$

$$MM_2 = (MM_1)\left(\frac{\text{Rate}_1}{\text{Rate}_2}\right)^2$$

Finally, we need to enter the values:

$$MM_2 = (16 \text{ g mol}^{-1})\left(\frac{10. \text{ mL s}^{-1}}{5.0 \text{ mL s}^{-1}}\right)^2 = 64 \text{ g mol}^{-1}$$

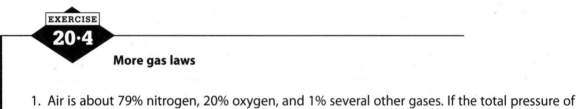

EXERCISE
20·4

More gas laws

1. Air is about 79% nitrogen, 20% oxygen, and 1% several other gases. If the total pressure of an air sample is 1.75 atm, what are the partial pressures of nitrogen and oxygen?

2. In a Graham's law experiment, a sample of chlorine gas was tested against an unknown gas. When the chlorine was used, 16.0 mL of the gas effused in 0.555 min. For the unknown gas, 20.1 mL effused in 1.00 min. What is the molar mass of the unknown gas?

Gas stoichiometry

Any stoichiometry problem may be done by finding the moles from $n = \dfrac{PV}{RT}$. Or the moles from some stoichiometry calculation may be plugged into $PV = nRT$. (Technically, both Avogadro's law and the combined gas law could be used instead of $PV = nRT$; however, neither of these is used very often.)

One of the most common errors is to do a gas law calculation when there is no gas involved. Gases laws work for gases and never for solids or liquids.

EXAMPLE 8

When $KClO_3(s)$ is heated, it decomposes according to the following equation:

$$2 \text{ KClO}_3(s) \rightarrow 2 \text{ KCl}(s) + 3 \text{ O}_2(g)$$

What volume of $O_2(g)$ in liters, at 745.0 torr and 25.00°C, may be formed from 15.00 g of $KClO_3(s)$ (molar mass = 122.548 g mol⁻¹)?

For oxygen, we have:

$V = ?$ L

$P = 745.0$ torr

$T = 25.00°C = 298.15$ K

$n = ?$

mass = 15.00 g $KClO_3(s)$

This table indicates that we will need to use: $V = \dfrac{nRT}{P}$. Unfortunately, there are two unknowns in this equation (V and n); therefore, we need to find one of these first. We still have not used the mass of the $KClO_3(s)$, the mass plus the molar mass will give us moles, albeit not the moles of $O_2(g)$ that we are seeking, so we need to use stoichiometry.

$$\text{Moles } O_2(g) = (15.00 \text{ g } KClO_3)\left(\frac{1 \text{ mol } KClO_3}{122.548 \text{ g } KClO_3}\right)\left(\frac{3 \text{ mol } O_2}{2 \text{ mol } KClO_3}\right) = n$$

Now that we have the moles of O_2, we can enter values into the rearranged ideal gas equation:

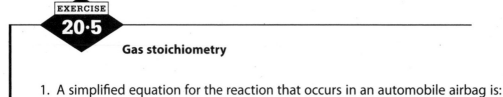

$$V = \frac{nRT}{P} = \frac{\left[(15.00 \text{ g } KClO_3)\left(\dfrac{1 \text{ mol } KClO_3}{122.548 \text{ g } KClO_3}\right)\left(\dfrac{3 \text{ mol } O_2}{2 \text{ mol } KClO_3}\right)\right]\left(0.08206 \dfrac{L \cdot atm}{mol \cdot K}\right)(298.15 \text{ K})}{(745.0 \text{ torr})}\left(\frac{760 \text{ torr}}{1 \text{ atm}}\right)$$

$= 4.582$ L $O_2(g)$

Notice that since all the given values had at least four significant figures, we needed to make any numbers we added to be either exact or have a minimum of four significant figures. These "added" values include the conversion from °C to K and the value of R.

EXERCISE

20·5

Gas stoichiometry

1. A simplified equation for the reaction that occurs in an automobile airbag is:

   $2 NaN_3(s) \rightarrow 2 Na(l) + 3 N_2(g)$

   What volume, in liters, of $N_2(g)$ at 20.0°C and 755.0 torr may be generated from 125.0 g of $NaN_3(s)$? The molar mass of sodium azide, NaN_3, is 65.011 g mol⁻¹.

2. The following reaction may be used to generate $NO_2(g)$:

   $Cu_2S(s) + 8 HNO_3(g) \rightarrow 2 Cu(NO_3)_2(aq) + 4 NO_2(g) + S(s) + 4 H_2O(s)$

   How many liters of $NO_2(g)$ at 298 K and 1.05 atm may be generated from the reaction of 15.2 g of $Cu_2S(s)$ with excess $HNO_3(aq)$? The molar mass of $Cu_2S(s)$ is 159.157 g mol⁻¹.

3. A 30.0% hydrogen peroxide solution has a density of 1.11 g mL⁻¹. When heated, the hydrogen peroxide decomposes according to the following equation:

   $2 H_2O_2(aq) \rightarrow 2 H_2O(l) + O_2(g)$

   What volume of oxygen gas, in liters, may be produced from the decomposition of the hydrogen peroxide in 175 mL of a 30.0% hydrogen peroxide solution at a temperature of 40.0°C under a pressure of 765 mm Hg? The molar mass of hydrogen peroxide is 34.0146 g mol⁻¹. Note, 30.0% means that 30.0% of the mass of the solution is H_2O_2.

4. Carbon dioxide gas may be generated by the following reaction:

$$3\ CaCO_3(s) + 2\ H_3PO_4(aq) \rightarrow Ca_3(PO_4)_2(s) + 3\ H_2O(l) + 3\ CO_2(g)$$

What pressure, in torr, will the $CO_2(g)$ generated by the reaction of a solution containing 0.3000 g of H_3PO_4 produce if it is collected in a 100.0 mL container at 50.25°C? The molar mass of H_3PO_4 is 97.9951 g mol^{-1}.

Kinetic molecular theory

Currently, the best explanation of the behavior of gases is kinetic molecular theory (KMT). Actually, KMT applies not only to gases, but also to solids and liquids. However, the KMT properties of solids and liquids are less obvious than those of gases.

There are various assumptions (postulates) that explain how KMT applies to gases. These are listed in various ways in a variety of orders, so do not be surprised to see variations to this list.

1. The particles in a gas have negligible volume.

2. The particles are in constant motion—colliding with each other and the walls of the container. Pressure is due to the force of the particles colliding with the walls of the container.

3. There are no attractive or repulsive interactions between the particles. This includes during collisions. This means that when the particles collide, they simple "bounce" off each other.

4. The average kinetic energy of the particles depends on the temperature.

The first assumption is that the particles (individual atoms or molecules) are extremely small when compared to the volume of the container. For example, the volume of an oxygen molecule is about 2×10^{-29} m^3, and when this is compared to the volume of a 1 L container (1×10^{-3} m^3), this is indeed negligible.

The second assumption comes from the fact that particles are moving at any temperature above absolute zero.

The third postulate assumes that if the attractive forces were greater, the gas would condense into a liquid.

Finally, we have the basis of Graham's law. The particles will have a range of kinetic energies; however, all gases have the same average at the same temperature. To achieve this average, lighter molecules must move faster than heavier molecules.

It is possible to tie all the gas laws directly to KMT. Ideal gases follow these assumptions exactly. For "real gases," the first and third assumptions break down.

KMT also explains observations such as why water evaporates from a glass at room temperature and why, over time, ice disappears in a freezer.

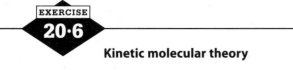

EXERCISE
20·6

Kinetic molecular theory

1. Which of the KMT assumptions helps with understanding Boyle's law? Why?

2. Which of the KMT assumptions helps with understanding Avogadro's law? Why?

Real gases

Gas laws rely on gases behaving ideally (or at least very close). There are two problems with assuming a gas is ideal. First, the gas particles do have volume (not negligible), and second, there are intermolecular forces between the gas particles (not "do not interact"). The first of these becomes more obvious when more gas particles are crammed into a small area. This occurs at high pressures. The second becomes more obvious when the gas molecules are moving more slowly and have time to interact (low temperatures). What high pressure and low temperature mean depends on the identity of the particles. Large particles and molecules with very strong intermolecular forces are the most likely candidates.

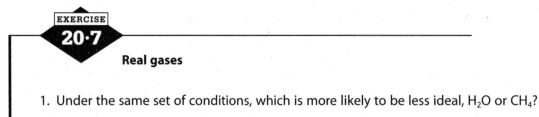

Real gases

1. Under the same set of conditions, which is more likely to be less ideal, H_2O or CH_4?

2. Under the same set of conditions, which is more likely to be less ideal, O_2 or O_3?

Solids and liquids

Solids and liquids both differ from gases because in gases the particles are widely separated with a great deal of space between the particles, while the particles are very close together in both liquids and solids (very little empty space). Solids and liquids may be grouped together and called condensed phases. Since liquid and solid particles are in contact, there are minimal pressure effects. Liquids and gases are similar in that the arrangement of the particles is random, which is why the two phases are grouped together and called fluids. The particles in a solid have a definite repeating arrangement called a lattice. Many of the properties of solids and liquids are strongly influenced by intermolecular forces. (Recall that for an ideal gas, according to kinetic molecular theory [KMT], there are no interactions between the particles.)

Some properties of liquids

Three commonly investigated properties of liquids are surface tension, viscosity, and capillary action.

As the name implies, **surface tension** occurs at the surface of a liquid. It occurs because the particles at the surface of the liquid are "different" from the particles in the bulk of the liquid. A particle in the bulk of the liquid is surrounded by other particles and for this reason is attracted by all the surrounding particles. For the most part, these surrounding attractions counter each other. However, particles at the surface of a liquid have other surface particles surrounding them on the surface and other particles in the bulk of the liquid attracting them. There are no particles above the surface of the liquid to attract them. The net result is that the particles have a net positive attraction in only one direction—down toward the bulk of the liquid. It is as if there is a tightly held "skin" forming the surface of the liquid. This skin is the consequence of surface tension. Surface tension allows certain insects to skitter along the surface of the water in a pond instead of breaking through and sinking. Surface tension allows you to fill a glass of water slightly above the rim of the glass.

Viscosity is the resistance to flow, that is, how easy it is to pour a liquid from one container into another. Let's compare water to, for example, honey. Honey does not pour as easily as water, because honey has a higher viscosity. Viscosity is temperature dependent. Observe how easy a sample of honey flows shortly after it is removed from a refrigerator. Then compare this rate to a sample of honey shortly after its container has been removed from a pan of warm water. The warmer honey pours more easily than the cold honey because warm honey has a lower viscosity. For some liquids the viscosity may be so high that it does not appear to flow at all.

Liquids of this type appear to be solids even though they are not. Such liquids are called amorphous solids. Another name for an amorphous solid is the same as that of the most common example – a glass. Glass and many plastics qualify as amorphous solids. To differentiate between an amorphous solid and a "true" solid, one must examine what happens when the substance is heated, because the viscosity of a liquid decreases on heating. True solids, such as ice, remain solids until reaching the melting point. However amorphous solids become softer as their viscosity decreases.

Unlike the other two properties discussed, **capillary action** depends upon a competition between intermolecular forces. The competition depends upon how the intermolecular forces within the liquid compare to the intermolecular forces between the liquid and the walls of the container. Capillary action is most obvious when most of the liquid particles are close to the walls of the container. This occurs when the liquid is in a narrow tube, a capillary. If the attraction between the liquid and the walls is greater than the attractions within the liquid, the liquid "climbs" the walls of the container. This may be seen in a piece of laboratory equipment called a burette. Water climbs the walls of a glass burette to form a meniscus. A meniscus is a concave upward surface of the liquid. If the glass were replaced with plastic, the attraction between the water and the walls is less than within the liquid and the surface is concave down.

EXERCISE
21·1

Some properties of liquids

1. Which of the following is expected to have the higher surface tension? Why?

   $CCl_4(l)$ $H_2O(l)$ $SO_2(l)$

2. Which of the following is expected to have the higher surface tension? Why? (Note, all have only London dispersion forces.)

   $C_5H_{12}(l)$ $C_{10}H_{22}(l)$ $C_{15}H_{32}(l)$

Phase changes

It is possible to convert one phase to another. Normally, this is done by changing the temperature; however, pressure changes will also work. There are six types of phase changes:

Solid to liquid	Melting (fusion)
Liquid to solid	Freezing (solidification)
Liquid to gas	Vaporization
Gas to liquid	Condensation
Solid to gas	Sublimation
Gas to solid	Deposition

Notice that each of these changes has a reverse. For example, freezing is the opposite of melting, and both occur at the same temperature; which one occurs depends upon whether heat is being added or removed.

Each of these changes has a quantity of energy associated with it. This energy depends upon the identity of the substance and which phase change is being considered. If the particles are atoms or molecules, intermolecular forces are the key; if the particles are ions, ionic bonds are the key. For other substances, covalent bonding or metallic bonding is the key. Note, in general, ionic, covalent, and metallic bonds are stronger than intermolecular forces (i.e., more energy is involved). Three of these changes (fusion, vaporization, and sublimation) require energy, while the remaining three (solidification, condensation, and deposition) release energy. If we consider the energy required for any change, the change that is associated with its reverse is numerically equal (but opposite in sign). Processes that require energy are endothermic (+), and processes that release energy are exothermic (−). Fusion and solidification usually involve less energy than the other four changes, while sublimation and deposition tend to involve more energy than the other four. These energies will be covered in more detail in the chapter on thermochemistry.

EXERCISE

21·2

Phase changes

1. Which of the following is expected to have the highest melting point? Why?
   H₂O(s)   NaCl(s)   N₂(s)

2. Which of the following is expected to have the lowest boiling point? Why?
   He(l)   Ar(l)   Xe(l)

Vapor pressure

If you leave an open glass of water on a table, it will slowly evaporate (evaporation is vaporization at temperatures below the boiling point). What happens here may be explained by KMT. According to KMT, the particles in a substance will have an average kinetic energy depending only on the temperature. As with any other average, some particles will be above average, and some will be below average. For particles within the bulk of the liquid, above/below average molecules cannot do much. However, an above energy particle moving away from the surface of the liquid may escape if its kinetic energy is above the average of a liquid at the same temperature.

If we change the earlier evaporation experiment by putting a lid on the container, the results begin the same way; however, eventually evaporation appears to stop. The lid traps the evaporated molecules inside the container. The trapped particles move around the container like other gas molecules, and when they hit the lid or walls, they bounce off; however, if they hit the surface of the liquid, they will be absorbed back into the liquid. When the lid is first placed on the container, liquid particles evaporate. The number of particles in the gas phase continues to increase while evaporation is occurring (minus the few that are reabsorbed by the liquid). Eventually there will be sufficient particles in the gas phase so that when one particle evaporates, another particle is reabsorbed. Then there is a balance, which makes it appear as if evaporation has stopped. When this balanced is achieved, the system is said to be in **equilibrium**. At equilibrium, the rate of evaporation equals the rate of condensation. However, both processes are continuing. The partial pressure of the liquid particles in the gas phase is the vapor pressure of the liquid. The vapor pressure at a given temperature will depend upon the nature of the liquid. Vapor pressures increase with temperature.

A liquid will boil whenever its vapor pressure equals the external pressure. If the vapor pressure equals 1 atm, then it is the normal boiling point.

The same behavior is observed for solids (sublimation and deposition); however, the vapor pressure for solids is normally much, much less than for liquids. This explains why an ice cube will slowly disappear while being stored in a freezer.

We will discuss vapor pressure in more detail in the chapter on solutions. Equilibrium will be discussed in more detail in several other chapters.

EXERCISE
21·3

Vapor pressure

1. Why does a pressure cooker cook food faster than an open container?

2. Why does a liquid evaporate faster at a higher temperature?

3. If you tip over a glass of water, it will evaporate faster. Why?

4. How will the temperature of the water change if you tip over a glass of water?

Phase diagrams

A phase diagram summarizes the phases present under a certain set of conditions. Such diagrams range from simple to complex. We will focus on phase diagrams involving a single substance. The general appearance of these diagrams follows:

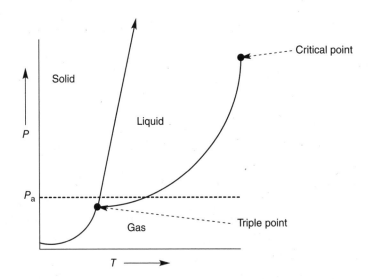

The numbers, length, and slope of the lines vary with the identity of the substance. The lines associated with the gas phase tend to be curved because gases, unlike solids and liquids, are more dependent on the pressure. The liquid-gas line stops a point known as the critical point. The lines come together at a point known as the triple point. In most one-component systems, the solid liquid line has a positive slope as seen here. Water is one of the few exceptions to this in that it has a negative slope. Complications appear in phase diagrams such as the diagram for carbon, which has two solid phases (graphite and diamond). All the points on the lines on the phase diagram represent an equilibrium at a particular temperature and pressure.

If we pick an arbitrary pressure, P_a, and start at a very low temperature (in the solid region) and begin heating, the system moves along the dotted line until it reaches the solid-liquid line where the solid melts (melting point). Once all the solid has melted, continued heating moves the system further to the right along the dotted line until the liquid-gas line is reached where the liquid vaporizes (boiling point). Once all the liquid has vaporized, the system continues along the dotted line. On the other hand, if the temperature were at the opposite end of the dotted line, and the system were cooled, the system would again move along the dotted line in the opposite direction, with the same phase changes occurring in reverse order. If P_a were equal to 1 atm, then the melting point and boiling point would be the normal melting point and the normal boiling point respectively. As P_a is varied, the melting and boiling points will vary accordingly. Note, if P_a were below the triple point, there is no liquid phase, and the phase transition would either be sublimation or deposition.

If the temperature were fixed, and the pressure were increased or decreased, the movement on the phase diagram would be vertical instead of horizontal, with the appropriate phase changes upon reaching one of the lines.

The **triple point** is a fixed point in a phase diagram when three phases are in equilibrium. The three phases are not necessarily solid, liquid, and gas. For example, in the phase diagram of carbon, there is a triple point relating graphite, diamond, and liquid carbon.

The critical point is at the high end of the liquid-gas line. The **critical point** is the highest temperature and pressure for which it is possible to distinguish between the liquid and gas phases. Recall that both liquids and gas have randomly arranged particles, and the only difference is that in a liquid, the particles are in contact, while in a gas they are widely separated. However, as the pressure on a gas is increased, the distance between the particles will decrease until the gas particles are in contact. Once the gas particles are in contact, there is no difference between a liquid and a gas. The critical point occurs at the critical pressure and critical temperature.

EXERCISE

21·4

Phase diagrams

1. What is the highest temperature where sublimation can occur?

2. The triple point for carbon dioxide is at −56.6°C and 5.1 atm. Solid carbon dioxide is also known as "dry ice," which may be purchased at some local grocery stores. Why do you think that solid carbon dioxide is also known as dry ice?

3. As mentioned in the discussion, the solid liquid line for water has a negative slope instead of the positive slope shown in the diagram. The slope in the diagram is the same as is seen in the phase diagram of CO_2. If you start with a solid sample of H_2O (ice) and CO_2 (dry ice), which is at the same pressure as the triple point for the respective substance and at a temperature slightly below that of the triple point, what will happen to each of the substances if the pressure is slowly increased while holding the temperature constant?

Types of solids

There are four basic types of solids. Each type depends upon the types of particles present along with the type of bonding. These four types are metallic solids, ionic solids, covalent (or covalent network solids), and molecular solids.

Metallic solids are solids containing only metal or metal plus metalloid atoms. Metallic solids are held together by metallic bonds.

Ionic solids normally involve a combination of metal ions and nonmetal ions, or polyatomic ions. Ionic solids are held together by ionic bonds. Note the interior bonds (covalent) of a polyatomic ion do not hold the solid together.

Covalent solids (or **covalent network solids**) are held together by covalent bonds. These are much less common than the other types of solids. Just because there are covalent bonds present does not make something a covalent solid. For example, H_2O contains covalent bonds; however, these bonds do not hold the water molecules together in ice. The water molecules in ice are held together by hydrogen bonds. Some examples of the few covalent solids are carbon (graphite and diamond) and silicon dioxide.

Molecular solids are held together by van der Waals forces. Recall that van der Waals forces include London dispersion forces, dipole-dipole forces, and hydrogen bonds. Molecular solids contain individual molecules.

The properties of the solids depend upon what holds them together (though in all cases, there are exceptions). Metallic solids are the only ones that conduct electricity; however, ionic solids will conduct electricity if they are first melted or if they are dissolved in water. Metallic solids have a typical shiny metallic appearance (as do a few metalloids). Metallic solids have the greatest variation in melting points ranging from very low (mercury –38.8°C) to very high (tungsten 3,422°C).

Molecular solids have the weakest forces holding them together (van der Waals); therefore, they have low melting points and are soft. A few will dissolve in water, but most are only soluble in a nonpolar solvent.

Covalent solids tend to be hard and have high melting points. Graphite is an exception as it is soft. Graphite is one of the few covalent solids that conducts electricity.

Ionic solids tend to be dull in appearance and have high melting points, and if they dissolve, they will dissolve in water. The smaller the ions and/or the greater the ionic charges, the stronger the ionic bonds, which leads to higher melting points and so on.

EXERCISE

21·5

Types of solids

1. Which of the following probably has the highest melting point? Why?

    CsI(s) SiO_2(s) CH_3OH(s)

2. Which of the following solids is most likely to conduct electricity if melted? Why?

    H_2O(s) CH_3OH(s)   NaCl(s)

3. A black crystalline solid has a high melting point. The solid does not conduct electricity; however, the molten material does. The substance is most likely to be which type of solid?

Structures of solids

True solids (crystalline solids), as opposed to amorphous solids, have an ordered arrangement of particles. This ordered arrangement is continued in three dimensions throughout the solid. This three-dimensional arrangement is known as a **crystal lattice,** or simply **lattice**. The following diagram gives a two-dimensional portion of a crystal. Each of the "dots" in the figure represents whatever the solid consists of. For example, the dots could represent individual atoms, groups of atoms, molecules, groups of molecules, ions, groups of ions, or whatever. To describe the lattice, it is necessary to describe the location of every lattice point. Unfortunately, even a small piece of a solid has close to Avogadro's number of lattice points, which would make describing each location an insurmountable challenge. To simplify the task of describing the lattice, it helps to take advantage of the repeating nature of the lattice. To do this, one describes the **unit cell**, which is the smallest repeating unit in the lattice. The number of particles in a unit cell is small enough to describe exactly. Consider two possible two-dimensional unit cells. Either of these, or others, may be used to construct the entire lattice by simple stacking identical unit cells together. In this case, the dotted unit cell is preferred since it is smaller. In three dimensions, unit cells are analogous to the bricks in a brick wall.

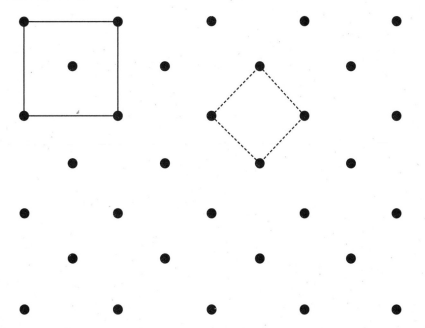

To describe a three-dimensional unit cell, it is necessary to know the length of the unit cell in each of the three dimensions; in addition, it is necessary to know the angles between the axes (which may or may not be 90°). There are six basic types of unit cells (seven in some classifications, as one of the six types may be divided into two). The simplest type is the cubic unit cell. There are three subtypes of cubic unit cells, all of which are cubes (all sides identical and all angles 90°). Other types of unit cells vary the length of one or more sides and/or one or more of the angles. The three types of cubic unit cells are as illustrated in the following figure:

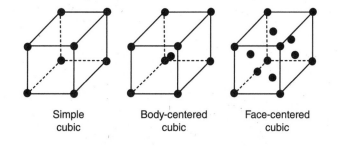

Simple cubic   Body-centered cubic   Face-centered cubic

In all cases, there are eight lattice points at the corners of the cubic unit cell. Only these eight points are present in a simple cubic unit cell. A body-centered unit cell has an additional lattice point in the very center of the unit cell. A face-centered unit cell has the corner points plus a lattice point at the center of each face. Do not forget, each of these lattice points must be absolutely identical. In addition, each lattice point may represent a wide variety of particles from atoms up to and including groups of molecules.

EXAMPLE 1

Assuming each lattice point in a simple cubic unit cell represents an atom, how many lattice points (atoms) are in the unit cell? One form of polonium adopts this type of structure.

The diagram of the simple cubic unit cell is misleading as it shows only part of the lattice. In the overall lattice, the corners are shared with the neighboring unit cells. A corner atom is at the corner of eight neighboring unit cells; therefore, only one-eighth of the corner atom is within the unit cell shown. Since there are eight corners, the total number of atoms present is (8 corners) × (1/8 atom per corner) = 1 atom. Note that if the corners are not 90°, the average contribution is usually 1/8.

EXAMPLE 2

Assuming each lattice point in a body-centered cubic unit cell represents an atom, how many lattice points (atoms) are in the unit cell? Many metals, such as sodium, adopt this type of structure.

The body-centered unit cell has the same type of corner atoms as a simple cubic unit cell; this accounts for one atom. Since there is an additional lattice point (atom) in the center of the unit cell, it counts as being entirely within the unit cell. This makes two total atoms in the unit cell.

EXAMPLE 3

Assuming each lattice point in a face-centered cubic unit cell represents an atom, how many lattice points (atoms) are in the unit cell? Many metals, such as aluminum, adopt this type of structure.

Just like a body-centered and cubic unit cell, there are eight points at the corners contributing one atom. The additional lattice points at the centers of each face are half in the cell shown and half in the neighboring unit cell. The lattice points at the centers of the faces contribute (6 faces) × (1/2 atom per face) = 3 atoms. The total in a face-centered cubic cell is four atoms.

There are additional ways of describing unit cells; however, they will not be discussed here.

Structures of solids

1. The following diagram represents a unit cell of CsCl. How many CsCl formula units are in the unit cell?

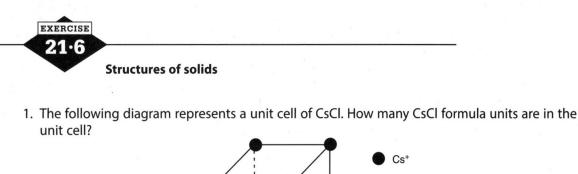

2. One form of phosphorus consists of P_4 molecules as the lattice points. This form is reported to be body-centered cubic. How many total phosphorus atoms are in each unit cell?

Metallic bonding

Most metals are body-center cubic, face-centered cubic (also known as cubic closest packed), or hexagonal closest packed. Such a small number of choices indicates that for metal atoms (metallic bonding), these structures must be particularly stable.

The best description of metallic bonding is in pure metals, where a combination of metal atoms, and a combination of metal and metalloid atoms are known as band theory. Unfortunately, band theory requires molecular orbital theory for its explanation. Molecular orbital theory is beyond the level of this book and will not be discussed here.

There is an older theory of metallic bonding involving what is known as the electron-sea or electron-gas model. According to this model, the metal atoms are present as ions consisting of the atomic nucleus and the core electrons. The valence electrons form a "sea" or "cloud" around these metal cores. The electrons are freely moving throughout the lattice, which facilitates the conductivity of electricity and thermal energy in solids. Since the metal cores are in a sea of electrons, no electrons are associated with any particular atom, this means that the structure may be distorted without breaking any bonds. The possible types of distortion include malleability and ductility. **Malleability** is the ability of a piece of metal to be flattened by using, for example, a hammer. **Ductility** is the ability of a piece of metal to be drawn to a wire.

Thermochemistry

All chemical processes involve energy. The quantity of energy depends upon the process and the type of energy. Energy is governed by the law of conservation of energy. This law states that "the total amount of energy in the universe is constant," or "Energy may change from one form to another, but the amount of energy does not change." Later, we will learn that this law is also known as the first law of thermodynamics. As the first law, it is one of three laws.

Energy

Energy is the capacity or ability to do work. This work may be the energy from gasoline or a battery moving a car to heat energy, making the molecules of air move faster as the temperature increases.

There are two general types of energy, each with many subtypes. **Kinetic energy** is the energy of motion, that is, energy is doing work. **Potential energy** is stored energy, that is the capacity to do work later.

There are a number of ways to measure energy. One example is calories. A **calorie** was originally defined as the energy required to warm exactly 1 g of pure water from 14.5°C to 15.5°C (though other temperature changes are sometimes used). Today it is defined as being exactly equal to 4.184 J. This is also known as the thermochemical calorie. Another energy is the Calorie with 1 Calorie = 1,000 calories. This **Calorie** (capitalized) is the nutritional calorie used to express the energy content of food. There are other units of energy, mostly used for specialized purposes, except for one, which is the joule. The **joule** is the SI unit of energy defined as:

$$E = \tfrac{1}{2}mv^2$$

In this definition, technically E is the kinetic energy, m is the mass of the object, and v is the velocity of the object, all measured in SI base units as:

$$E = \tfrac{1}{2}mv^2 = \frac{1}{2}(\text{kg})\left(\frac{\text{m}}{\text{s}}\right)^2 = \text{kg m}^2 \text{ s}^{-2} = \text{joules}$$

Do not forget, when using this relationship, mass must be in kilograms and velocity must be in meters per second.

Precise energy measurements are made with respect to the system, which is called a calorimeter. The actual measurement is how much energy enters or leaves the system from the surroundings. (More on calorimetry later.)

In some cases, it is important to know how much energy is involved in a process, while in other cases, it is more important to know if energy is being lost or gained, and in the remainder of the cases both of these need to be known. The terms *endothermic* and *exothermic* are used to describe the direction of energy flow (these are subsets of *endergonic* and *exergonic*, respectively, which we will discuss in another chapter).

An **endothermic** process is one that involves energy entering the system. Endothermic changes, with respect to the system, are positive (+). An **exothermic** process is one that involves energy leaving the system. Exothermic changes, with respect to the system, are negative (−). A fire releases energy, so you get warm setting next to a fire. Cold water absorbs energy, so you will get cold if you jump into a pool of cold water.

EXAMPLE 1

To date, the record speed for a downhill skier was set in 2006 by Klaus Kroell of Austria. During his run his speed was 96.6 mph. Assuming he weighed 91 kg, how much kinetic energy, in joules, was he expending?

We need to use the equation $E = \frac{1}{2}mv^2$. In this case, $m = 91$ kg and $v = 96.6$ mph. Entering these values into this equation gives:

$$E = \frac{1}{2}mv^2 = \frac{1}{2}(91\,\text{kg})\left(\frac{96.6\,\text{mi}}{\text{h}}\right)^2 =$$

This uses the given information; however, the units are incorrect. We need to convert miles to meters and hours to seconds. These may be done in either order. We will begin with miles. There are a number of acceptable conversions we could begin with, and we will choose 1,609 m = 1 mi as it is the simplest:

$$E = \frac{1}{2}mv^2 = \frac{1}{2}(91\,\text{kg})\left[\left(\frac{96.6\,\cancel{\text{mi}}}{\text{h}}\right)\left(\frac{1,609\,\text{m}}{1\,\cancel{\text{mi}}}\right)\right]^2$$

Now we will incorporate the hour conversion. As with the mile conversion, there is more than one way to make the conversion. For example, we could convert hours to minutes (60 min h^{-1}) and then convert the minutes to seconds (60 s min^{-1}) and get the correct answer, or we could use 3,600 s = 1 hr. We will use the later conversion because it is faster:

$$E = \frac{1}{2}mv^2 = \frac{1}{2}(91\,\text{kg})\left[\left(\frac{96.6\,\cancel{\text{mi}}}{\cancel{\text{h}}}\right)\left(\frac{1,609\,\text{m}}{1\,\cancel{\text{mi}}}\right)\left(\frac{1\,\cancel{\text{h}}}{3,600\,\text{s}}\right)\right]^2 =$$

To finish the problem, we will use the definition of a joule (= kg m² s⁻²) to finish the unit conversion:

$$E = \frac{1}{2}mv^2 = \frac{1}{2}(91\,\cancel{\text{kg}})\left[\left(\frac{96.6\,\cancel{\text{mi}}}{\cancel{\text{h}}}\right)\left(\frac{1,609\,\cancel{\text{m}}}{1\,\cancel{\text{mi}}}\right)\left(\frac{1\,\cancel{\text{h}}}{3,600\,\cancel{\text{s}}}\right)\right]^2\left(\frac{1\,\text{J}}{\frac{\cancel{\text{kg}}\,\cancel{\text{m}}^{-2}}{\cancel{\text{s}}^{-2}}}\right) = 8.5 \times 10^4\,\text{J}$$

A common error is to forget that all the units and numbers inside the square brackets are squared. (Note, the problem called for an answer in joules, not kilojoules.)

The final discussion in this section deals with energy entering or leaving the system. This change of energy is labeled ΔE and:

$$\Delta E = q + w$$

In this equation, q is the amount of heat involved and w is the amount of work involved. Heat (q) lost from the system is exothermic (−), and heat entering the system is endothermic (+). Work (w) done by the system is negative (−) and work done on the system is positive (+). Note that in both cases, what happens to the system is the key.

EXERCISE 22·1

Energy

1. Many different types of rounds may be used in a 30-06 rifle. One type is a 150 grain (10.0 g) round, which has an exit velocity of up to 2,910 ft/s. What is the energy of this type of bullet, in joules, as it exits the rifle?

2. A balloon (system) is warmed by 155 J, and it does 72 J of work as it expands. What is the change in energy (ΔE) of this system?

3. A balloon (system) is cooled by 155 J, and it has 72 J of work done on it as it contracts. What is the change in energy (ΔE) of this system?

Heat

At times, it is important to calculate the heat involved in a process. One way to do this is by a simple rearrangement of $\Delta E = q + w$ to $q = \Delta E - w$. However, a more commonly used method is to use the following equation:

$$q = mC\Delta T$$

Again, q is the heat, m is the mass, C is the heat capacity (specific heat), and ΔT is the change in temperature. There are many forms of C depending upon how and what are to be measured. Here, we will use the form at constant pressure, often written as C_P instead of just C. The typical units of C are the amount of energy required to warm one gram of a substance by one degree (either °C or K). If the other terms use different units, it will be necessary to make one or more conversions.

The following example points out one of the common errors when working this type of problem to determine ΔT.

EXAMPLE 2

Determine the temperature change (ΔT) when a sample of water is heated from 0°C to 25°C. Then determine the temperature change in Kelvin.

The change of temperature comes from the following equation:
$$\Delta T = T_{final} - T_{initial} = T_f - T_i$$

This gives us (in °C): $\Delta T = 25°C - 0°C \quad = 25°C$

This gives us (in K): $\Delta T = 298\ K - 273\ K = 25\ K$

It is important to note that no matter if the calculation is in °C or K, the numerical value of the change (ΔT) in either system is identical. Do not forget, this only works for ΔT.

EXAMPLE 3

What is the energy change that occurs when a 25.0 g sample of sand is heated from 25.0°C to 35.0°C? The specific heat of the sand is 0.830 J g^{-1} °C^{-1}.

We will begin by summarizing the information from the problem:

$$q = ? \quad m = 25.0\text{ g} \quad T_i = 25.0\text{ °C} \quad T_f = 35.0\text{ °C} \quad C = 0.830\text{ J g}^{-1}\text{ °C}^{-1}$$

Entering these values into $q = mC\Delta T$ gives:

$$q = mC\Delta T = (25.0\text{ g})(0.830\text{ J g}^{-1}\text{ °C}^{-1})(35.0 - 25.0)\text{ °C} = 208\text{ J}$$

No further unit conversions are necessary. The answer has the correct units and significant figures. Notice, that since the sample was heated, energy was absorbed (+).

EXAMPLE 4

What is the energy change that occurs when a 0.375 kg sample of sand is cooled from 335 K to 275 K? The specific heat of the sand is 0.830 J g^{-1} °C^{-1}.

We will begin by summarizing the information from the problem:

$$q = ? \quad m = 0.375\text{ kg} \quad T_i = 335\text{ K} \quad T_f = 275\text{ K} \quad C = 0.830\text{ J g}^{-1}\text{ °C}^{-1}$$

Entering these values into $q = mC\Delta T$ gives:

$$q = mC\Delta T = (0.375\text{ kg})(0.830\text{ J g}^{-1}\text{ °C}^{-1})(335 - 275)\text{ K}$$

In this case, there are some conversions necessary:

$$q = mC\Delta T = (0.375\text{ kg})(0.830\text{ J g}^{-1}\text{ °C}^{-1})[(335 - 275)\text{ K}]\left(\frac{1{,}000\text{ g}}{1\text{ kg}}\right)\left(\frac{\text{°C}}{\text{K}}\right)$$

$$= -1.87 \times 10^4\text{ J}$$

The conversions leave only joules as required. The answer has the correct units and significant figures. Notice, that since the sample was cooled, energy was lost (−).

In the relationship, $\Delta E = q + w$, if the change takes place under constant pressure, q becomes ΔH. H is the enthalpy. The **enthalpy** is the heat content of the system. We will see enthalpy and enthalpy changes many places in this chapter, and again in later chapters. We will investigate some ways of determining enthalpy later in this chapter.

EXERCISE
22·2

Heat

1. What is the energy change that occurs when a 0.375 kg sample of water is cooled from 335.0 K to 275.0 K? The specific heat of the water is 4.184 J g^{-1} °C^{-1}.

2. What is the temperature (K) change when a 0.375 kg sample of water is heated by 425 J? The specific heat of the water is 4.184 J g^{-1} °C^{-1}.

3. What is the mass (g) of iron that may be heated from 285.0 K to 305.0 K by 475 J? The specific heat of the iron is 0.462 J g^{-1} °C^{-1}.

4. What is the heat capacity of a sample of metal heated from 15.0°C to 27.0°C when 475 J are added to 35.0 g of metal?

Heats of reaction

Every chemical reaction has a certain quantity of energy associated with it. This energy is known as the heat of reaction. A more important name for the heat of reaction is the standard heat of reaction, which is the energy change associated with a reaction under a specific set of conditions. The **standard heat of reaction** is the energy change that occurs when all reactants and products are in their standard states (1 atm, and normally 25°C, 298 K) in terms of moles. This is also known as the standard enthalpy of reaction, $\Delta H°_{rxn}$ (note the addition of the degree symbol). Values for standard heats of reaction may be determined with a calorimeter (see the next section).

The chemical equation for the standard heat of reaction, a thermochemical equation, is different from the equation for other reactions in three ways. First, the equation must include information about the energy involved. Second, the reactants and products must be under "standard" conditions. This means that the reactants and products are in their standard states. The standard state is the most stable form of a substance under one atmosphere pressure at some specified temperature (normally 25°C). The second difference is that the reactions are always in terms of moles. This may not seem like much; however, let us continue the reaction:

$$2 H_2(g) + O_2(g) \rightarrow 2 H_2O(l)$$

In a general chemical equation, this reaction could mean 2 molecules H_2 will react with 1 molecule of O_2 to produce 2 molecules of H_2O, or 2 dozen molecules H_2 will react with 1 dozen molecules of O_2 to produce 2 dozen molecules of H_2O, or 2 mol H_2 will react with 1 mol O_2 to produce 2 mol H_2O, or a number of additional alternatives. However, for a thermochemical equation, only moles are acceptable. Finally, unlike "normal" chemical equations, the restriction to moles allows fractional coefficients. For example, in the water reaction above, a normal chemical equation can never involve a half oxygen molecule, but it is possible to have a half mole of oxygen.

The following are examples of correct thermochemical equations:

$$2 H_2(g) + O_2(g) \rightarrow 2 H_2O(l) \qquad \Delta H° = -484 \text{ kJ}$$

$$CH_4(g) + 2 O_2(g) \rightarrow CO_2(g) + 2 H_2O(l) \qquad \Delta H° = -891 \text{ kJ}$$

$$HCl(aq) + NaOH(aq) \rightarrow NaCl(aq) + H_2O(l) \qquad \Delta H° = -54 \text{ kJ}$$

$$2 CH_3OH(g) \rightarrow 2 CH_4(g) + O_2(g) \qquad \Delta H° = 253 \text{ kJ}$$

Note even though the units are given as kJ, for the first reaction the units are understood to be:

$$\frac{-484 \text{ kJ}}{2 \text{ mol } H_2} = \frac{-484 \text{ kJ}}{1 \text{ mol } O_2} = \frac{-484 \text{ kJ}}{2 \text{ mol } H_2O}$$

Here are some important properties of thermochemical equations (standard heats of reaction)

1. You can multiply or divide an equation by any value as long as all coefficients are treated the same way, and this includes the $\Delta H°$.

2. You can reverse a thermochemical equation as long as you reverse the sign on $\Delta H°$.

3. You can add any two or more thermochemical equations to produce a new thermochemical equation.

4. Any or all of these may be combined.

EXAMPLE 5

These are some thermochemical equations derived from the previous equations; determine what has been done in each case:

$$4\,H_2(g) + 2\,O_2(g) \rightarrow 4\,H_2O(l) \qquad \Delta H° = -968\text{ kJ}$$

Everything multiplied by 2.

$$0.5\,CH_4(g) + 1\,O_2(g) \rightarrow 0.5\,CO_2(g) + 1\,H_2O(l) \qquad \Delta H° = -446\text{ kJ}$$

Everything divided by 2.

$$3\,NaCl(aq) + 3\,H_2O(l) \rightarrow 3\,HCl(aq) + 3\,NaOH(aq) \qquad \Delta H° = +162\text{ kJ}$$

Reaction is reversed, $\Delta H°$ sign is reversed, everything is multiplied by 3.

$$4\,CH_3OH(g) \rightarrow 4\,CH_4(g) + 2\,O_2(g) \qquad \Delta H° = -506\text{ kJ}$$

Everything is multiplied by 2.

EXAMPLE 6

What is the energy change (in kJ) when 3.00 mol of $O_2(g)$ are reacted with excess $H_2(g)$ according to the following equation?

$$2\,H_2(g) + O_2(g) \rightarrow 2\,H_2O(l) \qquad \Delta H° = -484\text{ kJ}$$

Adding the given information to the chemical equation:

$$2\,H_2(g) + O_2(g) \rightarrow 2\,H_2O(l) \qquad \Delta H° = -484\text{ kJ}$$
$$\text{—} \qquad 3.00\text{ mol} \qquad \Delta H = ?$$

This is a one-step conversion problem:

$$\left(3.00 \; \cancel{\text{mol } O_2}\right)\left(\frac{-484\text{ kJ}}{1\;\cancel{\text{mol } O_2}}\right) = -1{,}450\text{ kJ}$$

We have the correct units and significant figures.

EXAMPLE 7

What is the energy change (in kJ) when 3.50 mol of $O_2(g)$ are reacted with 5.00 mol $H_2(g)$ according to the following equation?

$$2\,H_2(g) + O_2(g) \rightarrow 2\,H_2O(l) \qquad \Delta H° = -484\text{ kJ}$$

Adding the given information to the chemical equation:

$$2\,H_2(g) + O_2(g) \rightarrow 2\,H_2O(l) \qquad \Delta H° = -484\ kJ$$
$$5.00\ mol \quad 3.50\ mol \qquad\qquad \Delta H = ?$$

Since the amounts of two different reactants are given, we must first determine the limiting reagent. As always, this is done by dividing the mole of a substance by the coefficient of that substance in the balanced chemical equation.

$$\left(\frac{5.00\ mol\ H_2}{2}\right) = 2.50\ \text{(limiting reagent)}$$

$$\left(\frac{3.50\ mol\ O_2}{1}\right) = 3.50$$

Finishing the problem with the limiting reagent:

$$(5.00\ \cancel{mol\ H_2})\left(\frac{-484\ kJ}{2\ \cancel{mol\ H_2}}\right) = -1{,}210\ kJ$$

Notice that the second parentheses incorporates the coefficient of H_2 from the balanced chemical equation.

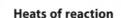

EXERCISE

22·3

Heats of reaction

1. What is the energy change (in kJ) when 3.85 mol of $CH_3OH(g)$ are decomposed according to the following equation?

   $$2\,CH_3OH(g) \rightarrow 2\,CH_4(g) + O_2(g) \qquad \Delta H° = 253\ kJ$$

2. What is the energy change (in kJ) when 7.00 mol of $O_2(g)$ are reacted with 3.75 $CH_4(g)$ according to the following equation?

   $$CH_4(g) + 2\,O_2(g) \rightarrow CO_2(g) + 2\,H_2O(l) \qquad \Delta H° = -891\ kJ$$

3. Consider the following thermochemical equation:

   $$H_2O(l) \rightarrow H_2O(g) \qquad \Delta H° = 44\ kJ$$

   What is the energy change when 3.5 mol of water vapor condense to liquid water?

Calorimetry

Calorimetry is the experimental method used to calculate energy changes involved in various processes. These processes include the heats of reaction. The equipment necessary ranges from a simple polystyrene cup (a coffee-cup calorimeter) with a thermometer through the lid to a complicated insulated steel vessel (a bomb calorimeter).

Whether simple or complicated equipment is used, the basic idea is to measure and add some reactants to the calorimeter, initiate the change, and measure the temperature change. Through other means, the heat capacities need to be known (often given in the problem, in accompanying tables, or by calibrating the calorimeter through some standard reaction). Every attempt is made to ensure that the measurements are done as precisely as possible.

The following example should let you see what is involved in a calorimetry problem. As with many "long" problems, there are a series of small simple steps not one gigantic challenging problem.

EXAMPLE 8

A sample of ethanol, $C_2H_5OH(l)$, weighing 2.130 g, was burned with excess $O_2(g)$ in a calorimeter. After completion of the reaction, the temperature of the calorimeter changed from 24.00°C to 30.54°C. The calorimeter also contained 0.754 kg of water. The heat capacity of the calorimeter was 11.31 kJ °C^{-1}. The specific heat of water is 4.184 J g^{-1} °C^{-1}. What was the heat of reaction in kJ mol^{-1} of C_2H_5OH? The reaction was:

$$C_2H_5OH(l) + 3\ O_2(g) \rightarrow 2\ CO_2(g) + 3\ H_2O(l)$$

As normal, we begin by adding the given information to the balanced chemical equation:

$$C_2H_5OH(l) + 3\ O_2(g) \qquad \rightarrow \qquad 2\ CO_2(g) + 3\ H_2O(l)$$

2.130 g

$T_i = 24.00°C$

$T_f = 30.54°C$

Mass $H_2O = 0.750$ kg

$C_{calorimeter} = 11.31$ kJ °C^{-1}

Specific heat of water $= 4.184$ J g^{-1} °C^{-1}.

$\Delta H = ?$

The numbers beneath the reaction arrow apply to the entire reaction.

To the given information, you need to calculate and add the molar mass of ethanol: 46.069 g mol^{-1}.

Some things to keep in mind:

1. Both the water originally in the calorimeter itself and the calorimeter are heated. (It is impossible to heat a pan full of water on the stove and heat only the pan or the water.)
2. Due to 1, it is necessary to apply the temperature change to both the calorimeter and the water.
3. It is necessary to combine the specific heat of water and the mass of water in the calorimeter.
4. It is necessary to convert the mass of ethanol to moles.

The calculation involves finding the energy and finding the moles. In the following calculation, the numerator is the energy, and the denominator is the moles. The energy is divided into two portions: the heat going to the calorimeter (first in the numerator) plus the heat going to the water (second in the numerator). To find the moles (denominator), multiply the mass by the inverse of the molar mass. Convert the final answer to negative:

$$\frac{\left[\left(\dfrac{11.31\,kJ}{°C}\right)(30.54-24.00)\,°C\right] + \left[\left(\dfrac{4.184\,J}{g°C}\right)\left(\dfrac{1{,}000\,g}{1\,kg}\right)\left(\dfrac{1\,kJ}{1{,}000\,J}\right)(0.500\,kg)(30.54-24.00)\,°C\right]}{\left(2.130\,g\,C_2H_5OH\right)\left(\dfrac{1\,mol\,C_2H_5OH}{46.069\,g\,C_2H_5OH}\right)}$$

$$= -1.90 \times 10^3\ kJ\ mol^{-1}$$

In the numerator, the first brackets enclose the energy absorbed by the calorimeter, while the second brackets enclose the energy absorbed by the water in the calorimeter. The plus sign between the two brackets shows that the two energies are to be added. The denominator contains the mole calculation. Note that the heat went from the calorimeter plus water out to the thermometer, which measured the temperature. For this reason, the process is exothermic, meaning that the heat of reaction *must* be negative.

The units and the significant figures are correct. This (burning) is an exothermic process; therefore, the final answer is negative.

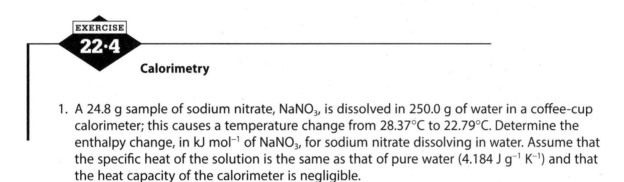

EXERCISE 22·4

Calorimetry

1. A 24.8 g sample of sodium nitrate, $NaNO_3$, is dissolved in 250.0 g of water in a coffee-cup calorimeter; this causes a temperature change from 28.37°C to 22.79°C. Determine the enthalpy change, in kJ mol^{-1} of $NaNO_3$, for sodium nitrate dissolving in water. Assume that the specific heat of the solution is the same as that of pure water (4.184 J g^{-1} K^{-1}) and that the heat capacity of the calorimeter is negligible.

2. A 1.5886 g sample of glucose, $C_6H_{12}O_6$, was combusted in a bomb calorimeter. The temperature increased from 20.000°C to 23.682°C. The heat capacity of the calorimeter was 3.562 kJ/°C, and the calorimeter contained 1.000 kg of water (4.184 J g^{-1} °C^{-1}). Find the molar heat of reaction for the reaction

$$C_6H_{12}O_6(s) + 6 O_2(g) \rightarrow 6 CO_2(g) + 6 H_2O(l)$$

3. Naphthalene is used in mothballs. In an experiment to measure naphthalene's heat of combustion, a 2.870 g sample was burned with an excess of O_2 and in a bomb calorimeter. After the reaction, the temperature of the calorimeter had increased from 23.00°C to 34.34°C. The calorimeter contained 2.000 kg of water (4.184 J g^{-1} °C^{-1}). The heat capacity of the empty calorimeter was 1.80 kJ/°C. Determine the heat of reaction in kJ mol^{-1} of naphthalene for the reaction

$$C_{10}H_8(s) + 12 O_2(g) \rightarrow 10 CO_2(g) + 4 H_2O(l)$$

Hess's law

It is important to be able to predict how much energy is involved in a reaction. There are three general ways to do this. One method is to use bonds energies, which is covered in another chapter. A second method is to use enthalpies of formation, which is covered in the next section. Finally, there is Hess's law (or Hess' law). **Hess's law** combines thermochemical equations to produce new thermochemical equations. It is necessary to apply Hess's law if the thermochemical equation of concern has never been performed for some experimental reason. In many cases, the reason why the thermochemical equation of concern has not been done is because as written the reaction will not take place in a calorimeter. For example, the reaction of sodium with nitrogen to form sodium azide:

$$6 Na(s) + N_2(g) \rightarrow 2 NaN_3(s)$$

This reaction will not proceed as written; however, the reverse reaction will occur as it does in an automobile airbag. Therefore, simply running the reverse reaction in a bomb calorimeter will give the following thermochemical equation:

$$2\,NaN_3(s) \rightarrow 6\,Na(s) + N_2(g) \qquad \Delta H° = +21.3 \text{ kJ mol}^{-1}$$

Simply reversing this thermochemical equation gives the equation being sought.

Hess's law works like the sodium azide example except that this law normally involves two or more thermochemical equations.

EXAMPLE 9

A student wishes to determine the enthalpy change for the following reaction:
$$CH_4(g) + 2\,O_2(g) \rightarrow CO_2(g) + 2\,H_2O(l) \qquad \Delta H° = ?$$

The student is given the following two thermochemical equations:
$$CH_4(g) + 2\,O_2(g) \rightarrow CO_2(g) + 2\,H_2O(g) \qquad \Delta H° = -799 \text{ kJ}$$
$$H_2O(l) \rightarrow H_2O(g) \qquad \Delta H° = +44 \text{ kJ}$$

We will use these two equations to demonstrate Hess's law.

We shall begin with a matching problem. We look at the equation we are seeking. The first thing that is present is $CH_4(g)$; therefore, we seek it in the two equations given. It is in the first equation, so we copy the equation down:
$$CH_4(g) + 2\,O_2(g) \rightarrow CO_2(g) + 2\,H_2O(g) \qquad \Delta H° = -799 \text{ kJ}$$

Continuing with the matching, the next thing in the equation being sought is $2\,O_2(g)$, which also appears as the second part of the equation we just wrote down. This is a piece of luck in that one equation does not give more than one component very often. There is another piece of luck when we move on to the third substance ($CO_2(g)$), which also balances. However, there is a problem when we move to the final substance ($2\,H_2O(l)$), the equation we are seeking requires $2\,H_2O(l)$ not $2\,H_2O(g)$. This requires us to move on to the second equation given.

The second equation does contain the $H_2O(l)$ we need. However, there are two problems: first it is on the wrong side of the reaction arrow, and second, the coefficient is 1 instead of 2. Since this is a thermochemical equation, it is possible to manipulate it. We need to compensate for the two problems. So we will reverse the second equation (reversing the sign on $\Delta H°$) and double the equation (both coefficients and $\Delta H°$ doubled). Adding this to what we have already written gives:
$$CH_4(g) + 2\,O_2(g) \rightarrow CO_2(g) + 2\,H_2O(g) \qquad \Delta H° = -799 \text{ kJ}$$
$$2\,H_2O(g) \rightarrow 2\,H_2O(l) \qquad \Delta H° = -88 \text{ kJ}$$

The next step is to cancel equal amounts of any substance appearing on opposite sides of the two equations as done here:
$$CH_4(g) + 2\,O_2(g) \rightarrow CO_2(g) + \cancel{2\,H_2O(g)} \qquad \Delta H° = -799 \text{ kJ}$$
$$\cancel{2\,H_2O(g)} \rightarrow 2\,H_2O(l) \qquad \Delta H° = -88 \text{ kJ}$$

Note, the substances *must* be identical to cancel.

Finally, add the two equations including the energies:
$$CH_4(g) + 2\,O_2(g) \rightarrow CO_2(g) + 2\,H_2O(l) \qquad \Delta H° = -799 \text{ kJ} - 88 \text{ kJ} = -887 \text{ kJ}$$

This is the correct final answer including appropriate units and significant figures.

EXAMPLE 10

Determine the $\Delta H°$ for:

$$N_2O(g) + NO_2(g) \rightarrow 3\,NO(g) \qquad \Delta H° = ?$$

To determine this value, use the following equations:

$$N_2(g) + O_2(g) \rightarrow 2\,NO(g) \qquad \Delta H° = +182\text{ kJ}$$
$$2\,NO(g) + O_2(g) \rightarrow 2\,NO_2(g) \qquad \Delta H° = -114\text{ kJ}$$
$$2\,N_2O(g) + \rightarrow 2\,N_2(g) + O_2(g) \qquad \Delta H° = -162\text{ kJ}$$

Again we will begin by matching. The equation sought begins with $N_2O(g)$; therefore, we must seek this substance in the given equations. This substance appears in the third equation; however, this equation needs to be divided by 2 to get the coefficient to match:

$$N_2O(g) + \rightarrow N_2(g) + 0.5\,O_2(g) \qquad \Delta H° = 0.5(-162\text{ kJ}) = -81\text{ kJ}$$

The next substance needed in the equation sought is $NO_2(g)$, which may be found in the second equation. Before the second equation may be used, it is necessary to reverse it (change sign) and divide by 2:

$$NO_2(g) \rightarrow NO(g) + 0.5\,O_2(g) \qquad \Delta H° = +57\text{ kJ}$$

Moving to the final substance in the equation we are seeking to match. We have $3\,NO(g)$; however, we already have $1\,NO(g)$, so we only require two more. $NO(g)$ appears in the first of the given equations, and since it is on the correct side, that is fine, and since it already has the correct coefficient, we can take it as given:

$$N_2(g) + O_2(g) \rightarrow 2\,NO(g) \qquad \Delta H° = +182\text{ kJ}$$

Next, we shall list the modified equations together:

$$N_2O(g) + \rightarrow N_2(g) + 0.5\,O_2(g) \qquad \Delta H° = -81\text{ kJ}$$
$$NO_2(g) \rightarrow NO(g) + 0.5\,O_2(g) \qquad \Delta H° = +57\text{ kJ}$$
$$N_2(g) + O_2(g) \rightarrow 2\,NO(g) \qquad \Delta H° = +182\text{ kJ}$$

Now, we will cancel:

$$N_2O(g) + \rightarrow \cancel{N_2(g)} + \cancel{0.5\,O_2(g)} \qquad \Delta H° = -81\text{ kJ}$$
$$NO_2(g) \rightarrow NO(g) + \cancel{0.5\,O_2(g)} \qquad \Delta H° = +57\text{ kJ}$$
$$\cancel{N_2(g)} + \cancel{O_2(g)} \rightarrow 2\,NO(g) \qquad \Delta H° = +182\text{ kJ}$$

Now add what remains, including the enthalpies, being careful to keep everything on the correct side of the reaction arrow:

$$N_2O(g) + NO_2(g) \rightarrow 3\,NO(g) \qquad \Delta H° = -81\text{ kJ} +57\text{ kJ} +182\text{ kJ} = 158\text{ kJ}$$

This is the correct final answer including appropriate units and significant figures.

Take your time in these problems. The most common error appears to be by students who try to answer the problem without writing out the "new" equations completely before combining them to get the final answer.

EXERCISE

22·5

Hess's law

1. Determine $\Delta H°$ for this reaction:

   $Pb(s) + S(s) + 2 O_2(g) \rightarrow PbSO_4(s)$

   Use the following information:

$Pb(s) + S(s) \rightarrow PbS(s)$	$\Delta H° = -94.31$ kJ
$PbS(s) + 2 O_2(g) \rightarrow PbSO_4(s)$	$\Delta H° = -824.0$ kJ

2. Determine $\Delta H°$ for this reaction:

   $2 Ca(s) + 2 C(s) + 3 O_2(g) \rightarrow 2 CaCO_3(s)$

   Use the following information:

$2 Ca(s) + O_2(g) \rightarrow 2 CaO(s)$	$\Delta H° = -635.3$ kJ
$C(s) + O_2(g) \rightarrow CO_2(g)$	$\Delta H° = -393.5$ kJ
$CaO(s) + CO_2(g) \rightarrow CaCO_3(s)$	$\Delta H° = -178.0$ kJ

3. Determine $\Delta H°$ for this reaction, which occurs when $C_6H_{12}O_6(s)$ undergoes fermentation:

   $C_6H_{12}O_6(s) \rightarrow 2 C_2H_5OH(l) + 2 CO_2(g)$

   Use the following information:

$C_6H_{12}O_6(s) + 6 O_2(g) \rightarrow 6 CO_2(g) + 6 H_2O(l)$	$\Delta H° = -2,817$ kJ
$C_2H_5OH(l) + 3 O_2(g) \rightarrow 2 CO_2(g) + 3 H_2O(l)$	$\Delta H° = -1,368$ kJ

Enthalpies of formation

Let us consider the following reaction:

$$2 Ca(s) + O_2(g) \rightarrow 2 CaO(s) \qquad \Delta H° = -635.3 \text{ kJ}$$

The degree symbol indicates that this reaction was conducted under standard conditions. If we now change the equation to:

$$Ca(s) + 0.5 O_2(g) \rightarrow CaO(s) \qquad \Delta H° = -317.6 \text{ kJ mol}^{-1} \text{ CaO}(s)$$

In this form, this equation is an example of a reaction to give the standard heat of formation. The **standard heat of formation** is the energy change that occurs when one mole of a substance is formed from the elements with all substances in their standard states. Note, there *must* be only *one* mole of a single substance on the product side, and everything on the reactant side *must* be an element. The standard heat of reaction is symbolized as $\Delta H_f°$. Values of standard heats of formation are found in tables. Note, for any element in its standard state $\Delta H_f° = 0$ (exactly).

Let's use standard heats of reaction to redo the last example in the previous section.

EXAMPLE 11

Determine the $\Delta H°$ for:

$$N_2O(g) + NO_2(g) \rightarrow 3\,NO(g) \qquad \Delta H° = ?$$

You should begin by locating the standard heat of formation for each substance in the balanced chemical equation and place the value beneath the substance in the equation:

$$N_2O(g) + \qquad NO_2(g) \rightarrow \qquad 3\,NO(g) \qquad \Delta H° = ?$$
$$+81.6 \qquad\qquad +33.84 \qquad +90.37 \qquad \text{All in kJ mol}^{-1}$$

The next step is to take the sum of the heats of reaction for the products and subtract the sum of the heats of reaction for the reactants. This may be expressed as:

$$\Delta H° = \sum \text{Products} - \sum \text{Reactants}$$

Using this relationship and the values we found:

$$\Delta H° = (3 \times 90.37\text{ kJ}) - (81.6\text{ kJ} + 33.84\text{ kJ}) = 155.7\text{ kJ}$$

This answer is slightly different from the previous answer due to rounding.

While this is a fairly simple problem, many students make errors extracting numbers from tables. For example, tables may contain standard heats of formation for $H_2O(s)$, $H_2O(l)$, and $H_2O(g)$, and if you pick the wrong one you will get the wrong answer. Also, in examples like this:

$$- (81.6\text{ kJ} + 33.84\text{ kJ})$$

some students forget that the minus sign applies to the sum of the two numbers and not just to the first number. Finally, do not forget that you are following the addition/subtraction rule for significant figures and not the multiplication/division rule.

There is a table of standard heats of formation at the end of this chapter. Since their standard heats of formation are 0, the values for elements in their standard states are omitted for the most part.

EXERCISE

22·6

Enthalpies of formation

1. Write a balanced chemical equation for the heat of formation of $Na_2CO_3(s)$.

2. Write a balanced chemical equation for the heat of formation of $HC_2H_3O_2(l)$.

3. Using standard heats of formation, determine the heat of reaction for the following:
   $$HC_2H_3O_2(l) + 2\,O_2(g) \rightarrow 2\,CO_2(g) + 2\,H_2O(l)$$

4. Using standard heats of formation, determine the heat of reaction for the following:
   $$H_2O(l) + SO_3(g) \rightarrow H_2SO_4(aq)$$

A simplified table of standard heats of formation:

	$\Delta H_f°$ (kJ mol^{-1})		$\Delta H_f°$ (kJ mol^{-1})
AgCl(s)	−127.03	C_2H_4(g)	52.47
Ag_2CrO_4(s)	−712	C_2H_6(g)	− 84.667
AgN_3(s)	378.5	C_4H_{10}(g)	−125
Ag_2O(s)	− 31.05	CH_3OH(g)	− 201.2
Al_2O_3(s)	−1,676	CH_3OH(l)	− 238.6
$Al_2(SO_4)_3$(s)	−3,441	C_2H_5OH(g)	− 235.1
B_2O_3(s)	−1,272	C_2H_5OH(l)	− 277.63
$BaCO_3$(s)	−1,219	$C_6H_{12}O_6$(s)	−1,273.3
BaO(s)	−548.1	CO(g)	−110.5
$BaSO_4$(s)	−1,465.4	CO_2(g)	−393.5
$BaCl_2 \cdot 2H_2O$(s)	−1,460	$COCl_2$(g)	−220
$Ba(OH)_2 \cdot 8H_2O$(s)	−3,350	$CaCO_3$(s)	−1,206.9
C (graphite)	0	CaH_2(s)	−45.1
C (diamond)	1.896	CaI_2(s)	−533.5
CH_4(g)	−74.87	CaO(s)	−635.1
C_2H_2(g)	227	$Ca(OH)_2$(s)	−986.09
Cu_2O(s)	−170.3	$CaSO_4$(s)	−1342.7
CuS(s)	−53.1	CuO(s)	−157.1
Cu_2S(s)	−79.5	NH_3(aq)	−80.83
$FeCl_3$(s)	−499.5	NH_4Cl(s)	−314.4
FeO(s)	−272.0	N_2O(g)	82.05
Fe_2O_3(s)	−825.5	NO(g)	90.29
HBr(g)	−36.3	NO_2(g)	33.2
HCl(g)	−92.31	NOCl(g)	51.71
HCl(aq)	−167.46	NaBr(s)	−361
HI(g)	25.9	NaCl(s)	−411.1
HF(g)	−273	NaI(s)	−288
$HC_2H_3O_2$(l)	−487.0	Na_2O_2(s)	−504.6
H_2O(g)	−241.826	NaOH(s)	−425.609
H_2O(l)	−285.840	Na_2CO_3(s)	−1,130.8
H_3PO_4(s)	−1,279	$NaHCO_3$(s)	−947.7
H_3PO_4(aq)	−1,277	P_4O_{10}(s)	−2,984
H_2S(g)	−20.2	PbI_2(s)	−175.5
H_2SO_4(l)	−813.989	$Pb(NO_3)_2$(s)	−452
H_2SO_4(aq)	−907.51	PbO(s)	−218
KCl(s)	−436.7	SO_2(g)	−296.9
$KClO_3$(s)	−397.7	SO_3(g)	−396
KI(s)	−327.9	$SiBr_4$(l)	−95.1
KNO_3(s)	−492.7	SiO_2(s)	−910.9
$MgCO_3$(s)	−1,112	$SrSO_4$(s)	−1,445
MgO(s)	−601.2	$TiCl_4$(g)	−763.2
MnO(s)	−385.2	TiO_2(s) (rutile)	−944.7
MnO_2(s)	−520.9	ZnO(s)	−348.0
NH_3(g)	−45.9	ZnS(s) (sphalerite)	−203

Solutions I

We have dealt with aqueous solutions many times. An **aqueous solution** is a homogeneous mixture where water is the solvent. Normally when we think of solutions we think of aqueous solutions, but there are a large number of solutions where water is not the solvent.

Solutions may be solid (steel and brass) or gases (air) in addition to liquids. Liquid solutions may exist with a wide variety of solvents other than water. In this chapter, we will consider not only water as a solvent but also other liquids.

There are additional types of intermolecular forces that may be present in solutions. These types of intermolecular forces include ion-dipole forces, ion-induced dipole forces, and dipole-induced dipole forces.

An **ion-dipole force** is the intermolecular force between an ion and a polar molecule (dipole). This is a very strong type of intermolecular force, which occurs in situations such as solutions of NaCl in water. Substances such as NaCl are electrolytes, which means that they may separate into individual Na^+ and Cl^- ions in solution. These ions will interact with any polar molecules in the vicinity. In the case of very polar molecules, such as water, this force is comparable in strength to a "normal" bond, be it ionic, covalent, or metallic. All electrolytes in polar solvents have some degree of ion-dipole forces.

An **ion-induced dipole force** occurs between an ion and a nonpolar molecule. This is not as strong as an ion-dipole force. Since most electrolytes do not ionize in nonpolar solvents, this type of intermolecular force is not very common except in cases where a solvent is actually a mixture of solvents, such as water containing a small amount of a nonpolar pollutant.

A **dipole-induced dipole force** is the intermolecular force between a polar molecule and a nonpolar molecule. Of the three types of intermolecular forces discussed here, this is the weakest, as it is only slightly stronger than the London dispersion forces present between nonpolar molecules.

Aqueous solutions

Before moving on, it is necessary to review a few of the properties of water and aqueous solutions.

Water is one of the best solvents known to most people. This is because water will dissolve a wide variety of substances that are common in the environment. This is good if you wish to make a cup of coffee, but not so good when one considers a pollutant. One of the reasons why water is such a good solvent is because it is a very polar molecule capable of hydrogen bonding. Other very polar substances capable of hydrogen bonding will dissolve most, if not all, the substances that are soluble in water. Molecules of lower polarity are less effective at dissolving water-soluble substances. On the other hand, substances that do not dissolve well in water tend to readily dissolve in each other.

What many people do not realize is that there are other solvents where their solutions behave very much like aqueous solutions.

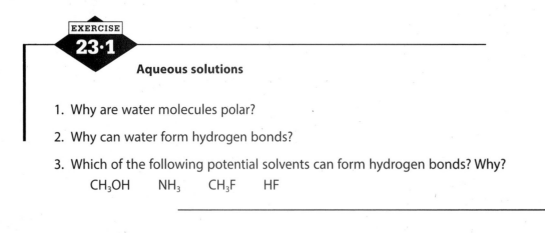

EXERCISE 23·1

Aqueous solutions

1. Why are water molecules polar?

2. Why can water form hydrogen bonds?

3. Which of the following potential solvents can form hydrogen bonds? Why?

CH_3OH NH_3 CH_3F HF

Electrolytes

Many important properties of solutions are due to the fact that solutes may behave as electrolytes in solution. An **electrolyte** is a substance that separates into ions in solution. This separation, dissociation, may be complete (strong electrolyte) or partial (weak electrolyte). Other substances, nonelectrolytes, do not separate into ions in solution.

Electrolytes may be acids, bases, or salts. Acids and bases will be covered later. The salts we will be considering here consist of a cation and an anion. A salt that behaves as an electrolyte in solution is soluble, while a salt that does not behave as an electrolyte in solution is insoluble (there are exceptions).

It is easy to determine if a solution contains an electrolyte. A solution containing an electrolyte will conduct electricity. How well the solution conducts depends upon how well the substance separates into ions and upon the concentrations of the ions.

In general, ions from an electrolyte in solution can exist only in the solution because they are surrounded by solvent molecules that interact with the ion through strong intermolecular forces. These forces must result in sufficient attraction that the tendency of the ions to come back together and reform ionic bonds is not energetically favorable. Since intermolecular forces are weaker than ionic bonds, this is accomplished by the number of intermolecular interactions formed and by the fact that solvents tend to insulate the charges of the ions from each other except at very high concentrations.

Let us examine what happens when you add an electrolyte to water.

We will consider solutions of three different electrolytes in aqueous solution. These electrolytes are NaOH, HCl, and HF. The first two are strong electrolytes and the third is a weak electrolyte. Nonelectrolytes are not included in this discussion because when they dissolve the molecules simply separate from each other in solution without breaking any bonds (they remain whole).

EXAMPLE 1

What happens when NaOH dissolves in water?

NaOH is a strong electrolyte that completely separates into Na^+ and OH^- ions when dissolved in water. This may be represented as:

$$NaOH(s) + H_2O(l) \rightarrow Na^+(aq) + OH^-(aq)$$

The "balancing" H_2O on the product side is included in the "aq." For this reason, this equation is usually simplified to:

$$NaOH(s) \rightarrow Na^+(aq) + OH^-(aq)$$

We will continue using this simplification.

The $Na^+(aq)$ and the $OH^-(aq)$ are examples of hydrated ions. A **hydrated ion** is an ion in solution with associated water molecules that surround the ion. The number of water molecules depends on the size and charge of the ion. Normally, there are four or more water molecules.

EXAMPLE 2

What happens when HCl dissolves in water?

HCl is a strong electrolyte that completely separates into H^+ and Cl^- ions when dissolved in water. This may be represented as:

$$HCl(g) \rightarrow H^+(aq) + Cl^-(aq)$$

The $H^+(aq)$ and the $Cl^-(aq)$ are additional examples of hydrated ions. Again the number of water molecules present depends upon the charge and the size of the ions. In the case of $H^+(aq)$, there are either four or five water molecules giving $[H(H_2O)_4]^+$ or $[H(H_2O)_5]^+$ which may be written as $H_9O_4^+$ or $H_{11}O_5^+$. These ions are often simplified to H_3O^+ even though this ion is not present in aqueous solution. The IUPAC name of H_3O^+ is the oxonium ion; however, many chemists use the common name hydronium ion.

EXAMPLE 3

What happens when HF dissolves in water?

HF is a weak electrolyte that incompletely separates into H^+ and F^- ions when dissolved in water. The question is how you indicate that some of the molecules have ionized and that some have not. The answer is to use a double arrow as:

$$HF(l) \leftrightarrows H^+(aq) + F^-(aq)$$

The arrow to the right indicates that some of the HF molecules have ionized. The arrow to the left indicates that some of the ions combine to reform HF molecules. In later chapters, we will learn that this is an equilibrium.

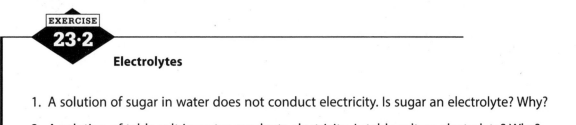

EXERCISE

23·2

Electrolytes

1. A solution of sugar in water does not conduct electricity. Is sugar an electrolyte? Why?

2. A solution of table salt in water conducts electricity. Is table salt an electrolyte? Why?

3. A solution of vinegar in water conducts electricity poorly. Is vinegar an electrolyte? Why?

Solubility

Water is a very polar solvent, and it will dissolve other polar substances in addition to electrolytes. Hexane, C_6H_{14}, is nonpolar and will dissolve other nonpolar substances. This behavior is often summarized by the statement "like dissolve like." While this statement is useful in a wide variety of situations, it does not explain the background for these observations.

There are three factors to consider, at this level, in the formation of a solution. All of these considerations deal with intermolecular forces or with chemical bonds. These three factors are:

1. The intermolecular forces or bonds between the solvent molecules must be overcome. This requires energy.

2. The intermolecular forces or bonds between the solute molecules must be overcome. This requires energy.

3. New interactions, intermolecular forces or bonds, form in the solution. This process releases energy.

The amounts of energy involved depend upon the strength of interactions involved. If the energy released does not compensate to the energy required, very little if any dissolution will occur.

(Later, we will see an additional factor known as entropy. Entropy always favors the formation of a solution. Entropy may make up for a small energy deficit.)

EXAMPLE 4

Describe the possible formation of an aqueous NaCl solution in terms of the previous three factors.

Going through each of the three factors:

1. It is necessary to overcome some of the hydrogen bonds in the water.

2. It is necessary to overcome the ionic bonds in NaCl.

3. Very strong ion-dipole forces form between the ions and the polar water molecules.

The charge on the ions in the compounds and their radii strongly influence the solubility of an ionic compound.

EXAMPLE 5

Describe the possible formation of a solution of Mn in Fe to form steel in terms of the previous three factors.

Going through each of the three factors:

1. It is necessary to overcome some of the metallic bonds in iron.

2. It is necessary to overcome the metallic bonds in manganese.

3. Strong metallic bonds form between the Mn and Fe.

EXAMPLE 6

Describe the possible formation of an aqueous O_2 solution in terms of the preceding three factors.

Going through each of the three factors:

1. It is necessary to overcome some of the hydrogen bonds in the water.

2. The intermolecular forces between gas molecules is negligible as long as the gas is close to ideal.

3. Weak dipole-induced dipole forces form.

Molecules of O_2 will dissolve in water primarily because step 2 is negligible, and entropy favors the dissolving. If O_2 was not soluble in H_2O, fish could not survive.

EXAMPLE 7

Describe the possible formation of a solution of I_2 (nonpolar) in CCl_4 (nonpolar) in terms of the given three factors.

Going through each of the three factors:

1. It is necessary to overcome some of the London dispersion forces in CCl_4.

2. It is necessary to overcome some of the London dispersion forces in I_2.

3. New London dispersion forces form.

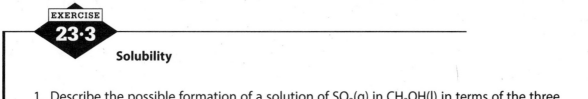

Solubility

1. Describe the possible formation of a solution of $SO_2(g)$ in $CH_3OH(l)$ in terms of the three factors discussed in this section.

Acids and bases

Throughout this book you will see different definitions of acids and bases. The best definition depends upon the circumstances. For example, two of the definitions are the Brønsted-Lowrey and the Lewis definitions. Each of these works well; however, they are normally not applied to the same reactions.

At this point, we will introduce you to the Arrhenius definition of an acid and a base. This theory was proposed by Svante Arrhenius in 1883. This theory is based upon hydrogen ions and hydroxide ions and ignores the fact that other species may behave as an acid or as a base, or even as both.

An **Arrhenius acid** is a substance that produces hydrogen ions in solution.

An **Arrhenius base** is a substance that produces hydroxide ions in solution.

In addition to the identity of the two ions, the process *must* be in solution, and it does not matter how the ions are produced.

The ions may be produced in solution through simple ionization:

$$HCl(aq) \rightarrow H^+(aq) + Cl^-(aq)$$

$$KOH(s) \rightarrow K^+(aq) + OH^-(aq)$$

$$HF(l) \leftrightarrows H^+(aq) + F^-(aq)$$

The ions may also be produced by a reaction with the solvent:

$$NH_3(aq) + H_2O(l) \leftrightarrows NH_4^+(aq) + OH^-(aq)$$

Both acids and bases may be either a strong or a weak electrolyte. The strong electrolytes are the strong acids and strong bases. The weak electrolytes are the weak acids and weak bases.

Only a few substances may be considered to be a strong acid, and only a few substances may be considered to be a strong base. Unless told otherwise, consider all other acids to be weak and all other bases to be weak.

The short list of strong acids are HCl, HBr, HI, H_2SO_4, HNO_3, $HClO_3$, and $HClO_4$. As strong acids they behave as strong electrolytes (100% ionized) in solution.

The short list of strong bases are (Li, Na, K, Rb, Cs)OH and (Ca, Sr, Ba)(OH)$_2$. As strong bases they behave as strong electrolytes (100% ionized) in solution.

Note there are compounds that will react with water to produce one of the strong acids or bases.

Acids have a sour taste as can be detected from the citric acid in lemonade and the acetic acid in vinegar. Bases taste bitter as do sodium bicarbonate and cocoa (not the drink, which has sugar added to cover the bitterness).

By all acid-base definitions, acids and bases are opposites. They will cancel (neutralize) each other.

The reaction of an acid with a base is a neutralization reaction. A neutralization reaction will always produce a type of compound called a salt. If the acid and base are both Arrhenius, water will usually form. (Note that substances that fall under other acid-base theories do not necessarily produce water in a neutralization reaction.) A **salt** is a compound containing the cation from a base and the anion from an acid.

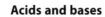

EXERCISE 23·4

Acids and bases

1. Identify the acid, the base, and the salt in the following neutralization reaction:
   $HCl(aq) + NaOH(aq) \rightarrow NaCl(aq) + H_2O(l)$

2. Identify the acid, the base, and the salt in the following neutralization reaction:
   $Ba(OH)_2(aq) + H_2SO_4(aq) \rightarrow BaSO_4(s) + 2\,H_2O(l)$

3. Identify the acid, the base, and the salt in the following neutralization reaction:
   $NH_3(aq) + HBr(aq) \rightarrow NH_4Br(aq)$

Net ionic equations

While net ionic equations may be applied to any solution containing an electrolyte (separate ions), they are most often applied to aqueous solutions, which we will do here. A net ionic equation shows what is happening when a reaction occurs between electrolytes in solution.

Let us consider the following reaction:

$$HCl(aq) + NaOH(aq) \rightarrow NaCl(aq) + H_2O(l)$$

This is a type of equation known as a molecular equation. A **molecular equation** treats all reactants and products as molecules regardless of whether they are molecules or not. Molecular equations are simple ways to describe reactions.

In some cases, it is more useful to consider what is happening in the solution. To do this we can use a net ionic or total ionic equation. A **total ionic equation** has all strong electrolytes separated into ions in solution. In the HCl/NaOH reaction, HCl, NaOH, and NaCl are strong electrolytes. Converting the previous molecular equation to a total ionic equation gives us:

$$H^+(aq) + Cl^-(aq) + Na^+(aq) + OH^-(aq) \rightarrow Na^+(aq) + Cl^-(aq) + H_2O(l)$$

You need to be able to recognize which substances are strong electrolytes and from your nomenclature, what ions they are likely to separate into. Do not forget, even though weak electrolytes also produce ions, they are not separated in ionic equations including total ionic equations. You should note that as an equation, a total ionic equation is a balanced equation; however, in balancing you now need to consider each element and the overall charge.

Look again at the total ionic equation. Note that this total ionic equation, like most total ionic equations, contains some ions, in this case the $Na^+(aq)$ and $Cl^-(aq)$ ions, which appear unchanged on each side of the reaction arrow. These are spectator ions. **Spectator ions** are ions that appear unchanged on both sides of the reaction arrow. Spectator ions are present to maintain the electrical neutrality of the solution. The spectator ions are not part of the actual reaction as their only purpose is to maintain neutrality.

A third type of equation is a net ionic equation. A **net ionic equation** is a total ionic equation with all the spectator ions removed. Converting the previous total ionic equation to a net ionic equation gives:

$$H^+(aq) + OH^-(aq) \rightarrow H_2O(l)$$

Again, the equation remains balanced (including the remaining charges). A net ionic equation summarizes the actual reaction that takes place.

Net ionic equations can also summarize other similar reactions. For example, if HBr, HI, H_2SO_4, HNO_3, $HClO_3$, or $HClO_4$ were substituted for the HCl, we would get six additional molecular equations, six additional total ionic equations, and zero additional net ionic equations.

A common student error is that when they are asked to write an ionic equation, total or net, they produce an equation with no charges. To be an ionic equation, there must be ions present.

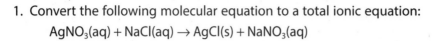

EXERCISE 23·5

Net ionic equations

1. Convert the following molecular equation to a total ionic equation:
$$AgNO_3(aq) + NaCl(aq) \rightarrow AgCl(s) + NaNO_3(aq)$$

2. Convert the following total ionic equation to a net ionic equation:
$$CaCO_3(s) + 2\,H^+(aq) + 2\,Cl^-(aq) \rightarrow Ca^{2+}(aq) + 2\,Cl^-(aq) + H_2CO_3(aq)$$

3. Convert the following molecular equation to a total ionic equation and then to a net ionic equation:
$$Ba(C_2H_3O_2)_2(aq) + H_2SO_4(aq) \rightarrow BaSO_4(s) + 2\,HC_2H_3O_2(aq)$$

4. Convert the following molecular equation to a total ionic equation and then to a net ionic equation:
$$CaCl_2(aq) + Na_2C_2O_4(aq) \rightarrow CaC_2O_4(s) + 2\,NaCl(aq)$$

Solutions II

Many chemical processes have been studied in solutions. Solutions often facilitate the control of a reaction to allow precise measurements to be made.

You should not be intimidated by the number of ways to express concentration. All you need to remember is their definitions and use these in the unit conversions needed.

Concentrations

There are many ways to determine how much solute is in a given amount of solution. Each of these ways is called a concentration. The reason why there are so many different ways to express the concentration of a solution is because there are so many different uses of solutions.

We have already visited one type of concentration unit, which is molarity. This was covered previously as it is one of the most versatile concentration units. As a reminder, **molarity**, M, is the moles of solute per liter (dm³) of solution. Part of molarity's usefulness is that it is easy to determine the moles present if you know the volume at a certain concentration.

Another common useful concentration unit is the mole fraction. The **mole fraction**, X, is the moles of a substance divided by the total moles present. Alternatively, for gaseous solutions, the mole fraction is the partial pressure of a component, P_A, divided by the total pressure, P_{Total}. Unlike the other concentration units, mole fractions have no units; therefore, in calculating mole fractions, you have made an error if all units do not cancel. Note, the maximum mole fraction is 1, which occurs only for a pure substance. In addition, the sum of all the mole fractions in a solution must be exactly 1.

EXAMPLE 1

Determine the molarity of a solution containing 1.00 mol of NaCl in 1.00 L of solution.

First, summarize the information from the problem, and then all that is needed is to write the definition of molarity and to insert the appropriate values:

$$M = ? \quad \text{Moles} = 1.00 \text{ mol NaCl} \quad \text{Volume} = 1.00 \text{ L}$$

$$M = \frac{\text{mol solute}}{\text{L solution}} = \frac{1.00 \text{ mol NaCl}}{1.00 \text{ L}} = 1.00 \text{ M NaCl}$$

Recall, the correct answer is 1.00 M and not 1 M, 1.0 M, or 1.000 M.

EXAMPLE 2

Determine the mole fraction of CH_3OH in a solution containing 3.00 mol of CH_3OH and 2.00 mol of H_2O.

First, summarize the information from the problem and then all that is needed is to write the definition of the mole fraction and to insert the appropriate values:

$X CH_3OH = ?$

Moles A = 3.00 mol CH_3OH Moles B = 2.00 mol H_2O

$$X CH_3OH = \frac{mol\ CH_3OH}{Total\ moles} = \frac{3.00\ \cancel{mol}\ CH_3OH}{(3.00+2.00)\ \cancel{mol}} = 0.600\ CH_3OH$$

Note, all mole fractions must be 1 or less.

EXAMPLE 3

A gas sample contains a mixture of noble gases. The partial pressures of the gases in this mixture are 0.500 atm argon, 0.250 atm neon, and 0.125 atm krypton. What is the mole fraction of krypton in this sample?

First, summarize the information from the problems, and then all that is needed is to write the definition of mole fraction and to insert the appropriate values:

$X Kr = ?$ Pressure A = 0.500 atm Ar

Pressure B = 0.250 atm Ne   Pressure C = 0.125 atm Kr

$$X Kr = \frac{Pressure\ CH_3OH}{Total\ pressure} = \frac{0.125\ \cancel{atm}\ Kr}{(0.500+0.250+0.125)\ \cancel{atm}} = 0.143\ Kr$$

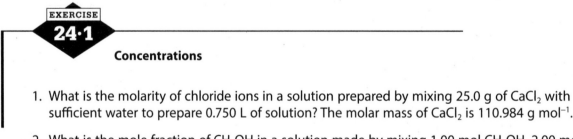

EXERCISE
24·1

Concentrations

1. What is the molarity of chloride ions in a solution prepared by mixing 25.0 g of $CaCl_2$ with sufficient water to prepare 0.750 L of solution? The molar mass of $CaCl_2$ is 110.984 g mol^{-1}.

2. What is the mole fraction of CH_3OH in a solution made by mixing 1.00 mol CH_3OH, 2.00 mol C_2H_5OH, and 2.00 mol of H_2O?

3. What is the mole fraction of O_2 in an air sample where the partial pressures of the components are 0.100 atm O_2, 0.250 atm N_2, and 0.150 atm Ar?

More concentrations

Some units involving percentages have a number of commercial uses; however, there are only a few applications to chemists. Two common ways of representing the percent concentration of a solution are mass-mass (weight-weight) percent and volume-volume percent. These may be expressed as:

$$\text{wt/wt\%} = \frac{\text{Mass of solute}}{\text{Mass of solution}} \times 100\%$$

$$\text{vol/vol\%} = \frac{\text{Volume of solute}}{\text{Volume of solution}} \times 100\%$$

Any units of mass or volume may be used as long as both masses or both volumes are in the same units. The vol/vol% is mainly for situations where both the solute and solution are in the liquid state.

There are additional concentration units that are of use in chemistry. The only one we will discuss here is the molality. **Molality**, *m*, is the moles of solute per kilogram of solvent. Note, unlike all the other concentrations mentioned, the denominator consists only of the solvent and not the entire solution. Be careful of the spelling difference between *molality* and *molarity* as well as similarities in the abbreviations, *m* and *M*.

EXAMPLE 4

Determine the molality of a solution containing 1.00 mol of NaCl in 500.0 g of water.

First, summarize the information from the problem and then all that is needed is to write the definition of molarity and to insert the appropriate values:

$m = ?$ Moles = 1.00 mol NaCl   Mass solvent = 500.0 g H_2O

$$m = \frac{\text{mol solute}}{\text{kg solvent}} = \left(\frac{1.00 \text{ mol NaCl}}{500.0 \text{ g } H_2O} \right)\left(\frac{1{,}000 \text{ g}}{\text{kg}} \right) = 2.00 \text{ m NaCl}$$

More concentrations

1. What is the molality of sodium ions in a solution made by adding 46.00 g Na_3PO_4 to 500.00 g of water? The molar mass of Na_3PO_4 is 212.266 g mol^{-1}.

2. What is the molality of a solution made by mixing 16.2500 g CH_3OH in *exactly* 1 kg of water? The molar mass of CH_3OH is 32.0417 g mol^{-1}.

Colligative properties

There are some properties of solutions that do not depend on the identity of the solute particles. For these properties only the number of these solute particles is important. These properties are known as colligative properties. A **colligative property** is a characteristic of a solution that depends only upon the concentration of the solute particles and not on the identity of the solute.

When dealing with colligative properties, it is necessary to deal with electrolytes in a slightly different manner than nonelectrolytes. Consider the following three cases:

Case 1. The solute is a nonelectrolyte like CH_3OH. What is the total concentration (molarity) of the particles when 1.00 mol of this substance is added to sufficient H_2O to produce 1.00 L of solution?

The addition of this solute to water may be symbolized as:

$$CH_3OH(l) \rightarrow CH_3OH(aq)$$

So adding 1.00 mol CH_3OH to enough water to make 1.00 L solution gives a solution that is 1.00 M solute particles.

Case 2. The solute is a strong electrolyte like K_2SO_4. What is the total concentration (molarity) of the particles when 1.00 mol of this substance is added to sufficient H_2O to produce 1.00 L of solution?

The addition of this solute to water may be symbolized as:

$$K_2SO_4(s) \rightarrow 2\,K^+(aq) + SO_4^{2-}(aq)$$

So adding 1.00 mol K_2SO_4 to 1.00 L solution gives a solution that is 2.00 M K^+ and 1.00 M SO_4^{2-} or 3.00 M solute total particles.

Case 3. The solute is a weak electrolyte like HNO_2. What is the total concentration (molarity) of the particles when 1.00 mol of this substance is added to sufficient H_2O to produce 1.00 L of solution?

The addition of this solute to water may be symbolized as:

$$HNO_2(l) \leftrightarrows H^+(aq) + NO_2^-(aq)$$

So adding 1.00 mol HNO_2 to 1.00 L solution gives a solution that is x M H^+, x M NO_2^-, and $1.00 - x$ M HNO_2 or $1.00 + x$ M solute total particles. The value of x depends on just how significant the partial ionization of the weak electrolyte is.

It is possible to simplify dealing with strong electrolytes by using a van't Hoff factor. The **van't Hoff factor**, i, is the number of ions generated by a strong electrolyte. For example, in the K_2SO_4 case $i = 3$. So if you have a 1.00 M K_2SO_4 solution, then the concentration of the particles would be $i \times 1.00$ M = 3.00 M. For strong electrolytes, the van't Hoff factor will be an integer. The value will not be an integer for weak electrolytes. At high concentrations, the van't Hoff factors tend to be higher than what is observed. The van't Hoff factor may be applied to other concentration units.

As noted in these three cases, adding equal moles of solute to form a certain volume of solution gives a different concentration of particles, and hence different colligative properties. This is a case where $i \neq 1$.

The colligative properties to be examined here are vapor pressure, freezing point depression, boiling point elevation, and osmotic pressure. Vapor pressure depends on the mole fraction. Both freezing point depression and boiling point elevation depend upon the molality, and for this reason will be treated together. Osmosis depends upon the molarity.

Vapor pressure

The vapor pressures of pure substances are discussed in the chapter on solids and liquids; therefore, we will not restate the theory and applications for pure substances.

For solutions, it is necessary to accommodate the presence of more than one substance. Since the vapor pressure involves particles "escaping" from the surface of an object, it is important to know what fraction of the surface a particular substance is. The fraction of the surface that is a certain substance is equal to the mole fraction of that substance. The following two examples will explain how finding that is done. One example deals with a situation where both the solvent and solute are volatile (have a vapor pressure), and the other example deals with a situation where the solvent is volatile, and the solute is nonvolatile.

EXAMPLE 5

What is the vapor pressure of a solution containing 3.00 mol of CCl_4 and 1.00 mol C_6H_6 at 20°C?

Two things are necessary to progress further. One is the mole fraction of each of the substances present in the solution. The other is the vapor pressure of each substance present.

Vapor pressures (at 20°C):

$$P°_{CCl_4} = 89.7 \text{ torr}$$

$$P°_{C_6H_6} = 95.8 \text{ torr}$$

Mole fractions:

$$X\,CCl_4 = \frac{3.00 \text{ mol } CCl_4}{(3.00+1.00)\text{ mol}} = 0.750\ X\,CCl_4$$

$$X\,C_6H_6 = = \frac{1.00 \text{ mol } C_6H_6}{(3.00+1.00)\text{ mol}} = 0.250\ X\,C_6H_6$$

The degree symbols indicate the values are for the pure substance. Vapor pressures vary with temperature, so make sure you use values from the appropriate temperature. If the pressure units do not match, one needs to be converted to the other. Any pressure units are acceptable.

Vapor pressures follow Raoult's law. For a solution, apply Raoult's law to both the solvent and solute separately and then sum the results. **Raoult's law** says that the vapor pressure of a substance is equal to the mole fraction of that substance times the vapor pressure of the pure substance.

Raoult's law:

$$P_{CCl_4} = (X\,CCl_4)\,(P°_{CCl_4}) = (0.750)\,(89.7 \text{ torr}) = 62.3 \text{ torr}$$
$$P_{C_6H_6} = (X\,C_6H_6)\,(P°_{C_6H_6}) = (0.250)\,(95.8 \text{ torr}) = 24.0 \text{ torr}$$

The vapor pressure of the solution is:

$$P_{CCl_4} + P_{C_6H_6} = 62.3 \text{ torr} + 24.0 \text{ torr} = 86.3 \text{ torr}$$

EXAMPLE 6

What is the vapor pressure of a solution containing 9.00 mol of methanol, CH_3OH, and 1.00 mol of glucose, $C_6H_{12}O_6$, at 25°C?

Two things are necessary to progress further. One is the mole fraction of each of the substances present in the solution. The other is the vapor pressure of each substance present.

Vapor pressures (at 25°C):

$P°_{CH_3OH} = 94.0$ torr

$P°_{C_6H_{12}O_6} = 0$ torr

Mole fractions:

$$X\,CH_3OH = \frac{9.00\;\text{mol}\;CH_3OH}{(9.00+1.00)\;\text{mol}} = 0.900\;X\,CH_3OH$$

$$X\,C_6H_{12}O_6 = = \frac{1.00\;\text{mol}\;C_6H_{12}O_6}{(9.00+1.00)\;\text{mol}} = 0.100\;X\,C_6H_{12}O_6$$

Actually it was not necessary to calculate $X\,C_6H_{12}O_6$ as $C_6H_{12}O_6$ was not volatile ($P°_{C_6H_{12}O_6} = 0$ torr).

Raoult's law:

$$P_{CH_3OH} = (X\,CH_3OH)\,(P°_{CH_3OH}) = (0.900)\,(94.0\;\text{torr}) = 84.6\;\text{torr}$$
$$P_{C_6H_{12}O_6} = (X\,C_6H_{12}O_6)\,(P°_{C_6H_{12}O_6}) = (0.100)\,(0\;\text{torr}) = 0\;\text{torr}$$

Again it is unnecessary to do the $C_6H_{12}O_6$ calculation.

The vapor pressure of the solution is:

$$P_{CH_3OH} + P_{C_6H_{12}O_6} = 84.6\;\text{torr} + 0\;\text{torr} = 84.6\;\text{torr}$$

An **ideal solution** follows Raoult's law exactly. Real solutions, however, often show some deviation from ideal.

Freezing point depression and boiling point elevation

The equations for calculating each of these are similar:

$$\Delta T_f = K_f\,m \qquad \Delta T_b = K_b\,m$$

If the solute contains a strong electrolyte, these equations become:

$$\Delta T_f = iK_f\,m \qquad \Delta T_b = iK_b\,m$$

The subscripts are "f" for freezing point and "b" for boiling point. The ΔT's are the change in the freezing points and boiling points. The K's are constants referring to either freezing or boiling. The m's are the molalities. The i's are van't Hoff factors.

The value of ΔT_f *must be subtracted* from the normal freezing point of the solvent. The value of ΔT_b *must be added* to the normal boiling point of the solvent. The values of the K's depend upon the identity of the solvent and must be given or looked up (unless the purpose of the problem is to determine K).

EXAMPLE 7

The normal freezing and boiling points of CCl_4 are −22.3°C and 76.8°C, respectively. The freezing and boiling points constants are $K_f = 29.8$°C m^{-1} and $K_b = 5.02$°C m^{-1}. Determine the freezing and boiling points of a 0.100 m $CHCl_3$ solution in CCl_4.

Freezing Point

Summarizing the freezing point information from the problem:

$\Delta T_f = ?$ Concentration = 0.100 m $K_f = 29.8$°C m^{-1} $T_f = -22.3$°C

$\Delta T_f = K_f m = (29.8$°C m$^{-1})(0.100$ m$) = 2.98$°C

Freezing point of solution = Normal freezing point $- \Delta T_f = -22.3$°C $- 2.98$°C $= -25.3$°C

Boiling Point

Summarizing the boiling point information from the problem:

$\Delta T_b = ?$ Concentration = 0.100 m $K_b = 5.02$°C m^{-1} $T_b = 76.8$°C

$\Delta T_b = K_b m = (5.02$°C m$^{-1})(0.100$ m$) = 0.502$°C

Boiling point of solution = Normal boiling point $+ \Delta T_b = 76.8$°C $+ 0.502$°C $= 77.3$°C

Common mistakes made when working boiling point elevation and freezing point depression problems are to stop after the ΔT calculation when the new freezing or boiling point is needed and not subtracting from the freezing point or adding to the boiling point. A less common error is to use values for the wrong solvent.

EXAMPLE 8

The normal boiling point of H_2O is 100.00°C. The boiling point constant is $K_b = 0.51$°C m^{-1}. Determine the boiling point of a 0.500 m KNO_3 solution in H_2O.

$\Delta T_b = iK_b m$

Summarizing the boiling point information from the problem:

$\Delta T_b = ?$ Concentration = 0.500 m $K_b = 0.51$°C m^{-1} $T_b = 100.00$°C   van't Hoff factor = 2

$\Delta T_b = iK_b m = 2 (0.51$°C m$^{-1})(0.500$ m$) = 0.51$°C

Boiling point of solution = Normal boiling point $+ \Delta T_b = 100.00$°C $+ 0.51$°C $= 100.51$°C

Since KNO_3 is a strong electrolyte, it is helpful to use the van't Hoff factor to indicate that there are two ions present.

Osmotic pressure

Osmotic pressure, Π, is the pressure generated through osmosis. The equation to determine the osmotic pressure is similar to the ideal gas equation:

$$\Pi V = nRT$$

This may be rearranged to:

$$\Pi = \left(\frac{n}{V}\right)RT = MRT$$

If there is a strong electrolyte present, the equation becomes:

$$\Pi = i\left(\frac{n}{V}\right)RT = iMRT$$

Note that in calculating the osmotic pressure you use the molarity (M) of the solution and not the molality (m).

EXAMPLE 9

What is the osmotic pressure of a 1.00 K_2SO_4 solution in water at 25°C?
   Summarizing the information from the problem:

   $\Pi = ?$ $M = 1.00$ M $T = 25°C = 298$ K   van't Hoff factor $= 3$

   $\Pi = iMRT = (3)\left(1.00 \ \frac{\text{mol}}{\text{L}}\right)\left(0.08206 \ \frac{\text{L atm}}{\text{mol K}}\right)(298 \ \text{K}) = 73.4$ atm

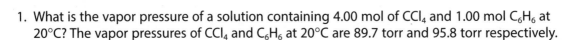

Colligative properties

1. What is the vapor pressure of a solution containing 4.00 mol of CCl_4 and 1.00 mol C_6H_6 at 20°C? The vapor pressures of CCl_4 and C_6H_6 at 20°C are 89.7 torr and 95.8 torr respectively.

2. What is the freezing point of a solution containing 0.250 mol of NaCl in 500.0 g of water? The normal freezing point of water is exactly 0°C and K_f for water is 1.86°C m^{-1}.

3. What is the boiling point of a solution containing 0.250 mol of CH_3OH in 500.0 g of C_2H_5OH? The normal freezing point of C_2H_5OH is 78.1°C and K_b for C_2H_5OH is 1.22°C m^{-1}.

4. What is the osmotic pressure of a 0.250 M $(NH_4)_2SO_4$ solution at 20.1°C?

Kinetics

Up until now we have considered a number of different reactions and made many different calculations based upon the balanced chemical equation. However, we have not considered that reactions take time to occur. Some reactions take place blindingly fast. These extremely fast reactions are known as explosions. Other reactions, such as the rusting of iron and steel, take place over a number of years. How fast a reaction occurs is the rate of the reaction. **Kinetics** is the study of reaction rates.

Introduction to kinetics

All reactions take time to occur. During this time, the concentration of the reactants decreases, and the concentration of the products increases. Concentrations are normally expressed as molarity and symbolized with square brackets, so the molarity of A is written as [A]. Timing how much a concentration changes, $\Delta[A]$, versus how much time has passed, Δt, indicates the rate of the reaction. This is normally expressed as:

$$\text{Reaction rate} = \frac{\Delta[A]}{\Delta t}$$

The rate may be determined for any substance in the reaction. Since the concentrations of the reactants are decreasing, the rate for a reactant is negative. On the other hand the concentration of the products is increasing, so the rate for a product is positive. The rate for any substance will change over time, because as the system begins to run out of reactant, the rate decreases until it reaches zero at the end of the reaction.

> Note, as clarification, that we will be using t to indicate time and T to indicate temperature.

What is taking place during a reaction may be symbolized as:

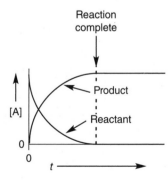

159

In addition to the nature of the reaction, there are additional factors that influence how fast a reaction will proceed. These factors include concentration, nature of the reactants, temperature, and the presence of a catalyst.

Reactions go faster when the concentration of the reactants is higher.
Reactions go faster if the reactants are finely dispersed especially if in the gas phase or in solution.
Reactions go faster at higher temperatures.
Catalysts (see later) make reactions go faster.

EXERCISE
25·1

Introduction to kinetics

1. What are the units of the reaction rate?

Reaction rates

We will begin our discussion of reaction rates by examining the following reaction:

$$H_2(g) + I_2(g) \rightarrow 2\ HI(g)$$

The iodine vapor is easy to see because it is purple. As the reaction proceeds, this purple color fades. Measuring the intensity of the purple color allows the concentration of the $I_2(g)$ to be determined. As the color fades, the concentration decreases. The change in concentration versus time for the iodine may be expressed as:

$$I_2\ \text{Rate} = -\frac{\Delta[I_2]}{\Delta t}$$

This rate has a negative value because the I_2 is decreasing.

The iodine was chosen because it is easy to measure its concentration due to the ability to see the purple color. But what if you wanted the rate for hydrogen? Do you need to do the entire experiment over? The answer is no because we have a balanced chemical equation. The balanced chemical equation says that every time an iodine molecule reacts, so does a hydrogen molecule; therefore, this one-to-one relationship makes the rate of H_2 equal to the rate for I_2 as:

$$-\frac{\Delta[I_2]}{\Delta t} = -\frac{\Delta[H_2]}{\Delta t}$$

Now, what about the hydrogen iodide? It would be possible to rerun the experiment to determine the HI rate. However, it is simpler to refer to the balanced chemical equation again. According to the balanced chemical equation, for every I_2 that reacts, two HI form; thus the HI rate is double the I_2 rate. Since HI is a product, its rate will be positive. Adding the HI rate the other rates gives:

$$-\frac{\Delta[I_2]}{\Delta t} = -\frac{\Delta[H_2]}{\Delta t} = \frac{1}{2}\frac{\Delta[HI]}{\Delta t}$$

The ½ multiplier is to compensate for the HI rate being double the I_2 rate.

EXAMPLE 25.1

Let us consider the following reaction:

$$2\,C_4H_{10}(g) + 13\,O_2(g) \rightarrow 8\,CO_2(g) + 10\,H_2O(g)$$

The rates are:

$$-\frac{1}{2}\frac{\Delta[C_4H_{10}]}{\Delta t} = -\frac{1}{13}\frac{\Delta[O_2]}{\Delta t} = \frac{1}{8}\frac{\Delta[CO_2]}{\Delta t} = \frac{1}{10}\frac{\Delta[H_2O]}{\Delta t}$$

EXERCISE
25·2

Reaction rates

1. Show how the reaction rates for the substances in the following reaction compare to each other.

$$C_3H_8(g) + 10\,F_2(g) \rightarrow 3\,CF_4(aq) + 8\,HF(g)$$

Rate laws

Writing a rate law helps summarize what is taking place in the reaction. A rate law ties together all of the separate rates in a reaction. The general form of a rate law is:

$$\text{Rate} = k\,[A]^m\,[B]^n....$$

The rate may be any of the rates for a reaction; however, the rate chosen is normally for a product. The k is the rate constant, which must be determined experimentally. (Note: *always* use a k not a K for the rate constant.) The terms in brackets represent the concentrations (molarities) of the various reactants. The exponents are the orders for each of the reactants. Both the rate constant and the order must be determined experimentally. Orders are usually small, whole numbers; however, the order may be fractional, 0, or even negative.

The following example illustrates how to determine the rate law for a reaction. This general procedure will work for simple rate laws.

EXAMPLE 25.2

The following reaction was investigated:

$$2\,NO(g) + 2\,H_2(g) \rightarrow N_2(g) + 2\,H_2O(g)$$

The following data were obtained on this reaction:

EXPERIMENT	INITIAL [NO]	INITIAL [H$_2$]	INITIAL RATE $-\dfrac{\Delta[NO]}{\Delta t}$ AS M s^{-1}
1	5.0×10^{-3}	2.0×10^{-3}	2.4×10^{-5}
2	15×10^{-3}	2.0×10^{-3}	2.2×10^{-4}
3	15×10^{-3}	4.0×10^{-3}	4.4×10^{-4}

What is the rate law for this reaction, and what is the value of k?

As one can see from the table, the experiment was run three times with different initial concentrations of the reactants, and the initial rate of disappearance of NO was measured in each case. (It does not matter which initial rate is used.) Note that nonstandard scientific notation is sometimes used to express number so that the values all use the same power of 10.

The first step in determining the rate law is to write the basic rate law for the reaction. In this case, we have:

$$\text{Rate} = k\,[NO]^m\,[H_2]^n$$

The table gives the rate, [NO], and [H$_2$], and it is our goal to determine k, m, and n.

We can begin by determining either m or n first. We will begin with n. We need to choose two of the experiments to determine n. These two experiments must meet two conditions. First, since n refers to H$_2$, the concentration of H$_2$ must be different in the two experiments, and second, none of the other concentrations must change. Experiments 2 and 3 meet these conditions. This means that for these two experiment, we can ignore the NO.

Comparing the data for experiments 2 and 3 shows that the H$_2$ is doubled, and the rate is doubled (or both are halved if 3 is compared to 2). The key here is that both the concentration and rate change by the same factor. This may be represented as:

$$\text{Rate} = [H_2]^n$$
$$2 = [2]^n$$

This means that $n = 1$; now we can move on to find m.

Experiments 1 and 2 meet the two requirements for determining m. Comparing the data for experiments 1 and 2 shows that the NO is tripled, and the rate is $\times 9$. This may be represented as:

$$\text{Rate} = [NO]^m$$
$$9 = [3]^m$$

This means that $m = 2$.

So far we have modified our rate law to:

$$\text{Rate} = k\,[NO]^2\,[H_2]^1$$

The "1" exponent is shown for you to see what happened; it is optional to show it in the final answer.

We can now move on to determining k. This is done by rearranging the rate law and entering the values from any of the three experiments. Since k is a constant, it does not matter which experiment is chosen. The rearranged rate law is, using the experiment 1 values:

$$k = \frac{\text{Rate}}{[NO]^2\,[H_2]} = \frac{2.4 \times 10^{-5}\ \cancel{M}\ s^{-1}}{[5.0 \times 10^{-3}\ M]^2 [2.0 \times 10^{-3}\ \cancel{M}]} = 480\ M^{-2}\ s^{-1}$$

Be careful of the units on k, as they may seem unusual at times.

This reaction is described as second-order in NO, first-order in H$_2$, and third-order overall. Note, the units of the rate are M s^{-1} and the units on k times the units from the concentration terms must cancel to give the rate units. Note, any resemblance of the orders to any coefficients in the balanced chemical equation is coincidental.

Rate laws where the orders are 0 or an integer work this way. If the order is not an integer, this procedure needs to be modified.

Rate laws

Here are some practice problems. For each, write the rate law for the reaction and determine the value of k.

1. The following reaction was investigated:

$$2\,NO(g) + Cl_2(g) \rightarrow 2\,NOCl(g)$$

The following data were obtained on this reaction:

EXPERIMENT	INITIAL [NO]	INITIAL [Cl$_2$]	INITIAL RATE $\dfrac{\Delta[NOCl]}{\Delta t}$ AS M s^{-1}
1	3.00×10^{-3}	2.25×10^{-3}	1.15×10^{-7}
2	6.00×10^{-3}	2.25×10^{-3}	4.60×10^{-7}
3	3.00×10^{-3}	4.50×10^{-3}	2.30×10^{-7}

What is the rate law for this reaction, and what is the value of k?

2. The following reaction was investigated:

$$S_2O_8^{2-}(aq) + 3\,I^-(aq) \rightarrow 2\,SO_4^{2-}(aq) + I_3^-(aq)$$

The following data were obtained on this reaction:

EXPERIMENT	INITIAL [S$_2$O$_8^{2-}$]	INITIAL [I$^-$]	INITIAL RATE $\dfrac{\Delta[I_3^-]}{\Delta t}$ AS M s^{-1}
1	0.040	0.12	2.9×10^{-5}
2	0.080	0.12	5.9×10^{-5}
3	0.080	0.24	1.2×10^{-4}

What is the rate law for this reaction, and what is the value of k?

Integrated rate laws

An integrated rate law ties the rate law to the time involved in a reaction. There are a variety of reasons why this is important. For example, it may be important to know when you need to order more material to keep the reaction going or maybe you need to know when you need to take another dose of your medication.

Our discussion of integrated rate laws will be limited to reactions involving only one reactant, that is, to rate laws such as:

$$Rate = k[A]^n$$

In addition, we will limit our rate laws to include only those with $n = 1$, 2, or 0.

Note, none of the equations in this section or the next will work if more than one reactant is involved.

There are several terms that will appear in all three of these integrated rate laws:

$[A]_0$ = initial concentration $[A]_t$ = concentration after some time t

t = time between $[A]_0$ and $[A]_t$ k = rate constant

First-order integrated rate law ($n = 1$):

$$\ln\frac{[A]_0}{[A]_t} = kt$$

Second-order integrated rate law ($n = 2$):

$$\frac{1}{[A]_t} - \frac{1}{[A]_0} = kt$$

Zero-order integrated rate law ($n = 0$):

$$[A]_0 - [A]_t = kt$$

The corresponding integrated rate laws in your textbook may appear different; however, they are just rearranged versions of these three. We report these integrated rates in this form so that what is to the right of the equal sign is always the same, so there is no worrying about sign changes or additional information being required.

Nuclear decay processes are excellent examples of first-order kinetics.

EXAMPLE 25.3

The following reaction is first order in $SO_2Cl_2(g)$:

$$SO_2Cl_2(g) \rightarrow SO_2(g) + Cl_2(g)$$

For this reaction, $k = 2.2 \times 10^{-5}$ s^{-1}. In one experiment, a sample with an initial concentration of 0.0050 M SO_2Cl_2 is allowed to decompose until $[SO_2Cl_2]$ = 0.0030 M. The sample was in a 2.0 L container. (a) How long did the experiment take? (b) How many moles of Cl_2 will have formed?

(a) The problem says the reaction is first order; therefore, we need to use the equation:

$$\ln\frac{[A]_0}{[A]_t} = kt$$

As normal, we will begin by extracting the data from the problem and labeling it.

$k = 2.2 \times 10^{-5}$ s^{-1} $[A]_0 = 0.0050$ M SO_2Cl_2 $[A]_t = 0.0030$ M SO_2Cl_2

$t = ?$

Rearranging the equation and entering the appropriate values:

$$t = \frac{\ln\frac{[A]_0}{[A]_t}}{k} = \frac{\ln\frac{[0.0050 \ \cancel{M}]_0}{[0.0030 \ \cancel{M}]_t}}{2.2 \times 10^{-5} \ \text{s}^{-1}} = 2.3 \times 10^4 \ \text{s}$$

Note, we were looking for time, so we need to end with time units. Time units include s (seconds) but never s^{-1}.

(b) If you attempted to do this part this way:

$$\text{Moles } SO_2Cl_2 \text{ original} = \left(\frac{0.0050 \text{ mol } SO_2Cl_2}{L}\right)(2.0 \text{ L}) = 0.0100 \text{ mol } SO_2Cl_2$$

$$\text{Moles } Cl_2 = (1.0 \times 10^{-2} \text{ mol } SO_2Cl_2)\left(\frac{1 \text{ mol } Cl_2}{1 \text{ mol } SO_2Cl_2}\right) = 1.0 \times 10^{-2} \text{ mol } Cl_2$$

You got the wrong answer, because this approach assumes that all the SO_2Cl_2 reacted not just part of it.

What you need to do is determine how many of the SO_2Cl_2 moles reacted in 2.3×10^4 seconds. To do this, we need to find the moles remaining when SO_2Cl_2 has dropped to 0.0030 M SO_2Cl_2.

$$\text{Moles } SO_2Cl_2 \text{ remaining} = \left(\frac{0.0030 \text{ mol } SO_2Cl_2}{L}\right)(2.0 \text{ L}) = 0.0060 \text{ mol } SO_2Cl_2$$

$$\text{Moles } SO_2Cl_2 \text{ reacted} = (0.0100 - 0.0060) \text{ mol } SO_2Cl_2 = 0.0040 \text{ mol } SO_2Cl_2$$

Now, we can calculate the moles of Cl_2 formed from the mole of SO_2Cl_2 that actually reacted:

$$\text{Moles } Cl_2 = (4.0 \times 10^{-3} \text{ mol } SO_2Cl_2)\left(\frac{1 \text{ mol } Cl_2}{1 \text{ mol } SO_2Cl_2}\right) = 4.0 \times 10^{-3} \text{ mol } Cl_2$$

EXAMPLE 25.4

At 400°C, the decomposition of HI(g) is second order in HI(g) with a rate constant $k = 1.19 \times 10^{-3} \text{ M}^{-1}\text{s}^{-1}$. How long will it take for 0.75 M HI to decompose to 0.25 M?

The problem states that the reaction is second order; therefore, the integrated rate law is:

$$\frac{1}{[A]_t} - \frac{1}{[A]_0} = kt$$

Summarizing the information from the problem gives:

$$T = 400°C \qquad k = 1.19 \times 10^{-3} \text{ M}^{-1}\text{s}^{-1} \qquad [A]_0 = 0.75 \text{ M HI} \qquad [A]_t = 0.25 \text{ M}$$
$$t = ?$$

Inspecting the integrated rate law shows that we can ignore the T.

We can now rearrange the integrated rate law to solve for t and then enter the given information:

$$t = \frac{\frac{1}{[A]_t} - \frac{1}{[A]_0}}{k} = \frac{\frac{1}{0.25 \text{ M}} - \frac{1}{0.75 \text{ M}}}{1.19 \times 10^{-3} \text{ M}^{-1}\text{s}^{-1}} = \frac{\left(\frac{1}{0.25} - \frac{1}{0.75}\right) \text{M}^{-1}}{1.19 \times 10^{-3} \text{ M}^{-1}\text{s}^{-1}} = 2.2 \times 10^3 \text{ s}$$

We end with time units and no other units, which is what the question is asking for.

EXAMPLE 25.5

Ammonia gas will decompose to the elements, through a zero-order reaction, when it comes in contact with a heated tungsten surface. A sealed container has a tungsten filament heated to 1,400 K. How much $NH_3(g)$ is left after 375 s if the initial ammonia concentration was 0.0475 M, and $k = 3.40 \times 10^{-6}$ M s^{-1}?

We will begin with the zero-order integrated rate law: $[A]_0 - [A]_t = kt$
Summarizing the information from the problem gives:

$$T = 1,400 \text{ K} \qquad k = 3.40 \times 10^{-6} \text{ M s}^{-1} \qquad [A]_0 = 0.0475 \text{ M NH}_3 \qquad [A]_t = ?$$
$$t = 375 \text{ s}$$

We can now rearrange the integrated rate law to solve for $[A]_t$ and then enter the given information:

$$[A]_t = [A]_0 - kt = 0.0475 \text{ M NH}_3 - (3.40 \times 10^{-6} \text{ M s}^{-1})(375 \text{ s}) = 0.0462 \text{ M NH}_3$$

We end with concentration units, and no other units, which is what the question is asking for.

EXERCISE

25·4

Integrated rate laws

1. The decomposition of NOCl(g) at 750°C is second order with a rate constant of 8.50×10^5 M^{-1}s^{-1}. How long will it take for the concentration of NOCl(g) to drop from 0.225 M to 0.100 M? The reaction is:

   $2 \text{ NOCl(g)} \rightarrow 2 \text{ NO(g)} + \text{Cl}_2(g)$

2. The decomposition of dimethyl ether $(CH_3)_2O$, at 777 K is a first-order reaction with $k = 4.28 \times 10^{-4}$ s^{-1}. How much dimethyl ether will remain for a 0.500 M sample after 1.00 h? The reaction is:

   $(CH_3)_2O(g) \rightarrow CH_4(g) + H_2(g) + CO(g)$

Half-life

A useful concept derived from the integrated rate laws is the half-life. The half-life, $t_{1/2}$, is the time required for one-half the reactant to be used. It is possible to determine the half-life by using the appropriate integrated rate law and entering $[A]_t = 1/2 [A]_0$ and solving for t. An alternative time-saving way is to use the appropriate equation from the following:

First-order: $\qquad t_{1/2} = \dfrac{\ln 2}{k} = \dfrac{0.693}{k}$

Second-order: $\qquad t_{1/2} = \dfrac{1}{k[A]_0}$

Zero-order: $\qquad t_{1/2} = \dfrac{[A]_0}{2k}$

There are cases where knowing the half-life of a process is more important than other kinetic information. For example, the half-life of a drug in your body can lead to a prediction of when you need to take another dose of that particular medication. Or the half-life of certain radioactive material lets the hospital know when they need to order additional quantities of material for either diagnostic or therapeutic radiation treatment.

Note that for each of the half-life equations given previously, the rate constant is important in every case, and in all except the first-order equation, the initial concentration is also needed.

EXAMPLE 25.6

Radioactive decay involves first-order kinetics. The half-life of tritium, ^3H, is 13.6 years. What is the rate constant for the radioactive decay of tritium?

We need to begin with the first-order half-life equation: $t_{1/2} = \dfrac{0.693}{k}$.

Summarizing the information from the problem gives: $t_{1/2} = 13.6$ years $\quad k = ?$

We need to rearrange the half-life equation to: $k = \dfrac{0.693}{t_{1/2}}$, and then entering the value of the half-life into the equation:

$$k = \frac{0.693}{t_{1/2}} = \frac{0.693}{13.6 \text{ y}} = 0.0510 \text{ y}^{-1}$$

Note that the units here are in terms of years instead of seconds as used in the other examples. Any time unit will always work, and it is only if you have mixed time units or are asked for a different time unit that you need to do any conversions.

The preceding example may be a stand-alone problem as in the example or part of a longer problem such as in the next example.

EXAMPLE 25.7

Radioactive carbon-14 is the substance used for dating most archeological samples. This process, known as radiocarbon dating, is based on the half-life of carbon-14, 5,730 years, and that living organisms have a constant level of carbon-14 present. When the organism dies, this level decreases with age. A sample of wood from an Egyptian tomb has a carbon-14 level that is 68.5% of the level found in living organisms. How old is the sample?

Since radioactive processes follow first-order kinetics, we need the following two equations:

$$\ln \frac{[A]_0}{[A]_t} = kt \quad \text{and} \quad t_{1/2} = \frac{0.693}{k}$$

Summarizing the information from the problem gives:

$$t_{1/2} = 5,730 \text{ years} \quad k = ? \quad [A]_0 = 100\% \quad [A]_t = 68.5\% \quad t = ?$$

Note, we are using percent instead of molarities for $[A]_0$ and $[A]_t$ as the actual molarities are not necessary (a percent of a molarity is still a molarity). This is because it is not necessary to know the exact concentration but only the relative concentration.

First, we need to determine the value of k by using the half-life equation:

$$k = \frac{0.693}{t_{1/2}} = \frac{0.693}{5,730 \text{ y}} = 1.21 \times 10^{-4} \text{ y}^{-1}$$

Now that we have the value of k, we can use the integrated rate law, rearranged, and with values entered:

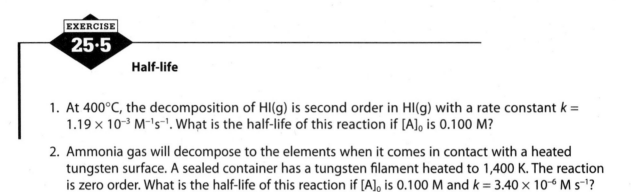

$$t = \frac{\ln \dfrac{[A]_0}{[A]_t}}{k} = \frac{\ln \dfrac{[100\%]_0}{[68.5\%]_t}}{1.21 \times 10^{-4} \, y^{-1}} = 3{,}130 \, y$$

1. At 400°C, the decomposition of HI(g) is second order in HI(g) with a rate constant $k = 1.19 \times 10^{-3} \, M^{-1}s^{-1}$. What is the half-life of this reaction if $[A]_0$ is 0.100 M?

2. Ammonia gas will decompose to the elements when it comes in contact with a heated tungsten surface. A sealed container has a tungsten filament heated to 1,400 K. The reaction is zero order. What is the half-life of this reaction if $[A]_0$ is 0.100 M and $k = 3.40 \times 10^{-6} \, M \, s^{-1}$?

3. The following reaction is first order in $SO_2Cl_2(g)$:

   $$SO_2Cl_2(g) \rightarrow SO_2(g) + Cl_2(g)$$

For this reaction, $k = 2.2 \times 10^{-5} \, s^{-1}$. What is the half-life of this reaction if $[A]_0$ is 0.100 M?

Energy profiles and activation energy

Why do some reactions happen very fast, while other reactions take place extremely slowly? Part of the answer is energy. It is possible to represent the progress of a reaction in terms of energy as:

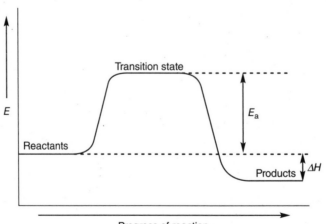

As shown by the diagram the reactants go through a transition state before they become products. The transition state is higher energy than the reactants by an amount known as the activation energy, E_a. In this case, the products are lower than the reactants by an amount known as the heat of reaction. This is an exothermic situation. If the process were endothermic, the products would be higher energy than the reactants. For kinetics the key is the activation energy. The **activation energy** is the minimum amount of energy necessary for a reaction to occur. The greater the activation energy, the slower the reaction.

To speed up a reaction, you could heat the reaction. On heating, the reactants will increase in energy, which would make it easier to get over the activation energy barrier. As a rule of thumb, a 10°C increase in temperature doubles the rate of a reaction.

Another way to increase the rate of a reaction is to use a catalyst. A catalyst, which is not consumed by the reaction, provides an alternate pathway for the reaction with a lower activation energy. A **catalyst** is a substance that alters the rate of a reaction without being consumed by the reaction.

Introduction to equilibrium

Stoichiometry is usually the first way that students examine calculations involving chemical reactions. However, stoichiometry has its limitations. For example, stoichiometry tells you how much product a reaction might produce; however, it is kinetics not stoichiometry that tells you it might take years for the reaction to produce the calculated amount of material. Kinetics is not the only adjustment necessary for stoichiometry calculations; equilibrium is an additional adjustment. Some reactions do not generate the calculated amount of product, because the reaction goes to equilibrium instead of completion.

In principle, all reactions are equilibrium processes; however, the equilibrium adjustment may be too small to be of concern. Equilibrium must be considered because, in general, every reaction is reversible. This reversibility means that the forward and reverse reactions proceed simultaneously, at the same rate to give a balance where the concentrations or partial pressures of the substances involved in the reactions do not appear to change.

The lack of change in the concentrations is a balance that allows a variety of calculations to be made, which leads to some important conclusions concerning everyday life.

Equilibria are described by using either a reaction quotient (Q) or an equilibrium constant (K). There may be subscripts to indicate a specific type of equilibrium. Q applies in all cases, while K applies only at equilibrium. Both Q and K require the activities of the substances involved in the reaction. (The determination of activities are a topic for an advanced chemistry course.) The activities may be approximated by either the molarity or the partial pressure (especially when the values are small). Activities are unitless quantities, so both K and Q are unitless. The setup for either a Q or K calculation is basically the same:

$$\frac{\text{Information on the products}}{\text{Information on the reactants}}$$

A "c" subscript, or sometimes no subscript, is used when the calculation is based on concentration (molarity), and a "p" subscript is used when the calculation is based on the partial pressures (normally atmospheres).

Equilibrium reactions

A chemical equation for an equilibrium reaction is indicted by the presence of a double arrow, $\leftrightarrows$, instead of a single arrow, $\rightarrow$. The double arrow reenforces the idea that both the forward and reverse reactions are occurring simultaneously.

Let us consider the following equilibrium reaction:

$$2\,C_2H_2\,(g) + 3\,O_2(g) \rightleftharpoons 4\,CO(g) + 2\,H_2O(g)$$

For this reaction, both Q and K will have information on the products $(4\,CO(g) + 2\,H_2O(g))$ over information on the reactants $(2\,C_2H_2\,(g) + 3\,O_2(g))$ as:

$$K_c = \frac{[CO]^4[H_2O]^2}{[C_2H_2]^2[O_2]^3} = 4.5 \qquad K_p = \frac{P_{CO}^4\,P_{H_2O}^2}{P_{C_2H_2}^2\,P_{O_2}^3} = 1.1 \times 10^2$$

The square brackets represent concentrations and the P's represent partial pressures. The coefficients from the balanced chemical equation are the exponents in the K (and Q) expressions, or mass action expressions. The values given (4.5 and 1.1×10^2) are experimentally determined values. This same general procedure is used, sometimes with minor adjustments, for every Q or K expression. The minor adjustments are that solids, liquids, and solvents are not included (because in the case of solids and liquids, their activities are defined as equal to 1, while the amount of solvent does not change significantly).

EXAMPLE 1

Write both a K_c and a K_p for each of the following. Try doing this before looking at the answers.

(a) $CO(g) + Cl_2(g) \rightleftharpoons COCl_2(g)$

(b) $4\,HCl(g) + O_2(g) \rightleftharpoons 2\,H_2O(g) + 2\,Cl_2(g)$

(c) $2\,NOCl(g) \rightleftharpoons 2\,NO(g) + Cl_2(g)$

(d) $N_2(g) + 3\,H_2(g) \rightleftharpoons 2\,NH_3(g)$

(e) $2\,NH_3(g) \rightleftharpoons N_2(g) + 3\,H_2(g)$

(a) $K_c = \dfrac{[COCl_2]}{[CO][Cl_2]}$ $\qquad$ $K_p = \dfrac{P_{COCl_2}}{P_{CO}\,P_{Cl_2}}$

(b) $K_c = \dfrac{[H_2O]^2[Cl_2]^2}{[HCl]^4[O_2]^2}$ $\qquad$ $K_p = \dfrac{P_{H_2O}^2\,P_{Cl_2}^2}{P_{HCl}^4\,P_{O_2}^2}$

(c) $K_c = \dfrac{[NO]^2\,[Cl_2]}{[NOCl]^2}$ $\qquad$ $K_p = \dfrac{P_{CO}^2\,P_{Cl_2}}{P_{NOCl}^2}$

(d) $K_c = \dfrac{[NH_3]^2}{[N_2][H_2]^3}$ $\qquad$ $K_p = \dfrac{P_{NH_3}^2}{P_{N_2}\,P_{H_2}^3}$

(e) $K_c = \dfrac{[N_2][H_2]^3}{[NH_3]^2}$ $\qquad$ $K_p = \dfrac{P_{N_2}\,P_{H_2}^3}{P_{NH_3}^2}$

If you got any different answers, try to understand why before moving on to the discussion.

Discussion

(a) Both of these equations are typical and yield a simple K_c and K_p because all of the coefficients are 1.

(b) These are like part (a) except that the coefficients are no longer ones; therefore, the superscripts are no longer understood ones.

(c) This is similar to part (b).

(d) and (e) These are treated together here as they illustrate an important point. If you reverse an equilibrium, the equilibrium expressions are the inverses of each other. In addition, if we rename the K_c's to $K_{c(d)}$ and $K_{c(e)}$, we find that the relationships between these two K's are:

$$K_{c(e)} = \frac{1}{K_{c(d)}}$$

This relationship allows you to convert from one to the other. This would also work if we use the K_p relationships for these two examples.

EXAMPLE 2

Write a K_c for each of the following. Try doing this before looking at the answers.

(a) $HNO_2(aq) \leftrightharpoons H^+(aq) + NO_2^-(aq)$
(b) $NH_3(aq) + H_2O(l) \leftrightharpoons OH^-(aq) + NH_4^+(aq)$
(c) $Cu^{2+}(aq) + 4\ Cl^-(aq) \leftrightharpoons [CuCl_4]^{2-}(aq)$
(d) $H_2O(l) \leftrightharpoons H^+(aq) + OH^-(aq)$
(e) $CaCO_3(s) \leftrightharpoons Ca^{2+}(aq) + CO_3^{2-}(aq)$

(a) $K_c = \dfrac{[H^+][NO_2^-]}{[HNO_2]}$

(b) $K_c = \dfrac{[OH^-][NH_4^+]}{[NH_3]}$

(c) $K_c = \dfrac{[CuCl_4^{2-}]}{[Cu^{2+}][Cl^-]^4}$

(d) $K_c = [H^+][OH^-]$

(e) $K_c = [Ca^{2+}][CO_3^{2-}]$

If you got any different answers, try to understand why before moving on to the discussion.

Discussion

This group of examples was chosen because they will appear repeatedly in upcoming equilibrium chapters. Note that these follow the same rules as all other equilibrium expressions.

(a) This type of equilibrium will appear whenever we examine equilibria involving weak acids. For weak acids, K_c is renamed to K_a with the subscript "a" being for "acid." When we see these again, we will see that they *always* have the form: $K_a = \dfrac{[H^+][\ ]}{[\ ]}$. Every weak acid has a K_a setup like this with *no* variations. For this reason, as soon as you see the subscript "a," you *know* exactly what the K expression will look like.

(b) This is similar to part (a) except that it is for weak bases. Note, the $H_2O(l)$ is left out of the K_c expression because water is the solvent. For weak bases, K_c is renamed to K_b with the subscript "b" being for base. When we see these again, we will see that they *always* have the form: $K_b = \dfrac{[OH^-][\ ]}{[\ ]}$. Every weak base has a K_b setup like this with *no* variations.

As with K_a, as soon as you see K_b, you *know* what the equilibrium expression looks like.

(c) This deals with the formation of a complex ion (defined later). When used, as here, for the formation of a complex ion, the K_c is renamed to K_f. A K_f always has only the complex on the product side and only the "pieces" on the reactant side. If this is reversed, it is a dissociation reaction with K_c renamed to K_d, a dissociation constant. K_d is the reciprocal of K_f.

(d) If you accidently wrote the incorrect K_c expression as: $K_c = \dfrac{[H^+][OH^-]}{[H_2O]}$, you forgot that liquids do not appear in K_c expressions. This equilibrium is present every time liquid water is involved, including when it is the solvent. When we see this equilibrium again, it will be renamed from K_c to K_w, and it also will never vary from how it appears here.

(e) This type of equilibrium applies to "insoluble" substances since nothing is completely insoluble. If you accidently wrote the incorrect K_c expression as: $K_c = \dfrac{[Ca^{2+}][CO_3^{2-}]}{[CaCO_3]}$, you forgot that solids do not appear in K_c expressions. This type of equilibrium reaction is a solubility reaction, and the equilibrium expression is a solubility product with K_c renamed to K_{sp}. These never have anything except the "insoluble" material on the reactant side and only the component ions or the product side, which means a K_{sp} never has a denominator since the denominator would only be a solid.

EXERCISE

26·1

Equilibrium reactions

1. Write a K_c for: $HCN(aq) \leftrightharpoons H^+(aq) + CN^-(aq)$

2. Write a K_p for: $2\,SO_2(g) + O_2(g) \leftrightharpoons 2\,SO_3(g)$

3. Write a K_c for: $Ag_2SO_4(s) \leftrightharpoons 2\,Ag^+(aq) + SO_4^{2-}(aq)$

4. Write a K_c for: $HC_2H_3O_2(aq) \leftrightharpoons C_2H_3O_2^-(aq) + H^+(aq)$

Equilibrium constants

There are situations where the simple K expression, of any type, is all that is needed; however, there are situations where the numerical value of K is needed. These need to be determined experimentally (either directly or indirectly) and represent a "constant" as long as the temperature is constant.

EXAMPLE 3

The following equilibrium is established in a sealed 10.0 L container at 180°C:

$$2\,NO_2(g) \leftrightharpoons 2\,NO(g) + O_2(g)$$

Analysis of the mixture found the following partial pressures:

$$NO_2 = 0.105 \text{ atm} \qquad NO = 0.0114 \text{ atm} \qquad O_2 = 0.0229 \text{ atm}$$

What is the value of K_p?

We begin this and all equilibrium calculations by writing the appropriate K expression:

$$K_p = \frac{P_{NO}^2 \, P_{O_2}}{P_{NO_2}^2}$$

We can now enter the appropriate partial pressures into this expression and determine K_p:

$$K_p = \frac{(0.0114)^2 (0.0229)}{(0.105)^2} = 2.70 \times 10^{-4}$$

Note that since K values are based upon unitless activities, there are no units needed anywhere in this calculation.

EXAMPLE 4

The following equilibrium is established at 25°C:

$$HC_2H_3O_2(aq) \leftrightarrows H^+(aq) + C_2H_3O_2^-(aq)$$

Analysis of the mixture found the following concentrations:

$$HC_2H_3O_2 = 0.9958 \text{ M} \qquad H^+ = 0.00424 \text{ M} \qquad C_2H_3O_2^- = 0.00424 \text{ M}$$

What is the value of K_c?
  First, write the appropriate K expression:

$$K_c = \frac{[H^+][C_2H_3O_2^-]}{[HC_2H_3O_2]}$$

We can now enter the appropriate concentrations into this expression and determine K_c:

$$K_c = \frac{[0.00424][0.00424]}{[0.9958]} = 1.81 \times 10^{-5}$$

Be careful about rounding, as minor intermediate rounding may lead to large errors.
  Other equilibrium constants may be determined as in these two examples; the only differences are how the concentrations and partial pressures are determined.

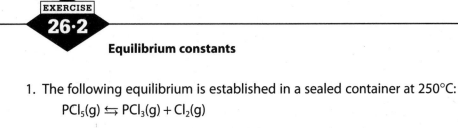

EXERCISE

26·2

Equilibrium constants

1. The following equilibrium is established in a sealed container at 250°C:
   $$PCl_5(g) \leftrightarrows PCl_3(g) + Cl_2(g)$$

   Analysis of the mixture found the following partial pressures:
   $$PCl_5 = 0.0662 \text{ atm} \qquad PCl_3 = 0.242 \text{ atm} \qquad Cl_2 = 0.348 \text{ atm}$$

   What is the value of K_p?

2. The following equilibrium is established at 25°C:
$$Ag^+(aq)^+ + 2\,NH_3(aq) \leftrightarrows [Ag(NH_3)_2]^+(aq)$$

Analysis of the mixture found the following concentrations:
$$Ag^+ = 8.0 \times 10^{-10}\,M \qquad NH_3 = 2.8\,M \qquad [Ag(NH_3)_2]^+ = 0.10\,M$$

What is the value of K_c?

3. The value of K_c for the following reaction is $K_c = 7.2 \times 10^{-4}$:
$$HNO_2(aq) \leftrightarrows H^+(aq) + NO_2^-(aq)$$

What is the value of K_c for this reaction?
$$H^+(aq) + NO_2^-(aq) \leftrightarrows HNO_2(aq)$$

Multiple equilibria

In some cases it is necessary to consider more than one related equilibrium. These multiple equilibria may be the same equilibrium occurring more than once or one equilibrium leading to another related equilibrium.

EXAMPLE 5

Consider the following equilibrium:
$$2\,SO_3(g) \leftrightarrows 2\,SO_2(g) + O_2(g) \qquad K_p = 0.023 \text{ at } 625°C$$

What is the value of K_p for each of the following two equilibria?
$$4\,SO_3(g) \leftrightarrows 4\,SO_2(g) + 2\,O_2(g)$$
$$SO_3(g) \leftrightarrows SO_2(g) + 1/2\,O_2(g)$$

The first of these two new equilibria is double the original equilibrium; however, its K_p is not double the original K_p.

To find the new K_p, let us first write the original K_p and compare it to the new K_p:

$$\text{Original } K_p = \frac{P^2_{SO_2}\,P_{O_2}}{P^2_{SO_3}} \qquad \text{New } K_p = \frac{P^4_{SO_2}\,P^2_{O_2}}{P^4_{SO_3}}$$

Doubling the original K_p would give: $2\,\dfrac{P^2_{SO_2}\,P_{O_2}}{P^2_{SO_3}}$, which is not equal to the new K_p. To get to the

new K_p from the old, we need to square the original K_p as:

$$\left(\frac{P^2_{SO_2}\,P_{O_2}}{P^2_{SO_3}}\right)^2 = \frac{P^4_{SO_2}\,P^2_{O_2}}{P^4_{SO_3}}$$

Therefore, K_p^2 means $(0.023)^2 = 5.3 \times 10^{-4} = $ the new K_p.

Now for the third K_p ($SO_3(g) \leftrightarrows SO_2(g) + 1/2\,O_2(g)$). This K_p is one-half the original, so its K_p is the square root of the original:

$$\sqrt{\frac{P^2_{SO_2}\,P_{O_2}}{P^2_{SO_3}}} = \frac{P_{SO_2}\,P^{1/2}_{O_2}}{P_{SO_3}}$$

So the third $K_p = (0.023)^{1/2} = 0.15$

This same procedure works for all K's, not just for K_p's and for all multiples or fractions (with appropriate exponents).

EXAMPLE 6

Aqueous solutions of H_2Se involve the following two equilibria:

$$H_2Se(aq) \leftrightharpoons H^+(aq) + HSe^-(aq) \qquad K_{c1} = 1.3 \times 10^{-4}$$
$$HSe^-(aq) \leftrightharpoons H^+(aq) + Se^{2-}(aq) \qquad K_{c2} = 1.0 \times 10^{-11}$$

(The 1 and 2 added to the subscripts are present to distinguish between the two.) We can add these to equilibria together as:

$$H_2Se(aq) + HSe^-(aq) \leftrightharpoons 2\,H^+(aq) + HSe^-(aq) + Se^{2-}(aq) \qquad K_c = ?$$

This may be simplified as follows:

$$H_2Se(aq) + \cancel{HSe^-(aq)} \leftrightharpoons 2\,H^+(aq) + \cancel{HSe^-(aq)} + Se^{2-}(aq) \qquad K_c =$$
$$H_2Se(aq) \leftrightharpoons 2\,H^+(aq) + Se^{2-}(aq) \qquad K_c = ?$$

What is the value of K for the new equilibrium?

Even though we added the two separate equilibria, the new K_c is not $K_{c1} + K_{2c}$. Let's write the individual K expressions:

$$K_{c1} = \frac{[H^+][HSe^-]}{[H_2Se]} \qquad K_{c2} = \frac{[H^+][Se^{2-}]}{[HSe^-]} \qquad K_c = \frac{[H^+]^2[Se^{2-}]}{[H_2Se]}$$

It is now obvious why $(K_{c1} + K_{2c})$ will not work. The addition of equilibria requires that the individual K's be multiplied as:

$$K_{c1} \times K_{c2} = \frac{[H^+][\cancel{HSe^-}]}{[H_2Se]} \times \frac{[H^+][Se^{2-}]}{[\cancel{HSe^-}]} = \frac{[H^+]^2[Se^{2-}]}{[HSe^-]}$$

Thus: $K_c = K_{c1} \times K_{c2} = (1.3 \times 10^{-4})(1.0 \times 10^{-11}) = 1.3 \times 10^{-15}$

Just as the preceding example, this works for all K's as long as they have something in common as is the HSe^- in this example.

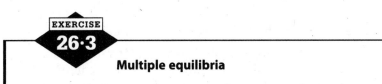

Multiple equilibria

1. The equilibrium constant for the following equilibrium is $K_p = 2.92 \times 10^{-6}$ at 575 K:

   $$N_2(g) + 3\,H_2(g) \leftrightharpoons 2\,NH_3(g)$$

   What is the equilibrium constant for the following?

   $$3\,N_2(g) + 9\,H_2(g) \leftrightharpoons 6\,NH_3(g)$$

2. Aqueous solutions of $H_2C_2O_4$ involve the following two equilibria:

   $$H_2C_2O_4(aq) \leftrightharpoons H^+(aq) + HC_2O_4^-(aq) \qquad K_{c1} = 1.3 \times 10^{-4}$$
   $$HC_2O_4^-(aq) \leftrightharpoons H^+(aq) + C_2O_4^{2-}(aq) \qquad K_{c2} = 1.0 \times 10^{-11}$$

   What is the value of K for the following equilibrium?

   $$H_2C_2O_4(aq) \leftrightharpoons 2\,H^+(aq) + C_2O_4^{2-}(aq) \qquad K = ?$$

Le Châtelier's principle

Equilibria such as we have been discussing are examples of dynamic equilibria. Dynamic equilibria are systems where there is a balance between two opposing processes going at equal rates. In the chemical systems we have been studying, the two opposing processes are the forward and reverse reactions. All dynamic equilibria are governed by Le Châtelier's principle.

Le Châtelier's principle says that if a stress is applied to a system at equilibrium, the system will respond to counter the stress and go to a new equilibrium.

The stress mentioned in the definition refers to any change made in the equilibrium system. Examples of stresses are adding or removing a reactant or product, changing the temperature, changing the pressure, or adding or removing a catalyst.

Changing the amount of a reactant or product may be countered by replacing anything removed or by subtracting anything added. These may be accomplished by temporarily increasing the rate of the forward or reverse reaction over that of the opposite reaction.

EXAMPLE 7

Let us use the following equilibrium as an example:

$$N_2(g) + 3\,H_2(g) \rightleftharpoons 2\,NH_3(g)$$

Case 1:	+
Case 2:	+
Case 3:	−
Case 4:	−

The plus sign indicates that a product or reactant has been added, and a minus sign indicates that a reactant or product has been removed. What side a substance is being added or removed from is indicated by the position of the sign.

Case 1:   The addition of a reactant (either or both reactants) speeds the forward reaction, using some of the added reactant (countering the stress) until the amount of new product formed increases the rate of the reverse reaction to the point where the rates become equal again.

Case 2:   The addition of a product speeds the reverse reaction, using some of the added product (countering the stress) until the amount of new reactant formed increases the rate of the forward reaction until the rates again become equal.

Case 3:   The removal of a reactant (either or both reactants) slows the forward reaction, making the reverse reaction relatively faster, allowing some of the reactant to be replaced (countering the stress) until the amount of new reactant formed increases the rate of the forward reaction until the rates become equal again.

Case 4:   The removal of a product slows the reverse reaction, making the forward reaction relatively faster, allowing some of the product to be replaced (countering the stress) until the amount of new product formed increases the rate of the reverse reaction until the rates again become equal.

Note: changes due to adding or removing a reactant or product are limited to substances appearing in the Q or K expression (but not solids, liquids, or solvents) unless the amount is reduced to 0, in which case, there is no equilibrium.

How a change in the temperature will shift a reaction depends on whether the reaction is endothermic or exothermic. For an exothermic reaction, heat may be treated as a product with an increase in temperature corresponding to the addition of a product and a decrease in temperature corresponding to a loss of product. For an endothermic reaction, heat may be treated as a reactant with an increase in temperature corresponding to the addition of a reactant and a decrease in temperature corresponding to a loss of reactant. This allows predictions of changes along the lines outlined in the preceding four cases.

Changes in pressure are only significant for gases. In addition, according to Boyle's gas law, pressure is inversely related to volume, which means that pressure changes are tied to volume changes. To examine pressure (volume) changes, it is necessary to compare the number of moles of gas on the product and reactant sides of the reaction. An increase in pressure (decrease in volume) leads to a shift toward the side with fewer moles of gas, while a decrease in pressure (increase in volume) leads to a shift to the side with more moles of gas.

The presence of a catalyst involves kinetics not equilibrium. A catalyst allows the reaction to reach equilibrium in less time; however, it catalyzes both the forward and reverse reactions, so the equilibrium amounts are the same.

EXAMPLE 8

We will begin with the equilibrium used in the preceding example:

$$N_2(g) + 3\,H_2(g) \leftrightarrows 2\,NH_3(g) \qquad \text{exothermic}$$

Which direction will this equilibrium shift if the following stresses are applied to the equilibrium:

(a) NH_3 is added.

(b) H_2 is removed.

(c) The pressure on the system is increased.

(d) The temperature is decreased.

(e) A catalyst is added.

(a) The system will shift to the left.

(b) The system will shift to the left.

(c) The system will shift to the right.

(d) The system will shift to the right.

(e) No change

It is also possible to determine the direction an equilibrium will shift by calculating a value for the reaction quotient, Q. Recall that the reaction quotient has the same form as an equilibrium constant. Unlike a K, the reaction quotient is not necessarily at equilibrium.

To predict the direction a reaction will shift depends on the relative values of Q and K.

If $K > Q$, the system will shift to the right to reach equilibrium.
If $K < Q$, the system will shift to the left to reach equilibrium.
If $K = Q$, the system is already at equilibrium, no shift is necessary.

EXAMPLE 9

The following equilibrium is established at 25°C:

$$HC_2H_3O_2(aq) \leftrightarrows H^+(aq) + C_2H_3O_2^-(aq) \qquad K_c = 1.81 \times 10^{-5}$$

Predict which direction the preceding equilibrium will shift to reach equilibrium:

(a) $[HC_2H_3O_2] = 1.002$ M $\quad$ $[H^+] = 0.00424$ M $\quad$ $[C_2H_3O_2^-] = 0.00424$ M

(b) $[HC_2H_3O_2] = 0.9958$ M $\quad$ $[H^+] = 0.00524$ M $\quad$ $[C_2H_3O_2^-] = 0.00524$ M

(c) $[HC_2H_3O_2] = 0.9958$ M $\quad$ $[H^+] = 0.00424$ M $\quad$ $[C_2H_3O_2^-] = 0.00424$ M

The reaction quotient for the preceding equilibrium is:

$$Q_c = \frac{[H^+][C_2H_3O_2^-]}{[HC_2H_3O_2]}$$

We can now enter the appropriate concentrations for each case into this expression and determine Q_c:

(a) $Q_c = \dfrac{[0.00420][0.00420]}{[1.002]} = 1.79 \times 10^{-5}$ $\quad K > Q \quad$ Shift right

(b) $Q_c = \dfrac{[0.00524][0.00524]}{[0.9958]} = 2.75 \times 10^{-5}$ $\quad K < Q \quad$ Shift left

(c) $Q_c = \dfrac{[0.00424][0.00424]}{[0.9958]} = 1.81 \times 10^{-5}$ $\quad K = Q \quad$ No shift (at equilibrium)

EXERCISE

26·4

Le Châtelier's principle

1. How will the following equilibrium shift if $CO_2(g)$ is added? Why?

    $CaCO_3(s) \leftrightarrows CaO(s) + CO_2(g)$

2. How will the following equilibrium shift if the following endothermic reaction is heated? Why?

    $CaCO_3(s) \leftrightarrows CaO(s) + CO_2(g)$

3. How will the following equilibrium shift if $CaCO_3(s)$ is added? Why?

    $CaCO_3(s) \leftrightarrows CaO(s) + CO_2(g)$

4. Which direction will the following reaction shift to reach equilibrium? Why? Initial concentrations are listed after the equilibrium equation.

    $PCl_5(g) \leftrightarrows PCl_3(g) + Cl_2(g)$ $\quad K_p = 1.27$

    $PCl_5 = 0.0862$ atm $\quad PCl_3 = 0.242$ atm $\quad Cl_2 = 0.348$ atm

5. Which direction will the following reaction shift to reach equilibrium? Why? Initial concentrations are listed after the equilibrium equation.

    $PCl_5(g) \leftrightarrows PCl_3(g) + Cl_2(g)$ $\quad K_p = 1.27$

    $PCl_5 = 0.0662$ atm $\quad PCl_3 = 0.250$ atm $\quad Cl_2 = 0.448$ atm

Equilibrium calculations

There are situations where it is helpful to be able to calculate either the concentration or partial pressure of a reactant or product in an equilibrium. In principle, it is always possible to calculate each of the substances in an equilibrium if the other substances are known. This process will not work for any substance not in the K expression.

EXAMPLE 10

We will use the following equilibrium as an example:

$$N_2(g) + 3\,H_2(g) \rightleftharpoons 2\,NH_3(g) \qquad K_p = 2.92 \times 10^{-6} \text{ at 575 K}$$

The question is what is the partial pressure of NH_3 if the partial pressure of N_2 is 0.375 atm and the partial pressure of H_2 is 0.850 atm?

As always, for an equilibrium calculation, we need the K expression:

$$K_p = \frac{p_{NH_3}^2}{p_{N_2}\,p_{H_2}^3} = 2.92 \times 10^{-6}$$

We can now enter the given partial pressures:

$$K_p = \frac{P_{NH_3}^2}{(0.375)(0.850)^3} = 2.92 \times 10^{-6}$$

Rearranging and taking the square root:

$$\sqrt{P_{NH_3}^2} = P_{NH_3} = \sqrt{K_p (0.375)(0.850)^3} = 8.20 \times 10^{-4} \text{ atm}$$

EXAMPLE 11

We will use the following equilibrium as an example:

$$HC_2H_3O_2(aq) \rightleftharpoons H^+(aq) + C_2H_3O_2^-(aq) \qquad K_c = 1.80 \times 10^{-5}$$

The question is what are the concentrations of $HC_2H_3O_2$, H^+, and $C_2H_3O_2^-$ when 1.000 mol of $HC_2H_3O_2$ is added to 1.000 L of solution?

As always, we need an equilibrium expression:

$$K_c = \frac{[H^+][C_2H_3O_2^-]}{[HC_2H_3O_2]}$$

We also need to know the initial concentration of $HC_2H_3O_2(aq) = \dfrac{1.000 \text{ mol } HC_2H_3O_2}{1.000 \text{ L}} = 1.000 \text{ M}.$

Next we will recopy the equilibrium into a table with the initial concentrations of the substances listed below each of the appropriate substances (—) if none is present. This line is labeled *Initial*.

	$HC_2H_3O_2(aq)$	$\rightleftharpoons$	$H^+(aq)$	+	$C_2H_3O_2^-(aq)$	$K_c = 1.80 \times 10^{-5}$
Initial	1.000 M		—		—	

In order to achieve equilibrium, some of the $HC_2H_3O_2$ must dissociate. We shall call this amount x. This amount will be subtracted from the initial $[HC_2H_3O_2]$, and since all the stoichiometric ratios are 1:1, x will be the amount of both H^+ and $C_2H_3O_2^-$ formed. We will add these changes to the table as a line labeled *Change*:

$$HC_2H_3O_2(aq) \leftrightarrows H^+(aq) + C_2H_3O_2^-(aq) \qquad K_c = 1.80 \times 10^{-5}$$

	$HC_2H_3O_2$	H^+	$C_2H_3O_2^-$
Initial	1.000 M	—	—
Change	$-x$	$+x$	$+x$

We will now add the values in each column to produce the entries for another line in the table called *Equilibrium*:

$$HC_2H_3O_2(aq) \leftrightarrows H^+(aq) + C_2H_3O_2^-(aq) \qquad K_c = 1.80 \times 10^{-5}$$

	$HC_2H_3O_2$	H^+	$C_2H_3O_2^-$
Initial	1.000 M	—	—
Change	$-x$	$+x$	$+x$
Equilibrium	$1.000 - x$	$+x$	$+x$

Taking the first letters of the words describing each line, we get ICE, which is why this type of table is commonly referred to as an ICE table. ICE tables are commonly used to solve equilibrium problems.

We can now enter the values from the Equilibrium line of the ICE table into the K expression:

$$K_c = \frac{[H^+][C_2H_3O_2^-]}{[HC_2H_3O_2]} = 1.80 \times 10^{-5} = \frac{[x][x]}{[1.000 - x]}$$

We can rearrange this equation to:

$$(1.80 \times 10^{-5})(1.000 - x) = x^2$$

Then:

$$x^2 + 1.80 \times 10^{-5}x - 1.80 \times 10^{-5} = 0$$

This is a quadratic equation. If you have not dealt with quadratic equations previously, skip the following paragraph.

This quadratic equation gives: $a = 1$, $b = +1.80 \times 10^{-5}$, and $c = -1.80 \times 10^{-5}$.

$$\text{Entering these into } x = \frac{-b \pm \sqrt{b^2 - 4ac}}{2a}$$

and solving gives $x = 4.23 \times 10^{-3}$ M or -4.25×10^{-3} M as results; however, negative concentrations are not possible, so the negative choice may be eliminated. We can now enter this x into the equilibrium line of the ICE table:

Equilibrium	$1.000 - 4.23 \times 10^{-3} = 0.996$ M	$+4.23 \times 10^{-3}$ M	$+4.23 \times 10^{-3}$ M

These are our answers.

Without the quadratic equation, we must refer to the value of K. When K is very large or very small, the calculation may be simplified. In this case, K is very small when compared to 1.000 M. This means that when the system reaches equilibrium 1.000 M $- x \approx 1.000$ M since x is so small. Using 1.000 M instead of 1.000 M $- x$ into the K expression gives us:

$$K_c = \frac{[H^+][C_2H_3O_2^-]}{[HC_2H_3O_2]} = 1.80 \times 10^{-5} = \frac{[x][x]}{[1.000]}$$

This rearranges to $x^2 = (1.000)(1.80 \times 10^{-5}) = 1.80 \times 10^{-5}$ with $x = 4.24 \times 10^{-3}$. This is indeed small compared to 1.000 M, which means this approach is acceptable. If it were not acceptable, this simplified procedure will not work. Since the answer is acceptable, we can enter it into the Equilibrium line of the ICE table:

Equilibrium	$1.000 - 4.24 \times 10^{-3} = 0.996$ M	$+4.24 \times 10^{-3}$ M	$+4.24 \times 10^{-3}$ M

EXAMPLE 12

We will use the following equilibrium as an example:

$$MgF_2(s) \leftrightarrows Mg^{2+}(aq) + 2\,F^-(aq) \qquad K_c = 3.7 \times 10^{-8}$$

The question is what are the concentrations of Mg^{2+}, and F^- when 1.000 mol of MgF_2 is added to 1.000 L of solution?

As always, we need an equilibrium expression:

$$K_c = [Mg^{2+}]\,[F^-]^2 = 3.7 \times 10^{-8}$$

MgF_2 is left out since it is a solid. This also means that we do not need its concentration.

The next step is to construct an ICE table:

$$MgF_2(s) \leftrightarrows Mg^{2+}(aq) + 2\,F^-(aq) \qquad K_c = 3.7 \times 10^{-8}$$

Initial	—	—	—
Change	—	$+x$	$+2x$
Equilibrium	—	$+x$	$+2x$

Note that since there is a 2:1 ratio for forming F^-, the amount formed is $2x$. All x values will have a multiplier equal to the coefficient in the equation.

Entering values from the equilibrium line into the K expression gives:

$$K_c = [Mg^{2+}][F^-]^2 = (x)(2x)^2 = 4x^3 = 3.7 \times 10^{-8}$$

Solving for x:

$$x^3 = \frac{3.7 \times 10^{-8}}{4} = 9.2 \times 10^{-9}$$

$$x = \sqrt[3]{x^3} = \sqrt[3]{9.2 \times 10^{-9}} = 2.0 \times 10^{-3}$$

Entering this value into the equilibrium line of the ICE table gives us our answers:

Equilibrium   — 2.0×10^{-3} M Mg^{2+} $+2\,(2.0 \times 10^{-3}) = 4.0 \times 10^{-3}$ M F^-

---

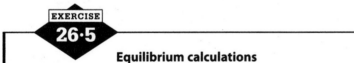

EXERCISE

26·5

Equilibrium calculations

1. For the following equilibrium, what is the partial pressure of CO if the partial pressure of H_2 is 266 atm, the partial pressure of CH_4 is 33.3 atm, and the partial pressure of H_2O is 28.4 atm?

   $$CO(g) + 3H_2(g) \leftrightarrows CH_4(g) + H_2O(g) \qquad K_p = 102 \text{ at } 500°C$$

2. For the following equilibrium, what are the concentrations of HOI, H^+, and OI^- when the initial concentration of HOI is 0.500 M?

   $$HOI(aq) \leftrightarrows H^+(aq) + OI^-(aq) \qquad K_c = 2.3 \times 10^{-11}$$

3. For the following equilibrium, what are the concentrations of Zn^{2+} and PO_4^{3-} when 1.000 mol of $Zn_3(PO_4)_2$ is added to 1.000 L of solution?

   $$Zn_3(PO_4)_2(s) \leftrightarrows 3Zn^{2+}(aq) + 2\,PO_4^{3-}(aq) \qquad K_c = 9.0 \times 10^{-33}$$

Acids and bases

Previously, we dealt with the Arrhenius definitions of acids and bases. An Arrhenius acid increases the hydrogen ion concentration in a solution. An Arrhenius base increases the hydroxide ion concentration in a solution. While the Arrhenius definitions work well in many cases, there are limitations that require slightly different definitions. The Arrhenius definitions do not cover processes that do not occur in solution, and these definitions neglect bases other than hydroxide ions. The new theory introduced in this chapter covers Brønsted-Lowry acids and bases.

Types of acids and bases

The Brønsted-Lowry theory was developed by Johannes Brønsted and Thomas Lowry in 1923. In this theory, the definitions of acids and bases follow the hydrogen ions. Accordingly, the definitions are:

A **Brønsted-Lowry acid** is a hydrogen ion donor.
A **Brønsted-Lowry base** is a hydrogen ion acceptor.

These definitions differ from the Arrhenius acids and bases in that a solution is not necessary and that hydroxide ions are not specifically isolated from all possible hydrogen ion acceptors.

Let us use the following equilibrium to illustrate this:

$$HC_2H_3O_2(aq) \leftrightarrows H^+(aq) + C_2H_3O_2^-(aq)$$

This equilibrium, like all equilibria, has two processes—the forward reaction and the reverse reaction. In the forward reaction, $HC_2H_3O_2(aq)$ donates a hydrogen ion to water, forming $H^+(aq)$; thus $HC_2H_3O_2(aq)$ is a Brønsted-Lowry acid. In the reverse reaction, the $C_2H_3O_2^-(aq)$ accepts a hydrogen to form $HC_2H_3O_2(aq)$; thus, $C_2H_3O_2^-(aq)$ is a Brønsted-Lowry base.

For acetic acids the species $HC_2H_3O_2(aq)$ and $C_2H_3O_2^-(aq)$ form a pair differing by only one hydrogen ion. Such pairs occur for every acid, even strong acids. This pair is an example of a conjugate acid-base pair with the acid form being the conjugate acid and the base form being the conjugate base. It is important to remember that the difference is one, and only one, hydrogen ion.

A **conjugate acid-base pair** are two species differing by only 1 H^+.
A **conjugate acid** is the member of a conjugate acid-base pair with one more H^+.
A **conjugate base** is a member of a conjugate acid-base pair with one less H^+.

Let us now compare acetic acid to hydrochloric acid:

$$HC_2H_3O_2(aq) \leftrightharpoons H^+(aq) + C_2H_3O_2^-(aq)$$

$$HCl(aq) \rightarrow H^+(aq) + Cl^-(aq)$$

Since HCl is a strong acid, its dissociation is not an equilibrium. Another way of looking at this is that $Cl^-(aq)$ is too weak a base to accept a H^+ like $C_2H_3O_2^-(aq)$ can; therefore, the acetate ion is a stronger base than the chloride ion. The strengths of the conjugate bases are important factors in comparing the strengths of the conjugate acids in that the stronger the conjugate base, the weaker the conjugate acid, and vice versa.

What about polyprotic acids like phosphoric acid, H_3PO_4? It is a mistake to assume that since H_3PO_4 can donate 3 H^+ leaving PO_4^{3-}, then H_3PO_4 and PO_4^{3-} are a conjugate acid-base pair. This assumption would be a major error since the two differ by 3 H^+ not 1 H^+. The conjugate acid-base pairs for H_3PO_4 are:

$$H_3PO_4(aq) \leftrightharpoons H^+(aq) + H_2PO_4^-(aq) \qquad \text{First pair}$$

$$H_2PO_4^-(aq) \leftrightharpoons H^+(aq) + HPO_4^{2-}(aq) \qquad \text{Second pair}$$

$$HPO_4^{2-}(aq) \leftrightharpoons H^+(aq) + PO_4^{3-}(aq) \qquad \text{Third pair}$$

You must be careful with phosphoric acid and all other polyprotic acids.

Note: for any conjugate acid-base pair, $K_a K_b = K_w = 1.00 \times 10^{-14}$.

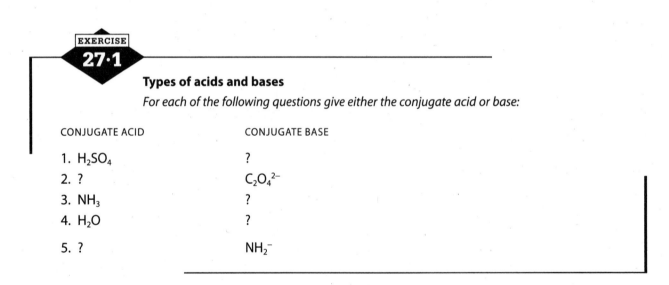

EXERCISE 27·1

Types of acids and bases

For each of the following questions give either the conjugate acid or base:

CONJUGATE ACID	CONJUGATE BASE
1. H_2SO_4	?
2. ?	$C_2O_4^{2-}$
3. NH_3	?
4. H_2O	?
5. ?	NH_2^-

Autoionization of water and pH

Water is an extremely important solvent; therefore, we need to examine its acid-base properties.

Water, like many solvents, undergoes autoionization. **Autoionization** occurs when a substance ionizes with itself. The autoionization of water may be represented by the following equilibrium:

$$2\, H_2O(l) \leftrightharpoons H_3O^+(aq) + OH^-(aq)$$

This reaction is normally represented as:

$$H_2O(l) \leftrightarrows H^+(aq) + OH^-(aq)$$

This equilibrium like all equilibria has an equilibrium constant defined as:

$$K_w = [H^+][OH^-]$$

As usual, the solvent, H_2O, is not included in the equilibrium expression. At 25°C, the value of this constant is 1.00×10^{-14}, which is what is normally used when studying aqueous systems. (Note: as with all equilibrium constants, K_w will vary with temperature.) The constant, K_w, applies to all systems involving liquid water; however, in most cases, it is too small to be significant.

From this equilibrium constant expression, we can determine the concentrations of the hydrogen and hydroxide ions in pure water. We will begin by creating an ICE table:

$$H_2O(l) \leftrightarrows H^+(aq) + OH^-(aq)$$

Initial	—	—	—
Change	—	$+x$	$+x$
Equilibrium	—	$+x$	$+x$

Entering this information into the equilibrium constant expression:

$$K_w = [H^+][OH^-] = [x][x] = x^2 = 1.00 \times 10^{-14}$$

Solving for x:

$$x = [H^+][OH^-] = 1.00 \times 10^{-7} \text{ M}$$

A situation such as this, where $[H^+] = [OH^-]$, is defined as neutral. If $[H^+] > [OH^-]$, it is defined as acidic. If $[H^+] < [OH^-]$, it is defined as basic. To produce an acidic solution, an acid needs to be added to water to increase the $[H^+]$, while adding a base will increase the $[OH^-]$.

The repeated writing of numbers such as 1.00×10^{-7} M $H^+(aq)$ can be cumbersome, so a simplified method of expressing the acidity/basicity of a solution has been devised. This simplified method is to express the acidity in terms of pH. The pH of a solution is calculated from the relationship:

$$pH = -\log[H^+]$$

Entering 1.00×10^{-7} M $H^+(aq)$ into this equation gives us $7.000 = pH$.

The rules for significant figures involving logarithms are different from the rules you have already seen. For example, entering 3.35×10^2 into your calculator and taking the logarithm displays 2.525044807, but how many of these digits are significant? Since the original number had three significant figures (3.35), the logarithm will also have three significant figures. However, the leading 2 (the characteristic) came from the exponent on 10^2, which was not significant in the original number and will not suddenly become significant in the logarithm. The part of the logarithm after the decimal point is the mantissa, and it is this part that includes the significant figures. In this case, the first three digits after the decimal point are the three significant figures. Therefore, reporting $\log 3.35 \times 10^2$ to three significant figures is 2.525, which has three not four significant figures. You can test this by taking and comparing the logarithms of each of the following three numbers: 3.35×10^1, 3.35×10^5, and 3.35×10^7.

A related but less commonly used term is the pOH, which is pOH = $-\log[OH^-]$. It is possible to relate the pH to the pOH of any aqueous solution as:

$$pH + pOH = pK_w = -\log K_w = -\log 1.00 \times 10^{-14} = 14.000$$

Note: the value of 14.000 assumes the solution is near room temperature (25°C).

EXAMPLE 1

Determine the pH of a 1.0 M HNO_3 solution.
   Since HNO_3 is a strong acid, it ionizes 100% as:

$$HNO_3(aq) \rightarrow H^+(aq) + NO_3^-(aq)$$

Since all the HNO_3 is converted to ions, the $H^+(aq)$ produced from a 1:1 mole ratio would be 1.0 M $H^+(aq)$, which when entered into the pH equation gives the pH of the solution:

$$pH = -\log[H^+] = -\log 1.0 = 0.00$$

A pH of 0.00 is strongly acidic, as expected from a solution containing a strong acid.

EXAMPLE 2

Determine the pH of a 1.0 M $Ca(OH)_2$ solution.
   Since $Ca(OH)_2$ is a strong base, it ionizes 100% as:

$$Ca(OH)_2(aq) \rightarrow Ca^{2+}(aq) + 2\,OH^-(aq)$$

The 2:1 ratio, or 2 OH^-: 1$Ca(OH)_2$, makes the $OH^-(aq)$ concentration 2.0 M OH^-. From here, there are two ways to reach the answer.

Approach 1

Use the K_w relationship to determine H^+ and use the pH definition to determine the final answer:

$$K_w = [H^+][OH^-] = [x][2.0] = 1.00 \times 10^{-14}\ M$$

$$[H^+] = x = \frac{1.00 \times 10^{-14}}{2.0} = 5.0 \times 10^{-15}\ M\ H^+(aq)$$

$$pH = -\log[H^+] = -\log 5.0 \times 10^{-15} = 14.30$$

A solution with pH = 14.30 is extremely basic.

Approach 2

We begin by finding the pOH:

$$pOH = -\log[OH^-] = -\log 2.0 = -0.30$$

Now we will use pH + pOH = pK_w = 14.000 to determine the answer:

$$pH = 14.000 - pOH = 14.000 - (-0.30) = 14.30$$

   It will be to your advantage to learn both approaches.
All strong acids and strong bases may be handled in a similar manner. Weak acids and weak bases require a different approach.

Autoionization of water and pH

1. What is the pH of a 0.0035 M $H^+(aq)$ solution?

2. What is the pH of a 0.0035 M $OH^-(aq)$ solution?

3. What is the pH of a 0.0135 M HCl(aq) solution?

4. What is the pH of a 0.0335 M $Ba(OH)_2(aq)$ solution?

K_a and K_b calculations

Whenever the pH or pOH of a weak acid or a weak base is needed, it is necessary to use an equilibrium constant.

$$HC_2H_3O_2(aq) \rightleftharpoons H^+(aq) + C_2H_3O_2^-(aq)$$

This is a weak acid with the following K expression:

$$K_c = \frac{[H^+][C_2H_2O_2^-]}{[HC_2H_3O_2]} = K_a = 1.8 \times 10^{-5}$$

The K_c for any weak acid is a K_a (acid equilibrium constant), which always has the form $\frac{[H^+][\text{Conjugate base}]}{[\text{Conjugate acid}]}$. The value (1.8×10^{-5}) is from a table or may be given in a problem. The K_a will be involved in some way in any weak acid equilibrium problem.

$$NH_3(aq) + H_2O(l) \rightleftharpoons OH^-(aq) + NH_4^+(aq)$$

This is an example of a weak base equilibrium. The equilibrium constant expression for this equilibrium is:

$$K_c = \frac{[OH^-][NH_4^+]}{[NH_3]} = K_b = 1.8 \times 10^{-5}$$

The K_c for any weak base is a K_b (base equilibrium constant), which *always* has the form $\frac{[OH^-][\text{Conjugate acid}]}{[\text{Conjugate base}]}$. The value (1.8×10^{-5}) is from a table or may be given in a problem. The K_b will be involved in some way in any weak base equilibrium problem.

EXAMPLE 3

What is the pH of a 0.500 M HNO_2 solution? The K_a for nitrous acid is 7.2×10^{-4}.

As with any equilibrium, we will begin with the equilibrium and the equilibrium constant expression:

$$HNO_2(aq) \rightleftharpoons H^+(aq) + NO_2^-(aq)$$

$$K_a = \frac{[H^+][NO_2^-]}{[HNO_2]} = 7.2 \times 10^{-4}$$

Here is the ICE table:

$$HNO2(aq) \leftrightarrows H^+(aq) + NO_2^-(aq)$$

Initial	0.500 M	—	—
Change	$-x$	$+x$	$+x$
Equilibrium	$0.500 - x$	$+x$	$+x$

Entering the equilibrium line into the K_a expression:

$$K_a = \frac{[H^+][NO_2^-]}{[HNO_2]} = \frac{[x][x]}{[0.500 - x]} = 7.2 \times 10^{-4}$$

We will assume that $0.500 - x \approx 0.500$ (an assumption that we will need to verify later). This assumption simplifies the calculation to:

$$\frac{[x][x]}{[0.500]} = 7.2 \times 10^{-4}$$

$$x^2 = (0.500)(7.2 \times 10^{-4})$$

$$x^2 = 3.6 \times 10^{-4} = [H^+]$$

$$x = 1.9 \times 10^{-4} = [H^+]$$

(Checking the assumption, $0.500 - 1.9 \times 10^{-4} = 0.500$ [based upon the significant figures from the addition-subtraction rule], which makes our simplifying assumption valid. If this had not been true, we would need to go back and redo the problems without the assumption and finish by using the quadratic formula.)

Finally:

$$pH = -\log [H^+] = -\log (1.9 \times 10^{-4}) = 1.72$$

Since we are dealing with an acid, the pH *must* be less than 7, which this is.

EXAMPLE 4

What is the pH of a 0.500 M NH_2OH, hydroxylamine, solution? The K_b for hydroxylamine is 9.1×10^{-9}. The K_b equilibrium equation is:

$$NH_2OH(aq) + H_2O(l) \leftrightarrows OH^-(aq) + NH_3OH^+(aq)$$

From here the calculations are similar to the K_a problem done previously except that we are looking for OH^- instead of H^+.

As with any equilibrium, we will begin with the equilibrium and the equilibrium constant expression:

$$NH_2OH(aq) + H_2O(l) \leftrightarrows OH^-(aq) + NH_3OH^+(aq)$$

$$K_b = \frac{[OH^-][NH_3OH^+]}{[NH_2OH]} = 9.1 \times 10^{-9}$$

Since H_2O is the solvent, it does not go in the K_b expression.

Here is the ICE table:

$$NH2OH(aq) + H_2O(l) \leftrightarrows OH^-(aq) + NH_3OH^+(aq)$$

Initial	0.500 M	—	—
Change	$-x$	$+x$	$+x$
Equilibrium	$0.500 - x$	$+x$	$+x$

Entering the equilibrium line into the K_b expression:

$$K_b = \frac{[OH^-][NH_3OH^+]}{[NH_2OH]} = \frac{[x][x]}{[0.500 - x]} = 9.1 \times 10^{-9}$$

We will assume that $0.500 - x \approx 0.500$ (an assumption that we will need to verify later). This assumption simplifies the calculation to:

$$\frac{[x][x]}{[0.500]} = 9.1 \times 10^{-9}$$

$$x^2 = (0.500)(9.1 \times 10^{-9})$$

$$x = 6.7 \times 10^{-5} = [OH^-]$$

(Checking the assumption, $0.500 - 6.7 \times 10^{-5} = 0.500$, which makes our simplifying assumption valid. If this had not been true, we would need to go back and redo the problems without the assumption, and finish by using the quadratic formula.)

Finally:

$$pOH = -\log [OH^-] = -\log 6.7 \times 10^{-5} = 4.17$$

$$pH = 14.000 - pOH = 14.000 - 4.17 = 9.83$$

Since we are dealing with a base, the pH *must* be greater than 7, which it is.
The smaller the K_a value, the weaker the acid (stronger the conjugate base).
The smaller the K_b value, the weaker the base (stronger the conjugate acid).

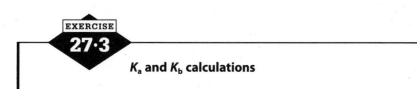

EXERCISE

27·3

K_a and K_b calculations

1. What is the pH of a 0.500 M HOCl solution? The K_a for hypochlorous acid is 2.9×10^{-8}.

2. What is the pH of a 0.500 M HN_3 solution? The K_a for hydrazoic acid is 1.9×10^{-5}.

3. What is the pH of a 0.500 M HF solution? The K_a for hydrofluoric acid is 6.6×10^{-4}.

4. What is the pH of a 0.500 M $C_2H_5NH_2$, ethylamine, solution? The K_b for ethylamine is 4.3×10^{-4}. The equilibrium equation for ethylamine is:

$$C_2H_5NH_2(aq) + H_2O(l) \leftrightarrows OH^-(aq) + C_2H_5NH_3^+(aq)$$

Multiple K's

Polyprotic acids can donate more than one H^+; therefore, they have more than one K_a. We will examine how to treat polyprotic acids in the next example.

EXAMPLE 5

What is the pH of a 0.500 M H_2CO_3 solution? The K_a's for carbonic acid are $K_{a1} = 4.4 \times 10^{-7}$ and $K_{a2} = 4.7 \times 10^{-11}$.

The K_{a1} is for the loss of the first H^+, and the K_{a2} is for the loss of the second H^+, as seen here:

$$H_2CO_3(aq) \leftrightarrows H^+(aq) + HCO_3^-(aq) \qquad K_{a1} = 4.4 \times 10^{-7}$$

$$HCO_3^-(aq) \leftrightarrows H^+(aq) + CO_3^{2-}(aq) \qquad K_{a2} = 4.7 \times 10^{-11}$$

(If there were more acidic hydrogens to be lost, there would be more K_a's.)

$$K_{a1} = \frac{[H^+][HCO_3^-]}{[H_2CO_3]} = 4.4 \times 10^{-7}$$

$$K_{a2} = \frac{[H^+][CO_3^{2-}]}{[HCO_3^-]} = 4.7 \times 10^{-11}$$

We will deal with the two K's separately.

The ICE table for K_{a1} is:

	$H_2CO_3(aq)$	$\leftrightarrows$	$H^+(aq)$	+	$HCO_3^-(aq)$	$K_{a1} = 4.4 \times 10^{-7}$
Initial	0.500 M		—		—	
Change	$-x$		$+x$		$+x$	
Equilibrium	$0.500 - x$		$+x$		$+x$	

Entering the equilibrium line into K_{a1}:

$$K_{a1} = \frac{[x][x]}{[0.500 - x]} = 4.4 \times 10^{-7}$$

Assuming $0.500 - x \approx 0.500$, and simplifying this relationship to:

$$K_{a1} = \frac{[x][x]}{[0.500]} = 4.4 \times 10^{-7}$$

Leads to $x = 4.69 \times 10^{-4} = [H^+] = [HCO_3^-]$ (we are keeping an extra significant figure to minimize intermediate rounding until after the next step).

Moving on to K_{a2}:

The ICE table for K_{a2} is:

	$HCO_3^-(aq)$	$\leftrightarrows$	$H^+(aq)$	+	$CO_3^{2-}(aq)$	$K_{a2} = 4.7 \times 10^{-11}$
Initial	4.69×10^{-4}		4.69×10^{-4}		—	
Change	$-x$		$+x$		$+x$	
Equilibrium	$4.69 \times 10^{-4} - x$		$4.69 \times 10^{-4} + x$		$+x$	

Note that the initial line is from the K_{a1} calculation (equilibrium line).

Entering the equilibrium line into K_{a2}:

$$K_{a2} = \frac{[H^+][CO_3^{2-}]}{[HCO_3^-]} = \frac{[4.69 \times 10^{-4} + x][x]}{[4.69 \times 10^{-4} - x]} = 4.7 \times 10^{-11}$$

Assuming that $(4.69 \times 10^{-4} - x) \approx (4.69 \times 10^{-4} + x)$, the preceding calculation simplifies to:

$$K_{a2} = \frac{[\cancel{4.69 \times 10^{-4}} + \cancel{x}][x]}{[\cancel{4.69 \times 10^{-4}} - \cancel{x}]} = 4.7 \times 10^{-11}$$

$$x = 4.7 \times 10^{-11}$$

$$[H^+] = 4.69 \times 10^{-4} + x = 4.69 \times 10^{-4} + 4.7 \times 10^{-11} \approx 4.69 \times 10^{-4} \text{ M } H^+(aq)$$

This is the total $[H^+]$ from the carbonic acid, and it will be used to determine the pH:

$$pH = -\log[H^+] = -\log(4.69 \times 10^{-4}) = 3.33 \text{ (rounded to proper significant figures).}$$

It is not always true that the H^+ from the second K_a will be too small to make a difference.

Acid-base properties of salts

Here is the dissociation equilibrium for acetic acid again:

$$\underset{\text{Conjugate Acid}}{HC_2H_3O_2(aq)} \leftrightarrows H^+(aq) + \underset{\text{Conjugate Base}}{C_2H_3O_2^-(aq)}$$

Now, let's look at the reaction of NaOH with acetic acid in total ionic form:

$$Na^+(aq) + OH^-(aq) + HC_2H_3O_2(aq) \rightarrow Na^+(aq) + C_2H_3O_2^-(aq) + H_2O(l)$$

Both reactions form the acetate ion, so both solutions contain a conjugate base. As a base, the acetate ion is involved in the following equilibrium:

$$\underset{\text{Conjugate Base}}{C_2H_3O_2^-(aq)} + H_2O(l) \leftrightarrows OH^-(aq) + \underset{\text{Conjugate Acid}}{HC_2H_3O_2(aq)}$$

This reaction with H_2O is an example of a hydrolysis reaction.

Now let's repeat this procedure for ammonia:

Here is the dissociation equilibrium for ammonia again:

$$\underset{\text{Conjugate Base}}{NH_3(aq)} + H_2O(l) \leftrightarrows OH^-(aq) + \underset{\text{Conjugate Acid}}{NH_4^+(aq)}$$

Now, here is the reaction for ammonia with hydrochloric acid in total ionic form:

$$NH_3(aq) + H^+(aq) + Cl^-(aq) \leftrightarrows NH_4^+(aq) + Cl^-(aq)$$

Both reactions form the ammonium ion, so both solutions contain a conjugate acid. As an acid, the ammonium ion is involved in the following equilibrium:

$$\underset{\text{Conjugate Acid}}{NH_4^+(aq)} \leftrightarrows H^+(aq) + \underset{\text{Conjugate Base}}{NH_3(aq)}$$

Even though there is no water shown, this is still considered a hydrolysis reaction.

For every acid and for every base there is a conjugate that may affect the acid-base properties of a solution. (The conjugates of strong acids and strong bases are too weak as conjugates to alter the acid-base properties of a solution.) If you have a weak acid or base, it will have a K_a or a K_b, and so will its conjugate. The relationship between the K's for a conjugate acid-base pair is:

$$K_a K_b = K_w = 1.00 \times 10^{-14}$$

For example:

$HC_2H_3O_2(aq) \leftrightarrows H^+(aq) + C_2H_3O_2^-(aq)$	$K_a = 1.8 \times 10^{-5}$
$C_2H_3O_2^-(aq) + H_2O(l) \leftrightarrows OH^-(aq) + HC_2H_3O_2(aq)$	$K_b = 5.6 \times 10^{-10}$

The K_a and K_b calculation procedure for any of these conjugates is no different from the "normal" K_a and K_b calculations.

The possibility of one or both ions present in a salt means that a salt may make a solution acidic or basic. The question becomes, how can we predict a salt is acidic or basic?

Let's look at four different reactions producing a salt:

$$HNO_3(aq) + KOH(aq) \rightarrow KNO_3(aq) + H_2O(l)$$
$$HNO_3(aq) + NH_3(aq) \rightarrow NH_4NO_3(aq)$$
$$HC_2H_3O_2(aq) + KOH(aq) \rightarrow KC_2H_3O_2(aq) + H_2O(l)$$
$$HC_2H_3O_2(aq) + NH_3(aq) \rightarrow NH_4C_2H_3O_2(aq)$$

Predictions are based upon the "parents" in the formation of the salt (even if the salt was made some other way).

The salts formed are:

KNO_3	Salt from a strong acid (SA) and a strong base (SB)
NH_4NO_3	Salt from a strong acid (SA) and a weak base (WB)
$KC_2H_3O_2$	Salt from a weak acid (WA) and a strong base (SB)
$NH_4C_2H_3O_2$	Salt from a weak acid (WA) and a weak base (WB)

All SA, SB, WA, and WB behave the same; the preceding one is only four examples from the large number of possibilities.

The generalizations are:

Salts from

SA-SB	Neutral (will not affect the pH)
SA-WB	Acidic (the pH will be lowered)
WA-SB	Basic (the pH will be increased)
WA-WB	Not easy to predict

As a reminder:

SA = HCl, HBr, HI, H_2SO_4, HNO_3, $HClO_3$, and $HClO_4$
WA = Every other acid (unless told otherwise)
SB = (Li, Na, K, Rb, Cs)(OH), (Ca, Sr, Ba)(OH)$_2$
WB = Every other base (unless told otherwise)

(Due to it being a diprotic acid, H_2SO_4-derived salts do not always work exactly as expected.)

If you are asked to determine the pH of a solution containing a SA-SB salt, you do not need to make any calculation as the pH will be 7.

If you are asked to determine the pH of a solution containing an SA-WB salt, you will need to do a K_a calculation on the cation, and the pH will be < 7. (Ignore the anion.)

If you are asked to determine the pH of a solution containing an WA-SB salt, you will need to do a K_b calculation on the anion, and the pH will be > 7. (Ignore the cation.)

If you are asked to determine the pH of a solution containing a WA-WB salt, you will need to do a K_a calculation on the cation and a K_b calculation on the anion, and decide which is more significant. The pH will probably be near 7; however, such situations normally appear only in advanced courses.

EXERCISE
27·4

Acid-base properties of salts

Predict whether each of the following solutions is acidic, basic, or neutral.

1. 1.0 M KCl(aq)

2. 1.0 M NH_4NO_2(aq)

3. 1.0 M LiF(aq)

4. 1.0 M $CH_3NH_2NO_3$(aq)

Other equilibria

Most processes involve some type of equilibrium process. In this chapter, we will investigate two additional types of equilibria: solubility and complex ion equilibria.

Solubility equilibria

When examining aqueous solutions, we find that many substances dissolve in water because water is such a good solvent. However, some substances are described as insoluble. These solubility options apply to other solvents, though there is limited data on the solubility of materials in solvents other than water.

When studying aqueous solutions, some potential solutes, such as CuS, were described as insoluble in older texts. Beginning about 100 years ago, scientific advancements in the analysis of solutions indicated that nothing is completely insoluble. In the case of CuS, this compound will dissolve to some extent, which leads to a copper(II) ion concentration of about 8×10^{-19} M.

Still more recently, environmental concerns have led to findings that there are substances that may cause health issues at extremely low levels. In many case, these levels were so low that they could not be detected until relatively recently. Since the levels could not be detected, there was no indication that the substance was causing health issues.

When we add an insoluble substance like barium carbonate, $BaCO_3$, to water, the solid $BaCO_3$ appears to sink to the bottom and simply sit there, doing nothing. However, what is really happening is:

$$BaCO_3(s) \leftrightarrows Ba^{2+}(aq) + CO_3^{2-}(aq)$$

As an equilibrium, this process is continually taking place. As this is an equilibrium reaction, it has an equilibrium constant. In this case, the equilibrium constant is $K_c = 5.1 \times 10^{-9}$. This solubility constant is called a solubility product constant, with K_c becoming K_{sp}. The form of the equilibrium constant expression is:

$$K_{sp} = [Ba^{2+}][SO_4^{2-}] = 5.1 \times 10^{-9}$$

As with all K_{sp} expressions, there is no term in the denominator, that is, the $BaCO_3(s)$ is not included. This is because the one and only reactant is a solid, and solids are never included in K expressions.

If we examine an insoluble substance like $Mg_3(PO_4)_2$, we see:

$$Mg_3(PO_4)_2(s) \leftrightarrows 3\,Mg^{2+}(aq) + 2\,PO_4^{3-}(aq) \qquad K_{sp} = 1 \times 10^{-25}$$

$$K_{sp} = [Mg^{2+}]^3\,[PO_4^{3-}]^2 = 1 \times 10^{-25}$$

Note, once the ions from an "insoluble" substance go into solution, they may undergo other processes such as hydrolysis (reaction with water); while such changes do occur, we will leave out these complications and treat K_{sp}'s as "simple" processes.

As with most equilibrium constants, K_{sp} values are found in tables or will be given in a problem.

Many students have trouble with K_{sp} problems because they are too easy, and students expect them to be more difficult.

EXERCISE
28·1

Solubility equilibria

1. Write the equation for the equilibrium reaction, and set up the K_{sp} expression for LiF(s), $K_{sp} = 3.8 \times 10^{-3}$.

2. Write the equation for the equilibrium reaction and set up the K_{sp} expression for PbSO₄(s), $K_{sp} = 1.6 \times 10^{-8}$.

3. Write the equation for the equilibrium reaction and set up the K_{sp} expression for Cr(OH)₃(s), $K_{sp} = 6.3 \times 10^{-31}$.

4. Write the equation for the equilibrium reaction and set up the K_{sp} expression for Cu₃(AsO₄)₂(s), $K_{sp} = 7.6 \times 10^{-36}$.

5. Write the equation for the equilibrium reaction and set up the K_{sp} expression for MgNH₄PO₄(s), $K_{sp} = 2.5 \times 10^{-13}$.

Complex ion equilibria

Many ions, usually metal ions, will interact with other substances to form what is known as a complex where the original ion plus the interacting substances form a group. The interacting substances are known as ligands. (In a few cases, the central species may be an atom instead of an ion.) Examples include the complex formed between the Fe^{2+} ions in hemoglobin with O_2 to allow the transport of oxygen throughout your body, and the complexing of Ca^{2+} in blood samples to prevent coagulation. Here we will examine simpler systems.

The following is an example showing the formation of a complex:

$$Cd^{2+}(aq) + 4\,NH_3(aq) \leftrightarrows [Cd(NH_3)_4]^{2+}(aq) \qquad K_c = 1.3 \times 10^7$$

The equilibrium constant, K_c, for the formation of a complex is known as a formation constant, K_f. The complex is normally enclosed in square brackets as shown here. The complex contains the central ion, Cd^{2+}, and the surrounding ligands, NH_3. Formation constants are always written with only the central ions and the ligands on the reactant side and only the complex on the product side. The reverse of a formation reaction is a dissociation reaction symbolized by a K_d:

$$[Cd(NH_3)_4]^{2+}(aq) \leftrightarrows Cd^{2+}(aq) + 4NH_3(aq) \qquad K_d = 7.7 \times 10^{-8} = 1/K_f$$

Notice that the K_d has only the complex on the reactant side and only the pieces of the complex on the product side.

The most common ligand is water, as all metal ions in aqueous solutions exist as complexes with water. In most cases, the formation of a complex is ignored and representations such as $Fe^{2+}(aq)$ are used instead of the "more" correct $[Fe(H_2O)_6]^{2+}$.

Here is a summary of the definitions:

A **complex ion** is an ion consisting of a central ion and one or more ligands.
A **formation constant**, K_f, is the equilibrium constant for the formation of a complex.
A **dissociation constant**, K_d, is the equilibrium constant for the breakdown of a complex.
A **ligand** is one species surrounding the central ion in a complex.
A **coordination compound** is a compound (neutral) containing a complex.

The formation of complexes, such as $[Cd(NH_3)_4]^{2+}(aq)$, is actually more involved than just shown. The ligands, in this case NH_3, add one at a time instead of all adding simultaneously. However, this complication need not concern us here, as we will only be using equilibria, which are the sum of all the steps.

Here is the formation of a different complex:

$$Fe^{3+}(aq) + 3\,C_2O_4^{2-}(aq) \leftrightarrows [Fe(C_2O_4)_3]^{3-}(aq) \qquad K_f = 2 \times 10^{20}$$

This shows that a complex ion may be an anion or a cation, which, even though they are larger, behave like other anions and cations, such as Cl^- and Na^+.

The K_f equations for the previously formed ions are:

$$K_f = \frac{[Cd(NH_3)_4^{2+}]}{[Cd^{2+}][NH_3]^4} = 1.3 \times 10^7 \qquad K_f = \frac{[Fe(C_2O_4)_3^{3-}]}{[Fe^{3+}][C_2O_4^{2-}]^3} = 2 \times 10^{20}$$

> Notice that in writing the K_f expressions, the ionic charges for the complex are moved from outside the square brackets to inside so as not to be mistaken for exponents. In addition, only the ligands have exponents other than an understood 1.

Here are examples of two coordination compounds containing these complexes:

$$[Cd(NH_3)_4]SO_4 \qquad K_3[Fe(C_2O_4)_3]$$

Complex ion equilibria

1. Write the equation for the equilibrium reaction, and set up the K_f expression for $[Al(OH)_4]^-(aq)$, $K_f = 1.1 \times 10^{33}$.

2. Write the equation for the equilibrium reaction, and set up the K_f expression for $[PbI_4]^{2-}(aq)$, $K_f = 3.0 \times 10^4$.

3. Write the equation for the equilibrium reaction, and set up the K_f expression for $[Ag(NO_2)_2]^-(aq)$, $K_f = 7 \times 10^2$.

4. Write the equation for the equilibrium reaction, and set up the K_d expression for $[Fe(CN)_6]^{3-}(aq)$, $K_d = 1 \times 10^{-31}$.

5. Write the equation for the equilibrium reaction, and set up the K_d expression for $[Zn(NH_3)_4]^{2+}(aq)$, $K_d = 2.0 \times 10^{-9}$.

Calculations

Other than being simpler, K_{sp} problems are worked like other equilibrium problems. K_f problems are much like other equilibrium problems. As long as you label and keep track of things and watch your units, the problems covered here should work fine.

EXAMPLE 1

What is the maximum silver ion concentration obtained by adding solid $Ag_2C_2O_4$ to water? The K_{sp} of silver oxalate is 3.5×10^{-11}.

As usual the first step is to write the appropriate reaction and equilibrium constant expression:

$$Ag_2C_2O_4(s) \leftrightarrows 2\,Ag^+(aq) + C_2O_4{}^{2-}(aq) \qquad K_{sp} = [Ag^+]^2\,[C_2O_4{}^{2-}] = 3.5 \times 10^{-11}$$

Normally, we would create an ICE table at this point; however, the simpler nature of K_{sp} problems allows us to skip straight to the last line in the ICE table. Which gives us:

$$Ag_2C_2O_4(s) \leftrightarrows 2\,Ag^+(aq) + C_2O_4{}^{2-}(aq) \qquad K_{sp} = [Ag^+]^2\,[C_2O_4{}^{2-}] = 3.5 \times 10^{-11}$$
$$\text{---} \qquad +2x \qquad +x$$

Entering this information into the K_{sp} expression gives us:

$$K_{sp} = [Ag^+]^2\,[C_2O_4{}^{2-}] = (2x)^2\,(x) = 4x^3 = 3.5 \times 10^{-11}$$

Solving for x:

$$x = 2.06 \times 10^{-4} \text{ (extra significant figure)}$$

Finally:

$$[Ag^+] = 2x = 2\,(2.06 \times 10^{-4}) = 4.1 \times 10^{-4} \text{ M (rounded to correct significant figures)}$$

Note: if you had rounded the 2.06×10^{-4} too soon, the final answer would be 4.2×10^{-4} M.

EXAMPLE 2

What is the maximum silver ion concentration obtained by adding solid $Ag_2C_2O_4$ to a 0.100 M $Na_2C_2O_4$ solution? The K_{sp} of silver oxalate is 3.5×10^{-11}.

We can start by recopying the ICE line from the previous example:

$$Ag_2C_2O_4(s) \leftrightarrows 2\,Ag^+(aq) + C_2O_4^{2-}(aq) \qquad K_{sp} = [Ag^+]^2\,[C_2O_4^{2-}] = 3.5 \times 10^{-11}$$

$$— \qquad +2x \qquad +x$$

Now, we deal with the change (presence of $Na_2C_2O_4(aq)$)
The sodium oxalate is a soluble strong electrolyte that behaves as follows:

$$Na_2C_2O_4(aq) \rightarrow 2\,Na^+(aq) + C_2O_4^{2-}(aq)$$

$$0.100\text{ M} \qquad 2(0.100) \qquad 0.100\text{ M}$$

The oxalate ion (0.100 M) appears in both the K_{sp} equation and the dissociation equation. This makes the oxalate ion, a common ion which we discuss in more detail later. Since this oxalate ion is in the same solution as the dissolving $Ag_2C_2O_4(s)$, it must be added to the ICE table. The sodium ions are not present in the K_{sp}, so they are not included. The new ICE table is:

$$Ag_2C_2O_4(s) \leftrightarrows 2\,Ag^+(aq) + C_2O_4^{2-}(aq) \qquad K_{sp} = [Ag^+]^2\,[C_2O_4^{2-}] = 3.5 \times 10^{-11}$$

$$— \qquad +2x \qquad 0.100\text{ M} + x$$

Entering this line into the K_{sp} gives:

$$K_{sp} = [2x]^2\,[0.100 + x] = 3.5 \times 10^{-11}$$

Since K_{sp} is so small, we will assume $[0.100 + x] \approx [0.100]$, and we can solve for x:

$$K_{sp} = 4x^2\,[0.100] = 3.5 \times 10^{-11}$$

Solving for x:

$$x = 9.35 \times 10^{-6} \text{ (extra significant figure)}$$

Finally:

$$[Ag^+] = 2x = 2\,(9.35 \times 10^{-6}) = 1.9 \times 10^{-5} \text{ M (rounded to correct significant figures)}$$

Note: the presence of the common ion significantly reduced the Ag^+ concentration.

EXAMPLE 3

What is the $Zn^{2+}(aq)$ concentration in a 0.100 M $[Zn(CN)_4]^{2-}(aq)$ solution? The K_d of $[Zn(CN)_4]^{2-}(aq)$ is 1.0×10^{-19}.

As usual the first step is to write the appropriate reaction and equilibrium constant expression:

$$[Zn(CN)_4]^{2-}(aq) \leftrightarrows Zn^{2+}(aq) + 4\,CN^-(aq) \qquad K_d = \frac{[Zn^{2+}]\,[CN^-]^4}{[Zn(CN)_4^{2-}]} = 1.0 \times 10^{-19}$$

Adding the last line of an ICE table:

$$[Zn(CN)_4]^{2-}(aq) \leftrightarrows Zn^{2+}(aq) + 4\,CN^-(aq) \qquad K_d = \frac{[Zn^{2+}][CN^-]^4}{[Zn(CN)_4^{2-}]} = 1.0 \times 10^{-19}$$

$$0.100 - x \qquad +x \qquad +4x$$

Entering this line into the K_d expression:

$$K_d = \frac{[Zn^{2+}][CN^-]^4}{[Zn(CN)_4^{2-}]} = \frac{[x][4x]^4}{[0.100-x]} = 1.0 \times 10^{-19}$$

Since K_d is so small, we will assume that $[0.100 - x] \approx [0.100]$, which simplifies the problem to allow us to determine x:

$$K_d = \frac{[Zn^{2+}][CN^-]^4}{[Zn(CN)_4^{2-}]} = \frac{[x][4x]^4}{[0.100]} = \frac{[x]\,256x^4}{[0.100]} = \frac{256x^5}{[0.100]} = 1.0 \times 10^{-19}$$

$$[Zn^{2+}] = x = 3.3 \times 10^{-5}\ M$$

EXAMPLE 4

At one time the dissolution of unreacted AgBr(s) from photographic film was important in photography. This was done by treating the film with a dilute $Na_2S_2O_3$ solution. The two equilibria involved were:

$$AgBr(s) \leftrightarrows Ag^+(aq) + Br^-(aq) \qquad\qquad K_{sp} = 5.0 \times 10^{-13}$$
$$Ag^+(aq) + 2\,S_2O_3^{2-}(aq) \leftrightarrows [Ag(S_2O_3)_2]^{3-}(aq) \qquad K_f = 2.9 \times 10^{13}$$

What is the equilibrium constant for AgBr(s) dissolving in a $Na_2S_2O_3$(aq) solution?
We can add these two related equilibria together:

$$
\begin{array}{lll}
AgBr(s) & \leftrightarrows \quad \cancel{Ag^+(aq)} + Br^-(aq) & K_{sp} = 5.0 \times 10^{-13} \\
\cancel{Ag^+(aq)} + 2\,S_2O_3^{2-}(aq) & \leftrightarrows \quad [Ag(S_2O_3)_2]^{3-}(aq) & K_f = 2.9 \times 10^{13}
\end{array}
$$

Combined: $\quad AgBr(s) + 2\,S_2O_3^{2-}(aq) \leftrightarrows [Ag(S_2O_3)_2]^{3-}(aq) + Br^-(aq)$

This is the equation for adding AgBr(s) to a $Na_2S_2O_3$(aq) solution, and the "new" K will be the product of the individual K's, or:

$$K = K_{sp}\,K_f = (5.0 \times 10^{-13})\,(2.9 \times 10^{13}) = 14$$

EXERCISE

28·3

Calculations

1. Calculate the Al^{3+}(aq) in a solution formed by the addition of $AlPO_4$(s) to water. The K_{sp} of $AlPO_4$(s) is 5.75×10^{-19}.

2. Calculate the Cl^-(aq) in a solution formed by the addition of $AuCl_3$(s) to water. The K_{sp} of $AuCl_3$(s) is 3.2×10^{-25}.

3. Calculate the Ba^{2+}(aq) in a solution formed by the addition of $Ba_3(AsO_4)_2$(s) to water. The K_{sp} of $Ba_3(AsO_4)_2$(s) is 7.8×10^{-51}.

4. Calculate the $Ag^+(aq)$ in a solution formed by the addition of $AgCl(s)$ to a 0.200 M NaCl solution. The K_{sp} of $AgCl(s)$ is 1.78×10^{-10}.

5. Calculate the $Ag^+(aq)$ in a solution formed by the addition of $Ag_2CrO_4(s)$ to a 0.200 M Na_2CrO_4 solution. The K_{sp} of $Ag_2CrO_4(s)$ is 1.1×10^{-12}.

6. Calculate the $Ca^{2+}(aq)$ in a solution formed by the addition of $Ca_3(PO_4)_2(s)$ to a 0.200 M PO_4^{3-} solution. The K_{sp} of $Ca_3(PO_4)_2(s)$ is 2.0×10^{-29}.

7. What is the $Zn^{3+}(aq)$ concentration in a 0.100 M $[Zn(NH_3)_4]^{2-}(aq)$ solution. The K_d of $[Zn(NH_3)_4]^{2-}(aq)$ is 2.0×10^{-9}.

Buffers and titrations

This chapter investigates acids and bases in more detail. However, the concepts covered here apply to other systems.

The common ion effect

Consider the following two examples. The first example should be a review.

EXAMPLE 1

Determine the pH of a 0.100 M acetic acid, $HC_2H_3O_2$, solution. The K_a of acetic acid is 1.74×10^{-5}.

We shall begin by writing the equilibrium equation and the K_a expression for acetic acid.

$$HC_2H_3O_2(aq) \rightleftharpoons H^+(aq) + C_2H_3O_2^-(aq)$$

$$K_a = \frac{[H^+][C_2H_3O_2^-]}{[HC_2H_3O_2]} = 1.74 \times 10^{-5}$$

Note, you should remember that this is the only way to write the reaction and K_a expression for a weak acid.

Next, we will add the equilibrium line from the ICE table below the equilibrium equation. (You may wish to do the entire ICE table for review.)

$$HC_2H_3O_2(aq) \rightleftharpoons H^+(aq) + C_2H_3O_2^-(aq)$$

$$0.100 - x \qquad +x \qquad +x$$

Entering this information into the K_a expression:

$$K_a = \frac{[H^+][C_2H_3O_2^-]}{[HC_2H_3O_2]} = \frac{[x][x]}{[0.100 - x]} = 1.74 \times 10^{-5}$$

The K_a value is sufficiently low to assume $[0.100 - x] \approx [0.100]$, which simplifies the calculation:

$$K_a = \frac{[H^+][C_2H_3O_2^-]}{[HC_2H_3O_2]} = \frac{[x][x]}{[0.100]} = 1.74 \times 10^{-5}$$

Solving for x:

$$x = \sqrt{(0.100)(1.74 \times 10^{-5})} = 1.32 \times 10^{-3} \, M = [H^+] = [C_2H_3O_2^-]$$

The value of x is sufficiently small to justify our assumption.
Finally, using the equation for the definition of pH:

$$pH = -\log[H^+] = -\log[1.32 \times 10^{-3}] = 2.879$$

EXAMPLE 2

Determine the pH of a 0.100 M acetic acid, $HC_2H_3O_2$, solution, which is also 0.100 M in $NaC_2H_3O_2$. The K_a of acetic acid is 1.74×10^{-5}.

We shall begin by writing the equilibrium equation and the K_a expression for acetic acid (exactly as before).

$$HC_2H_3O_2(aq) \leftrightarrows H^+(aq) + C_2H_3O_2^-(aq)$$

$$K_a = \frac{[H^+][C_2H_3O_2^-]}{[HC_2H_3O_2]} = 1.74 \times 10^{-5}$$

For the sodium acetate, we have:

$$NaC_2H_3O_2(aq) \rightarrow Na^+(aq) + C_2H_3O_2^-(aq)$$

$$\text{—} \qquad 0.100 \qquad 0.100$$

Sodium acetate is a strong electrolyte; therefore, this is not an equilibrium (100% ionized).

The acetate ion appears in both chemical equations, so when constructing our ICE table, both sources of acetate ion need to be accounted for. The sodium ion does not appear in the K_a expression; therefore, it is a spectator ion and does not belong in the ICE table.

$$HC_2H_3O_2(aq) \leftrightarrows H^+(aq) + C_2H_3O_2^-(aq)$$

$$0.100 - x \qquad +x \qquad 0.100 + x$$

Entering this information into the K_a expression:

$$K_a = \frac{[H^+][C_2H_3O_2^-]}{[HC_2H_3O_2]} = \frac{[x][0.100 + x]}{[0.100 - x]} = 1.74 \times 10^{-5}$$

The K_a value is sufficiently low to assume $[0.100 - x] \approx [0.100 + x] \approx [0.100]$, which simplifies the calculation:

$$K_a = \frac{[H^+][C_2H_3O_2^-]}{[HC_2H_3O_2]} = \frac{[x][\cancel{0.100}]}{[\cancel{0.100}]} = 1.74 \times 10^{-5}$$

Solving for x:

$$x = 1.74 \times 10^{-5} \, M = [H^+] = [C_2H_3O_2^-]$$

The value of x is sufficiently small to justify our assumption.
Finally, using the equation for the definition of pH:

$$pH = -\log[H^+] = -\log[1.74 \times 10^{-5}] = 4.757$$

Both solutions in these two examples have the same starting acetic acid concentration, the sodium ion is a spectator ion, so the difference in pH must be due to the acetate ion. The acetate ion appears both in the acetic acid equilibrium equation and in the sodium acetate ionization equation. For this reason, the acetate ion is considered to be a common ion, and as such affects the equilibrium calculation. A **common ion** is a species added to an equilibrium from an outside source. The presence of a common ion affects the equilibrium by what is called the common ion effect. The **common ion effect** is the alteration of an equilibrium by the addition of a common ion. The common effect applies to all solution equilibria.

The common ion effect

1. Determine the pH of a 0.100 M nitrous acid, HNO_2, solution, which is also 0.100 M in KNO_2. The K_a of nitrous acid is 5.1×10^{-4}.

2. Determine the pH of a 0.100 M hypochlorous acid, HClO, solution, which is also 0.200 M in NaClO. The K_a of acetic acid is 5.0×10^{-8}.

Buffer solutions

A common application of the common ion effect is their use as buffer solutions. A **buffer solution** is a solution that resists changes in pH. For acid-base systems, a buffer solution contains a weak acid or weak base *plus* its conjugate. If you examine the last example or either of the preceding practice problems, you will see that the solutions involved qualify as buffer solutions.

There is a shorter method for solving the K_a equation for buffer solutions. The general K_a equation is:

$$K_a = \frac{[H^+] \, [\text{Conjugate base}]}{[\text{Conjugate acid}]} = \frac{[H^+] \, [CB]}{[CA]}$$

Here we are substituting CA for conjugate acid and CB for conjugate base. If we take the −log of each side of the equation, we get:

$$-\log K_a = -\log [H^+] - \log \frac{[CB]}{[CA]}$$

Substituting $pK_a = -\log K_a$ and $pH = -\log [H^+]$ into this equation gives:

$$pK_a = pH - \log \frac{[CB]}{[CA]}$$

This equation is usually rearranged to:

$$pH = pK_a + \log \frac{[CB]}{[CA]}$$

In this form, this equation is known as the Henderson-Hasselbalch equation, which provides a short-cut for determining the pH of a buffer solution. There is a related form for determining the pOH:

$$pOH = pK_b + \log \frac{[CA]}{[CB]}$$

These two forms of the Henderson-Hasselbalch equation plus the relationship $pK_w = pH + pOH$ allows the two forms of the Henderson-Hasselbalch equation to be related.

Again, the Henderson-Hasselbalch equation, in either form, applies only to buffer solutions, for only in buffer solutions is it possible for neither [CA] nor [CB] to equal just x. There must be another number, other than 0, present in the concentration terms.

EXAMPLE 3

Determine the pK_a for acetic acid and the pK_b for methylamine, CH_3NH_2. The K_a of acetic acid is 1.74×10^{-5}, and the K_b for methylamine is 5.25×10^{-3}.

The definitions of the pK_a and the pK_b are:
$$pK_a = -\log K_a \qquad pK_b = -\log K_b$$

Entering the appropriate K into each of these definitions gives:
$$pK_a = -\log (1.74 \times 10^{-5}) = 4.759 \qquad pK_b = -\log (5.25 \times 10^{-3}) = 2.281$$

EXAMPLE 4

Using the Henderson-Hasselbalch equation, determine the pH of a 0.100 M acetic acid, $HC_2H_3O_2$, solution, which is also 0.200 M in $NaC_2H_3O_2$. The pK_a of acetic acid is 4.759.

Beginning with the Henderson-Hasselbalch equation:
$$pH = pK_a + \log \frac{[CB]}{[CA]}$$

We have [CA] = 0.100 M $HC_2H_3O_2$, [CB] = 0.200 M $C_2H_3O_2^-$, and $pK_a = 4.759$.

Entering the given values into the equation gives:
$$pH = pK_a + \log \frac{[CB]}{[CA]} = pH = 4.759 + \log \frac{[0.200]}{[0.100]} = 5.060$$

As a special case, when [CA] = [CB], $pH = pK_a$ (or $pOH = pK_b$).

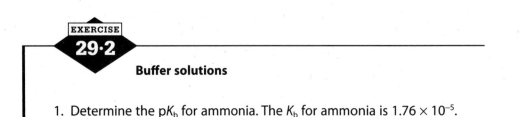

EXERCISE

29·2

Buffer solutions

1. Determine the pK_b for ammonia. The K_b for ammonia is 1.76×10^{-5}.

2. Determine the pH of a 0.200 M nitrous acid, HNO_2, solution, which is also 0.100 M in KNO_2. The K_a of nitrous acid is 5.1×10^{-4}.

3. Determine the pH of a 0.100 M ammonia, NH_3, solution, which is also 0.100 M in ammonium sulfate, $(NH_4)_2SO_4$. The K_b of ammonia is 1.76×10^{-5}.

Calculating the pH of a buffer solution

A buffer solution is a solution that resists changes in pH. How does this happen?

We shall answer this questions by using the example we gave previously.

Using the Henderson-Hasselbalch equation, determine the pH of a 0.100 M acetic acid, $HC_2H_3O_2$, solution, which is also 0.200 M in $NaC_2H_3O_2$. The pK_a of acetic acid is 4.759.

We saw that this buffer solution had a pH of 4.757. If this is truly a buffer solution, then the addition of some acid or base to this solution will not produce a large pH change. We shall do this with the following two examples.

EXAMPLE 5

We shall begin with 1.000 L of the buffer solution. If we have 1.000 L solution, then (MV = moles) we have 0.100 mol of $HC_2H_3O_2$ and 0.200 mol of $NaC_2H_3O_2$.

What happens if 0.010 mol of NaOH is added to the buffer solution? We will ignore any change in volume of the solution. (Note: if this were not a buffer, adding that quantity of NaOH would raise the pH to 12.000.)

We will begin with the buffer, which has:

$$HC_2H_3O_2(aq) \leftrightarrows H^+(aq) + C_2H_3O_2^-(aq)$$

0.100 mol   + x   0.200 mol

A base, in this case NaOH, will react with an acid. Since the sodium ion is a spectator ion, we will ignore it. The reaction of hydroxide ion with the acid gives this equation:

$$OH^-(aq) + HC_2H_3O_2(aq) \rightarrow C_2H_3O_2^-(aq) + H_2O(l)$$

Adding the moles of each substance given:

$$OH^-(aq) + HC_2H_3O_2(aq) \rightarrow C_2H_3O_2^-(aq) + H_2O(l)$$

0.010   0.100   0.200   mol

Examination of moles present shows that the hydroxide ion is the limiting reagent. Since all mole ratios here are 1:1, the stoichiometry gives:

$$OH^-(aq) + HC_2H_3O_2(aq) \rightarrow C_2H_3O_2^-(aq) + H_2O(l)$$

0.010   0.100   0.200   mol

−0.010   −0.010   +0.010   mol

0.000   0.090   0.210   mol after reaction.

Dividing these moles by the volume of the solution (1.000 L) gives the molarities.

[CA] = 0.090 M $HC_2H_3O_2$, [CB] = 0.210 M $C_2H_3O_2^-$

Entering these molarities into the Henderson-Hasselbalch equation:

$$pH = pK_a + \log \frac{[CB]}{[CA]} = pH = 4.759 + \log \frac{[0.210]}{[0.090]} = 5.127$$

The pH did change slightly, but it is nowhere near the pH = 12 that the solution would have reached if the solution were not a buffer.

EXAMPLE 6

We shall again begin with 1.000 L of the buffer solution. If we have 1.000 L solution, then ($MV =$ moles) we have 0.100 mol of $HC_2H_3O_2$ and 0.200 mol of $NaC_2H_3O_2$.

What happens if 0.010 mol of HCl is added to the buffer solution? Again, we are ignoring any change in the volume of the solution. (Note: if this were not a buffer, adding that quantity of HCl would lower the pH to 2.00.)

We will begin with the buffer which has:

$$HC_2H_3O_2(aq) \leftrightharpoons H^+(aq) + C_2H_3O_2^-(aq)$$

$$0.100 \text{ mol} \qquad +x \qquad 0.200 \text{ mol}$$

An acid, in this case HCl, will react with a base. Since the chloride ion is a spectator ion, we will ignore it. The reaction of hydrogen ion with the base gives this equation:

$$H^+(aq) + C_2H_3O_2^-(aq) \rightarrow HC_2H_3O_2(aq)$$

Adding the moles of each substance given:

$$H^+(aq) + C_2H_3O_2^-(aq) \rightarrow HC_2H_3O_2(aq)$$

$$0.010 \qquad 0.200 \qquad 0.100 \qquad \text{mol}$$

Examination of moles present shows that the hydrogen ion is the limiting reagent. Since all mole ratios here are 1:1, the stoichiometry gives:

$$H^+(aq) + C_2H_3O_2^-(aq) \rightarrow HC_2H_3O_2(aq)$$

0.010	0.200	0.100	mol
−0.010	−0.010	+0.010	mol
0.000	0.190	0.110	mol after reaction.

Dividing these moles by the volume of the solution (1.000 L) gives the molarities.

$$[CA] = 0.110 \text{ M } HC_2H_3O_2, [CB] = 0.190 \text{ M } C_2H_3O_2^-$$

Entering these molarities into the Henderson-Hasselbalch equation:

$$pH = pK_a + \log \frac{[CB]}{[CA]} = pH = 4.759 + \log \frac{[0.190]}{[0.110]} = 4.996$$

The pH did change slightly, but it is nowhere near the pH = 2 that the solution would have reached if the solution were not a buffer.

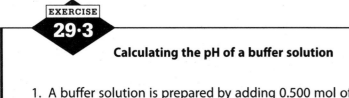

EXERCISE
29·3

Calculating the pH of a buffer solution

1. A buffer solution is prepared by adding 0.500 mol of KNO_2 and 0.400 mol of HNO_2 to sufficient water to produce 0.500 L of solution. The K_a of HNO_2 is 5.1×10^{-4}. What is the pH of this solution after 0.020 mol of NaOH are added to this solution?

2. A buffer solution is prepared by adding 0.500 mol of KNO_2 and 0.400 mol of HNO_2 to sufficient water to produce 0.500 L of solution. The K_a of HNO_2 is 5.1×10^{-4}. What is the pH of this solution after 0.020 mol of HNO_3 are added to this solution?

Buffer capacity

There is a limit to just how much acid or base may be added to a buffer before the solution ceases to behave as a buffer. Looking at the two examples in the preceding section, these limits are 0.100 mol of NaOH or 0.200 mol of HCl. When either of these quantities is added, both the base or acid and the substance it reacts with are limiting, which means that either [CA] or [CB] is 0 and the solution is no longer a buffer. The **buffer capacity** of a solution is the maximum amount of acid or base that may be added to a buffer before the solution ceases to be a buffer.

Types of acid-base titrations

Acid-base titrations involve the reaction between an acid and a base with the goal of learning some unknown information about either the acid or the base. There are four possible types of acid-base titrations. These different types depend upon whether the acid is strong (SA) or weak (WA) and if the base is strong (SB) or weak (WB). The four possible types are:

SA-SB
SA-WB
WA-SB
WA-WB

The SA-SB is the simplest and WA-WB is the most difficult. Difficulties with WA-WB titrations lead to such titrations not being done; therefore, there are only the other three to consider. An SA-SB titration is only concerned with stoichiometry as there are no significant equilibria involved. Both SA-WB and WA-SB titrations are concerned with stoichiometry and equilibria. In many cases, it is important to keep track of the pH during the titration.

In all types of acid-base titrations, either the acid is added to the base or the base is added to the acid. The addition is from a burette into a beaker or a flask, with the pH in the beaker or flask being tracked, perhaps with a pH meter.

There are four regions that are normally involved in the calculations: (1) the initial pH in the flask before the addition of any of the other reactant; (2) the region between the initial and the equivalence point; (3) the equivalence point where the acid and base both, through stoichiometry, become limiting reagents; (4) and finally, the region after the equivalence point. All the calculations needed are identical to the ones that you have done previously. However, here you will need to keep track of where you are in the titration.

SA-SB titrations

The initial pH is simply based on the concentration of the SA or SB. For example, 0.10 M HNO_3 gives 0.10 M H^+(aq), which leads to pH = $-\log$ (0.10) = 1.00. For a strong base, such as 0.10 M NaOH gives 0.10 M OH^-(aq), which leads to pOH = $-\log$ (0.10) = 1.00, and pH = 14.000 $-$ pOH = 13.00.

In the region between the initial and the equivalence points, you do a limiting reagent calculation, and the excess acid or base remaining is used in a pH calculation similar to what was done initially.

At the equivalence point, the pH = 7.

After the equivalence point, the calculations are similar to those between the initial and equivalence points. However, the excess moles being added will be used to calculate the pH.

WA-SB or SA-WB titrations

If the solution in the beaker or flask is weak, then either a K_a or a K_b calculation is needed. If the SA or SB is in the container, the calculation is the same as for the initial calculation in an SA-SB titration.

Before the equivalence point there are two ways to go. If the SA or SB is in the container, the calculations are the same as in an SA-SB titration. If the WA or WB is in the container, the stoichiometry is followed by either a K_a or K_b calculation using the stoichiometry results. (Note, if the WA or WB is in the container, the solution is a buffer up to the equivalence point.

At the equivalence point, the calculation is based upon the K_b of the conjugate base of the WA or the K_a of the conjugate acid of the WB.

After the equivalence point, it is similar to before the equivalence point, with the identities of the reactants and products reversed.

Note that doing all the calculations *is not difficult*; however, it is time consuming. In most cases, if a titration problem appears on an exam, it is only one point during the titration instead of all points.

We will now do an example of a WA-SB titrations.

EXAMPLE 7

The initial conditions are:

A sample consisting of 50.00 mL of 0.1000 M acetic acid, $HC_2H_3O_2$, solution is placed into a flask and 0.2000 M sodium hydroxide solution is added from a burette. The K_a of acetic acid is 1.74×10^{-5} ($pK_a = 4.759$). Determine the pH in the flask after the addition of 0.00 mL of NaOH, 12.50 mL of NaOH, 25.00 mL of NaOH, and 50.00 mL of NaOH.

Note, there is initially 0.005000 mol of acetic acid.

0.00 mL of NaOH

This is the initial pH, which means, in this case, this is a straight K_a calculation.

$$K_a = \frac{[H^+][C_2H_3O_2^-]}{[HC_2H_3O_2]} = \frac{[x][x]}{[0.1000-x]} = 1.74 \times 10^{-5}$$

The K_a value is sufficiently low to assume $[0.1000 - x] \approx [0.1000]$, which simplifies the calculation:

$$K_a = \frac{[H^+][C_2H_3O_2^-]}{[HC_2H_3O_2]} = \frac{[x][x]}{[0.1000]} = 1.74 \times 10^{-5}$$

Solving for x:

$$x = \sqrt{(0.100)(1.74 \times 10^{-5})} = 1.32 \times 10^{-3}\ M = [H^+] = [C_2H_3O_2^-]$$

The value of x is sufficiently small to justify our assumption.
Finally, using the equation for the definition of pH:
$$pH = -\log [H^+] = -\log [1.32 \times 10^{-3}] = 2.879$$

12.50 mL of NaOH,

It is necessary to do a stoichiometry calculation. We need to know the moles of NaOH (MV = 0.002500 mol NaOH).

The balanced chemical equation along with the moles of reactants are:

$$NaOH(aq) + HC_2H_3O_2(aq) \rightarrow C_2H_3O_2^-(aq) + Na^+(aq) + H_2O(l)$$

| 0.002500 | 0.005000 | — | | mol |

The NaOH is the limiting reagent, and the stoichiometric ratios are 1:1.

$$NaOH(aq) + HC_2H_3O_2(aq) \rightarrow C_2H_3O_2^-(aq) + Na^+(aq) + H_2O(l)$$

0.002500	0.005000	—	mol
−0.002500	−0.002500	+0.002500	mol
0	0.002500	0.002500	mol

Next, we need to convert the moles to molarities by dividing by the new solution volume (50.00 + 12.50) mL = 62.50 mL = 0.06250 L. Dividing the moles of each substance remaining by the liters of solution gives the molarities: 0.04000 M = $[HC_2H_3O_2]$ = $[C_2H_3O_2^-]$

This may be solved as a K_a problem, or since we have both the CA and CB of a WA, we have a buffer where we can use the Henderson-Hasselbalch equation. We will use the Henderson-Hasselbalch equation.

Entering the given values into the equation gives:

$$pH = pK_a + \log \frac{[CB]}{[CA]} = pH = 4.759 + \log \frac{[0.04000]}{[0.04000]} = 4.759$$

This is a special case where [CA] = [CB].

25.00 mL of NaOH

It is necessary to do a stoichiometry calculation. We need to know the moles of NaOH (MV = 0.005000 mol NaOH.

The balanced chemical equation along with the moles of reactants are:

$$NaOH(aq) + HC_2H_3O_2(aq) \rightarrow C_2H_3O_2^-(aq) + Na^+(aq) + H_2O(l)$$

0.005000	0.005000	—	mol

The NaOH is the limiting reagent, and the stoichiometric ratios are 1:1.

$$NaOH(aq) + HC_2H_3O_2(aq) \rightarrow C_2H_3O_2^-(aq) + Na^+(aq) + H_2O(l)$$

0.005000	0.005000	—	mol
−0.005000	−0.005000	+0.005000	mol
0	0	0.005000	mol

Since both reactants have 0 moles, this is the equivalence point. The only thing remaining to alter the pH is the $C_2H_3O_2^-(aq)$, which is the CB of the acetic acid. As a CB, a K_b is necessary. The K_b can be determined from the relationship $K_aK_b = K_w = 1.00 \times 10^{-14}$.

$$K_b = \frac{K_w}{K_a} = \frac{1.00 \times 10^{-14}}{1.74 \times 10^{-5}} = 5.75 \times 10^{-10}$$

Next, we need to convert the moles to molarity by dividing by the new solution volume (50.00 + 25.00) mL = 75.00 mL = 0.07500 L. Dividing the moles of each substance remaining by the liters of solution gives the molarities: 0.06667 M = $[C_2H_3O_2^-]$.

The K_b reaction and equilibrium constant expression are:

$$C_2H_3O_2^-(aq) + H_2O(l) \leftrightharpoons OH^-(aq) + HC_2H_3O_2(aq)$$

$$K_b = \frac{[OH^-][HC_2H_3O_2]}{[C_2H_3O_2^-]} = 5.75 \times 10^{-10}$$

Adding the results from the ICE table and entering them into the equilibrium constant expression, then setting $[0.06667 - x] \approx [0.06667]$ because K_b is so small:

$$C_2H_3O_2^-(aq) + H_2O(l) \leftrightharpoons OH^-(aq) + HC_2H_3O_2(aq)$$

$$\begin{array}{ccc} 0.06667 \text{ M} & +x & +x \end{array}$$

$$K_b = \frac{[OH^-][HC_2H_3O_2]}{[C_2H_3O_2^-]} = \frac{[x][x]}{[0.06667-x]} = 5.75 \times 10^{-10}$$

Solving for x:

$$x = \sqrt{(0.06667)(5.75 \times 10^{-10})} = 6.19 \ 10^{-6} \text{ M} = [OH^-]$$

$$pOH = -\log[OH^-] = -\log[6.19 \times 10^{-6}] = 5.208$$

$$pH = pK_w - pOH = 14.000 - 5.208 = 8.792$$

50.00 mL of NaOH

This amount of 0.2000 M NaOH solution supplies 0.0100 mol of NaOH (MV calculation). The balanced chemical equation along with the moles of reactants are:

$$NaOH(aq) + HC_2H_3O_2(aq) \rightarrow C_2H_3O_2^-(aq) + Na^+(aq) + H_2O(l)$$

$$\begin{array}{cccc} 0.01000 & 0.005000 & — & \text{mol} \end{array}$$

The $HC_2H_3O_2$ is the limiting reagent, and the stoichiometric ratios are 1:1.

$$NaOH(aq) + HC_2H_3O_2(aq) \rightarrow C_2H_3O_2^-(aq) + Na^+(aq) + H_2O(l)$$

0.01000	0.005000	—		mol
−0.005000	−0.005000	+0.005000		mol
0.005000	0	0.005000		mol

There are two bases present, NaOH(aq) and $C_2H_3O_2^-$(aq). The strong base is the dominant factor leading to the pH, with a minor (ignored) contribution from the $C_2H_3O_2^-$(aq)
The volume of the solution at this point is (50.00 + 50.00) mL = 100.00 mL = 0.10000 L
The molarity of the hydroxide ion from the NaOH is:

$$M \ OH^-(aq) = \frac{0.005000 \text{ mol NaOH}}{0.10000 \text{ L}} = 0.05000 \text{ M}$$

The pOH and pH are:

$$pOH = -\log[OH^-] = -\log(0.05000) = 1.3010$$

$$pH = pK_w - pOH = 14.000 - 1.3010 = 12.699$$

Titration calculations

1. A 25.00 mL sample of 0.1000 M NaOH solution is titrated with 0.5000 M HCl. What is the pH at the equivalence point?

2. A 25.00 mL sample of 0.1000 M NH_3 solution is titrated with 0.5000 M HCl. What is the pH at the equivalence point? The K_b of NH_3 is 1.76×10^{-5}.

3. A 25.00 mL sample of 0.1000 M $HC_2H_3O_2$ solution is titrated with 0.5000 M NaOH. What is the pH at the equivalence point? The K_a of $HC_2H_3O_2$ is 1.74×10^{-5}.

4. A 25.00 mL sample of 0.1000 M NaOH solution is titrated with 0.5000 M HCl. What is the pH after 4.00 mL of HCl has been added?

5. A 25.00 mL sample of 0.1000 M NH_3 solution is titrated with 0.5000 M HCl. What is the pH after 4.00 mL of HCl has been added? The K_b of NH_3 is 1.76×10^{-5}.

6. A 25.00 mL sample of 0.1000 M $HC_2H_3O_2$ solution is titrated with 0.5000 M NaOH. What is the pH after 4.00 mL of NaOH has been added? The K_a of $HC_2H_3O_2$ is 1.74×10^{-5}.

7. A 25.00 mL sample of 0.1000 M NaOH solution is titrated with 0.5000 M HCl. What is the pH after 15.00 mL of HCl has been added?

8. A 25.00 mL sample of 0.1000 M NH_3 solution is titrated with 0.5000 M HCl. What is the pH after 15.00 mL of HCl has been added? The K_b of NH_3 is 1.76×10^{-5}.

9. A 25.00 mL sample of 0.1000 M $HC_2H_3O_2$ solution is titrated with 0.5000 M NaOH. What is the pH after 15.00 mL of NaOH has been added? The K_a of $HC_2H_3O_2$ is 1.74×10^{-5}.

Thermodynamics

Thermodynamics is the study of energy transfer. Previously, we encountered a subset of thermodynamics known as thermochemistry. When considering thermochemistry, we dealt with the first law of thermodynamics (law of conservation of energy), enthalpy, and calorimetry. Here, we will see two more laws of thermodynamics, entropy, and free energy.

The problems will be postponed until the end of the chapter, as students benefit from seeing an overview first.

Considerations

As a reminder, the first law of thermodynamics may be expressed as:

$$\Delta E = q + w$$

In this equation, ΔE is the change in energy, q is the heat involved, and w is the work involved. We learned that under conditions of constant pressure, q was the enthalpy change represented as ΔH. The enthalpy is the heat content of the system.

We saw that thermochemical equations could be manipulated through Hess's law to find new thermochemical equations with new enthalpy values. In addition, for any substance there is a standard heat of formation to express how much energy was released (exothermic) or absorbed (endothermic) when a mole of a substance was formed under standard conditions. The standard heats of formation could be combined to predict the enthalpy change for other reactions. This is expressed as:

$$\Delta H = \text{Products} - \text{Reactants}$$

where "products" includes the sum of all the heats of formation for the products in the reaction and "reactants" is the sum of all the heats of formation for the reactants. The individual heats of formation are collected into tables.

Recall that thermochemical equations always deal with moles. For this reason, fractional coefficients are allowed, because while you cannot have a fraction of a molecule, you can have a fraction of a mole.

If the reaction takes place under standard conditions, a degree symbol is attached to the variable as $\Delta H°$.

We will build on these ideas in this chapter.

Entropy

One of the concepts to be introduced here is entropy symbolized with an *S*. **Entropy** is a thermodynamic quantity representing the system's energy that is unavailable for conversion to work. Some sources consider entropy to be a measure of the disorder of a system; however, this definition may lead to some erroneous conclusions.

As with standard heats of formation, there are standard entropy values symbolized as $S°$. These standard entropy values are also collected in tables; however, at any temperature above absolute zero, the entropy values are greater than zero. These values may be applied to reactions in the form:

$$\Delta S = \text{Products} - \text{Reactants}$$

Note that the entropy change in a system must be combined with the entropy change of the surroundings to get the entropy change required for the Second Law. This is sometimes expressed as:

$$\Delta S_{\text{Universe}} = \Delta S_{\text{System}} + \Delta S_{\text{Surroundings}}$$

Note, it is possible to calculate $\Delta S_{\text{Surroundings}}$ from:

$$\Delta S_{\text{Surroundings}} = \frac{-\Delta H_{\text{System}}}{T}$$

In general the entropy of a gas is much greater than that of a liquid, and the entropy of a liquid is greater than the entropy of a solid.

Laws of thermodynamics

As stated previously, there are three laws of thermodynamics. These laws may be stated as:

> **First law of thermodynamics** states that the total amount of energy in the Universe is constant.
>
> **Second law of thermodynamics** says that for any spontaneous process, the entropy change is positive.
>
> **Third law of thermodynamics** states that any pure crystalline substance has its entropy equal to zero at absolute zero.

Each of these may be stated differently; however, their meaning is very precise.

We have already examined the first law, so we will not cover it further here.

The second law brings up the term *spontaneous*. A **spontaneous process** is a change that occurs without the input of any outside energy. A **nonspontaneous process** is a change that requires an input of energy from some other source for the process to take place.

The third law allows the calculation of absolute entropies starting with 0 at 0 K. Other thermodynamic terms, such as *enthalpy*, are relative terms based on a zero value being set at some arbitrary set of conditions.

The second law leads us to the question "How can we predict if a process is spontaneous or nonspontaneous?" Such predictions depend upon both the entropy change and the enthalpy change. We do not need to know the exact values of either the entropy or the enthalpy, we only need to know the signs. There are four possibilities:

ΔH	ΔS	Prediction
–	+	Always spontaneous
+	–	Never spontaneous
–	–	Spontaneous at low temperatures
+	+	Spontaneous at high temperatures

Spontaneous under high or low temperatures means that the process must be heated or cooled for the process to occur. Just how much heating or cooling is necessary depends upon the numeric values instead of only the signs.

Gibb's free energy

While it is useful to predict the spontaneity of a process by examining the signs of the enthalpy and entropy changes, it would be useful to narrow such predictions to one variable. Fortunately, there is a thermodynamic function that fits this requirement. It is the Gibb's free energy.

The Gibb's free energy is concerned with the work available to or required by a system. The change in the Gibb's free energy, ΔG, is the maximum amount of work that may be extracted from a system or the minimum amount of required work to force a process to occur.

A process with a negative free-energy change is exergonic, while a process with a positive free-energy change is endergonic.

Like entropy and enthalpy values, Gibb's free energies are tabulated in tables. The Gibb's free energies of formation, like the $\Delta H_f°$ values for any element in its standard state, are defined as being exactly zero.

The Gibb's free energy may be tied to the entropy and enthalpy through the relationship:

$$\Delta G = \Delta H - T\Delta S$$

under standard conditions:

$$\Delta G° = \Delta H° - T\Delta S°$$

> Note, both Gibb's free energies and enthalpy values in tables tend to be in kJ mol^{-1}, while entropy values in the same tables tend to be in J mol^{-1} K^{-1}. For this reason when using relationships such as the preceding two equations, care should be taken to convert J to kJ or vice versa.

The sign of the Gibb's free energy allows one to predict the spontaneity of a process. The three possibilities are

$\Delta G < 0$	Spontaneous
$\Delta G > 0$	Nonspontaneous
$\Delta G = 0$	Equilibrium

We will return to the $\Delta G = 0$ situation later.

Nonstandard conditions

As mentioned previously, the tabulated values in tables are determined under standard conditions (normally at 1 atm, 298 K, and molar quantities). However, what if the process under consideration is occurring under nonstandard conditions? In general, only adjustments in the Gibb's free energy are important. The adjustment normally involves adding or subtracting a value from the standard free energy, $\Delta G°$, to produce a nonstandard free energy, ΔG. One way to make such an adjustment is to use the following relationship:

$$\Delta G = \Delta G° + RT \ln Q$$

In this equation, ΔG and $\Delta G°$ are the nonstandard and standard Gibb's free energies, respectively. R is the gas constant in energy terms, $R = 8.3145$ J mol^{-1} K^{-1}, T is the Kelvin temperature. The ln is the natural logarithm. Q is the reaction quotient, which we saw in the equilibrium chapters.

Note that taking this equation and setting $\Delta G = 0$, makes this an equilibrium process, which leads to $Q = K$. Making these changes and rearranging gives:

$$\Delta G° = -RT \ln K$$

This allows a means of predicting the value of the equilibrium constant from the Gibb's free energy (and vice versa). Values determined by this means are questionable as there are many other factors leading to erroneous values. However, sometimes this is the only choice.

Thermodynamic calculations

Note that standard $\Delta H°$, $S°$, and $\Delta G°$ are in the table at the end of this chapter.

EXAMPLE 1

Calculate $\Delta S°$ for the following reaction:
$$CaCO_3(s) + H_2SO_4(l) \rightarrow CaSO_4(s) + H_2O(g) + CO_2(g)$$

To begin, we need to locate the individual $S°$ values in a table, such as the one at the end of this chapter. Once done, enter the values below the substances in the balanced chemical equation:

CaCO$_3$(s)	+	H$_2$SO$_4$(l)	$\rightarrow$	CaSO$_4$(s)	+	H$_2$O(g)	+	CO$_2$(g)
22.2		156.90		107		188.7		213.7 J mol^{-1} K^{-1}

Enter these values into: $\Delta S° =$ Products − Reactants as:
$$([107 + 188.7 + 213.7] - [22.2 + 156.90]) \text{ J/mol K} = 330 \text{ J mol}^{-1} \text{ K}^{-1}$$

EXAMPLE 2

Calculate the standard entropy change for the formation reaction of $CaSO_4 \cdot 2\,H_2O$(s).

We need a balanced chemical equation first. A standard formation reaction is a thermochemical equation for preparing one mole of a substance from the elements with all substances being in their standard states. The reactant side of the equation will only have the appropriate elements. The product side will only have one mole of the substance being prepared. Next, the equation is balanced with the requirement that the coefficient of the one and only product <u>must</u> remain 1. Once balanced, it is necessary to locate the appropriate $S°$ values in a table and add this information to the balanced thermochemical equation.

Ca(s)	+	S(s)	+	3 O$_2$(g)	+	2 H$_2$(g)	→	CaSO$_4$·2 H$_2$O(s)
41.63		31.88		205		130.6		194.0 J mol^{-1} K^{-1}

Enter these values into: $\Delta S° = $ Products − Reactants as:

$$([194.0] - [41.63 + 31.88 + 3(205) + 2(130.6)])\ \text{J mol}^{-1}\ \text{K}^{-1} = -756\ \text{J mol}^{-1}\ \text{K}^{-1}$$

EXAMPLE 3

Calculate $\Delta G°$ for the following reaction:

$$2\ NH_4Cl(s) + CaO(s) \rightarrow CaCl_2(s) + H_2O(l) + 2\ NH_3(g)$$

It is necessary to locate the appropriate $\Delta G°$ values in a table and add this information to the balanced thermochemical equation.

2 NH$_4$Cl(s)	+	CaO(s)	→	CaCl$_2$(s)	+	H$_2$O(l)	+	2 NH$_3$(g)
−203.9		−603.5		−750.2		−237.2		−16.7 kJ mol^{-1}

Enter these values into: $\Delta G° = $ Products − Reactants as:

$$([-750.2 - 237.2 + 2(-16.7)] - [2(-203.9) -603.5])\ \text{kJ mol}^{-1} = -9.5\ \text{kJ mol}^{-1}$$

EXAMPLE 4

At the melting point of aluminum, this equilibrium is present.

$$Al(s) \leftrightarrows Al(l)$$

The ΔH for this process is 10.0 kJ mol^{-1} and the ΔS for this process is 9.50 J mol^{-1} K^{-1}. Calculate the melting point of Al.

Since the melting process is an equilibrium process, $\Delta G = 0$, which leads to:

$$\Delta G = 0 = \Delta H - T\Delta S$$

Rearranging this equation and entering the appropriate vales gives:

$$T = \frac{\Delta H}{\Delta S} = \frac{10.0\ \cancel{\text{kJ mol}}^{-1}}{9.50\ \cancel{\text{J mol}}^{-1}\ \text{K}^{-1}} \left(\frac{1{,}000\ \cancel{\text{J}}}{1\ \cancel{\text{kJ}}} \right) = 1.05 \times 10^3\ \text{K}$$

EXAMPLE 5

Calculate $\Delta G°$ for the following equilibrium:

$$2\ O_3(g) \leftrightarrows 3\ O_2(g) \qquad K_p = 4.17 \times 10^{14}$$

Since this is under standard conditions (as indicated by the °), the temperature is 298 K.

The relationship between $\Delta G°$ and K_p is:
$$\Delta G° = -RT \ln K$$

Entering the values gives:
$$\Delta G° = \frac{-(8.314 \text{ J})(298 \text{ K})}{\text{mol K}} \ln 4.17 \times 10^{14} = -8.34 \times 10^4 \text{ J/mol}$$

EXAMPLE 6

Calculate K_p for the following equilibrium:
$$I_2(g) \rightleftharpoons 2 \text{ I}(g) \qquad \Delta G° = 2.52 \times 10^4 \text{ J/mol}$$

Again we need the relationship between $\Delta G°$ and K:
$$\Delta G° = -RT \ln K$$

This may be rearranged to:
$$\ln K = \frac{\Delta G°}{-RT}$$

Entering the given values into this equation gives:
$$\ln K = \frac{\Delta G°}{-RT} = \frac{\left(2.52 \times 10^4 \, \cancel{\text{J}}/\cancel{\text{mol}}\right)}{-\left(8.314 \, \cancel{\text{J}}/\cancel{\text{mol K}}\right)(298 \, \cancel{\text{K}})} = -10.171$$

Now we need to take the inverse ln:
$$K = e^{-10.171} = 3.82 \times 10^{-5} = 4 \times 10^{-5} \text{ (after rounding to the appropriate significant figures)}$$

EXAMPLE 7

Calculate ΔG for the following reaction at 500. K:

2 NO(g)	+	O$_2$(g)	→	2 NO$_2$(g)
2.00 M		0.500 M		1.00 M

The molarities beneath each formula is the concentration of that substance.
We will need to use the equation:
$$\Delta G = \Delta G° + RT \ln Q$$

We are looking for ΔG, and we will assume that the correction takes care of the difference between ΔG and $\Delta G°$.
We can determine the ΔG_{rxn} by Products − Reactants by looking up the appropriate values:
$$\Delta G_{rxn} = [2(51.84)] - [2(86.71) + 0.000] = -69.74 \text{ kJ mol}^{-1}$$

The reaction quotient is:

$$Q = \frac{[NO_2]^2}{[NO]^2[O_2]}$$

Entering these values into:

$$\Delta G^{500} = \Delta G_{rxn} + RT \ln Q$$

gives:

$$\Delta G^{500} = (-69.74 \text{ kJ mol}^{-1})\left(\frac{1,000 \text{ J}}{1 \text{ kJ}}\right) + (8.314 \text{ J mol}^{-1} \text{ K}^{-1})(500. \text{ K}) \ln \frac{[1.00]^2}{[2.00]^2[0.500]}$$

The kJ to J conversion is necessary to make sure the units between the two terms match ($J \text{ mol}^{-1}$)

$$\Delta G^{500} = (-69.74)(1000 \text{ J mol}^{-1}) + (4157 \text{ J mol}^{-1})(-0.693147) == -7.262 \times 10^4 \text{ J mol}^{-1}$$

EXERCISE 30·1

Thermodynamic calculations

1. Calculate $\Delta S°$ for the following reaction:
   $$H_2(g) + 1/2 \, O_2(g) \rightarrow H_2O(g)$$

2. Calculate $\Delta S°$ for the following reaction:
   $$H_2(g) + 1/2 \, O_2(g) \rightarrow H_2O(l)$$

3. Calculate the standard entropy change for the formation reaction of $Al_2O_3(s)$.

4. Calculate the standard entropy change for the formation reaction of $N_2O_4(g)$.

5. Calculate $\Delta G°$ for the following reaction:
   $$C_2H_5OH(l) \rightarrow C_2H_4(g) + H_2O(g)$$

6. Calculate $\Delta G°$ for the following reaction:
   $$Ca(s) + 2 \, H_2SO_4(l) \rightarrow CaSO_4(s) + SO_2(g) + 2 \, H_2O(l)$$

7. Acetone boils at 56.2°C, at this temperature the ΔH is 31.9 kJ mol^{-1}. Calculate the entropy for this process. At the boiling point, liquid and gaseous acetone are in equilibrium.

8. Do you expect the entropy change for the following reaction to be positive or negative?
   $$N_2(g) + 3 \, H_2(g) \rightarrow 2NH_3(g)$$

9. Do you expect the entropy change for the following reaction to be positive or negative?
   $$2 \, KClO_3(s) \rightarrow 2 \, KCl(s) + 3 \, O_2(g)$$

10. What is the equilibrium constant for the vaporization of water under standard conditions? Hint: use the table at the end of the chapter to calculate $\Delta G°$.

Thermodynamic Properties of Some Elements and Compounds.

	$\Delta H_f°$ kJ mol^{-1}	$S°$ J mol^{-1} K^{-1}	$\Delta G_f°$ kJ mol^{-1}
Ag(s)	0	42.7	0
Ag$^+$(aq)	105.9	73.9	77.1
AgCl(s)	−109.7	96.11	−127
Ag$_2$CrO$_4$(s)	−712	214	−622
AgN$_3$(s)	378.5	99.2	310.3
Ag$_2$O(s)	−31.05	121.3	−11.20
Al(s)	0	28.32	0
Al$_2$O$_3$(s)	−1,669.8	50.99	−1,576.4
Al$_2$(SO$_4$)$_3$(s)	−3,441	239	−3,100.
B$_2$O$_3$(s)	−1,272	53.8	−1,193
BaCO$_3$(s)	−1,218.8	112.1	−1,138.9
BaO(s)	−558.2	70.3	−528.4
BaSO$_4$(s)	−1,464.4	132.2	−1,353.1
BaCl$_2 \cdot$ 2H$_2$O(s)	−1,460	202.9	−1,296
Ba(OH)$_2 \cdot$ 8H$_2$O(s)	−3,350	400.	−1,366
Br$_2$(g)	−30.91	245.38	−3.13
Br$_2$(l)	0	152.3	0
C(s) (graphite)	0	5.686	0
C(s) (diamond)	1.896	2.439	2.866
CH$_4$(g)	−74.9	186.19	−50.8
C$_2$H$_2$(g)	54.19	48.00	50.0
C$_2$H$_4$(g)	52.47	219.22	68.36
C$_2$H$_6$(g)	−84.7	229.49	−32.89
C$_4$H$_{10}$(g)	−125	310.0	−16
C$_6$H$_{12}$(l) (cyclohexane)	−123.1	298.2	31.76
C$_8$H$_{18}$(l) n-octane	−250.3	361.2	−358
CH$_3$OH(g)	−201.2	237.6	−161.9
CH$_3$OH(l)	−238.6	126.8	−166.23
C$_2$H$_5$OH(l)	−277.63	161.04	−174.8
C$_2$H$_5$OH(g)	−235.1	282.6	−168.6
C$_6$H$_{12}$O$_6$(s)	−1,273.3	212.1	−910.56
CO(g)	−110.5	197.5	−137.2
CO$_2$(g)	−393.5	213.7	−394.4
COCl$_2$(g)	−223.0	289.2	−210.5
Ca(s)	0	41.63	0
CaCO$_3$(s)	−288.45	22.2	−269.78
CaCl$_2$(s)	−796	104.6	−750.2
CaH$_2$(s)	−45.1	10	−35.8
CaI$_2$(s)	−533.5	142	−528.9
Ca(OH)$_2$(s)	−986.6	76.1	−896.76
CaO(s)	−635.1	38.2	−603.5
CaSO$_4$(s)	−1,434	107	−1,320.3
CaSO$_4 \cdot$2H$_2$O(s)	?	194.0	?
Cl$_2$(g)	0	222.96	0
Cr$_2$O$_3$(s)	−1,140	81	−1,058
Cu(s)	0	33.1	0
CuO(s)	−157.3	42.6	−129.7
CuS(s)	−48.5	66.5	−49.0

	$\Delta H_f°$ kJ mol^{-1}	$S°$ J mol^{-1} K^{-1}	$\Delta G_f°$ kJ mol^{-1}
Cu$_2$S(s)	−79.5	120.9	−86.2
FeCl$_3$(s)	−499.5	142	−334
FeO(s)	−271.9	60.75	−255.2
Fe$_2$O$_3$(s)	−822.16	89.96	−740.98
H$_2$(g)	0	130.6	0
HBr(g)	−36.23	198.49	−53.22
HCl(aq)	−167.46	55.06	−131.17
HCl(g)	−95.3	186.8	−92.3
HI(g)	26	206	1.30
HF(g)	−268.61	173.51	−270.70
HC$_2$H$_3$O$_2$(l)	−487.0	159.8	−392.4
H$_2$O(g)	−241.8	188.7	−228.6
H$_2$O(l)	−285.8	69.94	−237.2
H$_3$PO$_4$(aq)	−1288.3	158	−1,142.7
H$_3$PO$_4$(s)	−1279	110.	−1,119
H$_2$S(g)	−20.2	205.6	−33
H$_2$SO$_4$(aq)	−907.51	17	−741.99
H$_2$SO$_4$(l)	−813.989	156.90	−690.059
KCl(s)	−436.7	82.59	−409.2
KClO$_3$(s)	−397.7	143.1	−296.3
KI(s)	−327.9	106.3	−325
KNO$_3$(s)	−492.7	133	−393
MgCO$_3$(s)	−1,112.9	65.69	−1,029.3
MgO(s)	−601.8	26.78	−569.6
Mn(s)	0	32.0	0
MnO(s)	−385.2	59.7	−362.9
MnO$_2$(s)	−520.9	53.1	−466.1
N$_2$(g)	0	191.5	0
NH$_3$(g)	−46.1	192.2	−16.7
NH$_3$(aq)	−80.8	110.	−26.7
NH$_4$Cl(s)	−315	94.6	−203.9
N$_2$O(g)	81.6	220.0	51.84
NO(g)	90.37	210.62	86.71
NO$_2$(g)	33.84	240.45	51.84
NOCl(g)	51.71	261.6	66.07
N$_2$O$_4$(g)	9.66	304.3	98.28
NaBr(s)	−86.30	20.75	−83.409
NaCl(s)	−411.0	72.38	−384.0
NaI(s)	−288	91.2	−286.1
Na$_2$O$_2$(s)	−504.6	94.98	−447.69
NaOH(s)	−426.8	64.454	−379.53
Na$_2$CO$_3$(s)	−1,131	136	−1,044.49
NaHCO$_3$(s)	−947.7	155	−851.0
O(g)	249.2	160.95	231.7
O$_2$(g)	0	205	0
O$_3$(g)	143	238.82	163
P$_4$O$_{10}$(s)	−3,110.	229	−2,698
Pb(s)	0	68.85	0
PbI$_2$(s)	−175.5	175	−173.6

	$\Delta H_f°$ kJ mol^{-1}	$S°$ J mol^{-1} K^{-1}	$\Delta G_f°$ kJ mol^{-1}
Pb(NO$_3$)$_2$(s)	−452		
PbO(s)	−217.3	68.70	−187.9
S(s) [S$_8$(s)]	0	31.88	0
SO$_2$(g)	−296.9	248.5	−300.4
SO$_3$(g)	−396	256.22	−370.4
SO$_4^{2-}$(aq)	−907.51	17	−741.99
SiBr$_4$(l)	−95.1	66.4	−106.1
SiO$_2$(s)	−914.4	41.3	−856.0
Sr^{2+}(aq)	−545.51	−39	−557.3
SrSO$_4$(s)	−1,445	122	−1,334
TiCl$_4$(g)	−763.2	354.9	−726.8
TiO$_2$(s) (rutile)	−944.7	50.29	−889.4
ZnO(s)	−83.24	10.43	−76.08
ZnS(s)	−203	57.7	−198

Electrochemistry

Electrochemistry involves the chemical application of electricity. Some chemical processes produce electricity, while other chemical process require electricity. For example, batteries use chemical reactions to produce electricity, while recharging the battery uses electricity to reverse the discharge process.

Electrochemistry involves oxidation and reduction. **Oxidation** is the loss of electrons. **Reduction** is the gain of electrons. These are complimentary processes, which *must always* occur together. The number electrons lost during oxidation are the same as the number of electrons gained during reduction. There will never be an excess or deficit of electrons nor will some of the electrons come from or go to some other place.

Review of redox reactions

While oxidation and reduction must occur together, they may be examined separately. One reason we examine the two separately is in the balancing of electrochemical equations, commonly referred to as redox equations.

The balancing of redox equations may be simple or involved. We introduced simple redox equations previously, and we will review those in this section. We will see how to deal with more involved redox equations in the next section. If you carefully follow the rules, you can balance any redox equation.

EXAMPLE 1

We will begin by examining the reaction of silver ions with copper metal to produce silver metal and copper(II) ions. The unbalanced net ionic equation is:

$$Ag^+(aq) + Cu(s) \rightarrow Ag(s) + Cu^{2+}(aq)$$

> Net ionic equations are easier to balance, as the spectator ions are no longer present to camouflage the reaction.

The atoms in this equation are balanced (one of each on each side); however, we must also consider the charges on the ions. As it is, the total charge on the left is +1, while that on the right is +2, and $1 \neq 2$. While we could fix this by inspection, that will not help you with the more involved equations that you will see later.

What we will do is separate the given unbalanced equation into half-reactions. A **half-reaction** is a simplified equation dealing with only oxidation or only reduction. In this case, one half-reaction will involve silver, and the other half-reaction will involve copper. We start this pair of half-reactions by extracting the appropriate species from our unbalanced equation:

$$Ag^+(aq) \rightarrow Ag(s)$$

$$Cu(s) \rightarrow Cu^{2+}(aq)$$

These are our starting half-reactions, which we will balance separately. Since the atoms are balanced, the only thing unbalanced are the charges. We do this balancing with electrons, e^-, (since half-reactions are not actually taking place separately, having "loose" electrons as acceptable).

The charges are balanced by adding electrons as:

$$1\ e^- + Ag^+(aq) \rightarrow Ag(s)$$

$$Cu(s) \rightarrow Cu^{2+}(aq) + 2\ e^-$$

Note that the $Ag^+(aq)$ is gaining an electron; therefore, this half-reaction is a reduction half-reaction, while the $Cu(s)$ is losing two electrons, which makes the copper half-reaction the oxidation half-reaction.

The next step is to adjust the half-reaction so that the number of electrons in each match. Each half-reaction is to be multiplied by a value to achieve the lowest common multiple of the charges. In this case, the lowest common multiple is 2, which means that the silver half-reaction is multiplied by 2 and the copper half-reaction is multiplied by 1:

$$2 \times (1\ e^- + Ag^+(aq) \rightarrow Ag(s)$$

$$1 \times (Cu(s) \rightarrow Cu^{2+}(aq) + 2\ e^-)$$

giving:

$$2\ e^- + 2\ Ag^+(aq) \rightarrow 2\ Ag(s)$$

$$Cu(s) \rightarrow Cu^{2+}(aq) + 2\ e^-$$

Now we need to add the two halves together to make a whole. We also need to cancel any spectators (only the electrons in this example).

$$\cancel{2\ e^-} + 2\ Ag^+(aq) + Cu(s) \rightarrow 2\ Ag(s) + Cu^{2+}(aq) + \cancel{2\ e^-}$$

$$2\ Ag^+(aq) + Cu(s) \rightarrow 2\ Ag(s) + Cu^{2+}(aq)$$

Finally, we need to double-check to make sure the final equation is balanced. The most common error is accidentally skipping one of the species when multiplying.

> Note that when combining the half-reactions, the electrons *must* cancel; if they do not, you have made an error.

This procedure works for other redox equations of this type; however, more involved redox equations, such as the next one, require additional steps.

$$Cr_2O_7^{2-}(aq) + Fe^{2+}(aq) + H^+(aq) \rightarrow Cr^{3+}(aq) + Fe^{3+}(aq) + H_2O(l) \qquad \text{Unbalanced}$$

Review of redox reactions

1. We will begin by examining the reaction of iron(III) ions with aluminum metal to produce iron(II) ions and aluminum ions. The unbalanced net ionic equation is:

$$Fe^{3+}(aq) + Al(s) \rightarrow Fe^{2+}(s) + Al^{3+}(aq)$$

Balance the redox reaction.

Balancing redox equations

More involved redox equations require additional rules. It is important to follow each rule carefully. In addition the rules *must* be followed in order. Not doing the rules in order is the most common problem that people have and transforms a simple balancing into a complicated mess.

We will use the equation presented at the end of the previous section to illustrate the rules and their sequence. The equation presented was:

$$Cr_2O_7{}^{2-}(aq) + Fe^{2+}(aq) + H^+(aq) \rightarrow Cr^{3+}(aq) + Fe^{3+}(aq) + H_2O(l) \qquad \text{Unbalanced}$$

It is important to note that the reaction is taking place in an acidic solution (based upon the presence of $H^+(aq)$ on either side of the reaction arrow). Two rules need to be modified in bases ($OH^-(aq)$ present). If neither hydroxide nor hydrogen ions are shown, look for an acid or base, or a designation that the solution is acidic or basic. Neutral solutions may be treated as acidic.

This example will be for acidic solutions. The next example is for basic solutions.

EXAMPLE 2

Rule 1

Separate the equation into two half-reactions, one for oxidation and one for reduction.

In the previous example, this was obvious; as there were only two pairs of species given, whereas in this reaction and others, more species are present. However, the goal of producing two separate half-reactions is the same.

In some cases, one pair may be obvious, this occurs for $Fe^{2+}(aq) \rightarrow Fe^{3+}(aq)$. By implication, the other species must be in the other half-reaction. Unfortunately, assuming what belongs in a specific half-reaction may lead to errors.

A safe procedure is to assign an oxidation number to each species in the redox equation. (You may need to review how to do this.) Adding oxidation numbers to our reaction gives:

$$Cr_2O_7{}^{2-}(aq) + Fe^{2+}(aq) + H^+(aq) \rightarrow Cr^{3+}(aq) + Fe^{3+}(aq) + H_2O(l)$$
$$\phantom{Cr_2O_7{}^{2-}(aq)}\;+6\;-2\qquad\;+2\qquad\;+1\qquad\;+3\qquad\;+3\quad\;+1\;-2$$

Note the oxidation numbers are placed under the symbols to minimize the possibility that they will be mistaken for ionic charges.

The two changes are Cr^{6+} to Cr^{3+}, and Fe^{2+} to Fe^{3+}. These two changes are the basis for selecting a beginning to each half-reaction. Since the Cr^{6+} is part of a group, the group must be kept together. Our preliminary two half-reactions are:

$$Fe^{2+}(aq) \rightarrow Fe^{3+}(aq)$$

$$Cr_2O_7^{2-}(aq) \rightarrow Cr^{3+}(aq)$$

Other materials from the original equation may be ignored at this point.

Rule 2

Balance all atoms in each half-reaction except H and O.
    The iron atoms are already balanced, and a 2 coefficient for the $Cr^{3+}(aq)$ balances the chromium atoms:

$$Fe^{2+}(aq) \rightarrow Fe^{3+}(aq)$$

$$Cr_2O_7^{2-}(aq) \rightarrow 2\ Cr^{3+}(aq)$$

Rule 3

Since we are in an acid, we balance the oxygen atoms by adding $H_2O(l)$ to the side needing oxygen.
    Oxygen appears only in the chromium half-reaction, and there are 7 oxygen atoms on the reactant side; therefore, we need 7 water molecules on the product side:

$$Fe^{2+}(aq) \rightarrow Fe^{3+}(aq)$$

$$Cr_2O_7^{2-}(aq) \rightarrow 2\ Cr^{3+}(aq) + 7\ H_2O(l)$$

At this point all atoms, except H, should be balanced.

Rule 4

Since we are in an acid, we balance the hydrogen atoms by adding $H^+(aq)$ to the side needing hydrogen.
    Since only the chromium half-reaction contains hydrogen, and there are 14 hydrogen atoms on the product side, we need to add 14 $H^+(aq)$ to the reactant side:

$$Fe^{2+}(aq) \rightarrow Fe^{3+}(aq)$$

$$14\ H^+(aq) + Cr_2O_7^{2-}(aq) \rightarrow 2\ Cr^{3+}(aq) + 7\ H_2O(l)$$

At this point, all atoms should be balanced. If not, you miscounted something.
It is time to move to the charges.

Rule 5

Balance the charges by adding electrons.
    To do this, simply add up the charges actually shown on each side of each half reaction and use electrons (–1 each) to balance the charges. For the iron half-reaction, the charges are 2+ and 3+ (differing by 1), while for the chromium half-reaction, the charges are +12 and +6 (differing by 6). Adding the appropriate numbers of electrons gives:

$$Fe^{2+}(aq) \rightarrow Fe^{3+}(aq) + 1\ e^-$$

$$6\ e^- + 14\ H^+(aq) + Cr_2O_7^{2-}(aq) \rightarrow 2\ Cr^{3+}(aq) + 7\ H_2O(l)$$

While this is a more involved situation, this is the same step are we saw in the preceding example.

<u>Rule 6</u>

Multiply each half-reaction so that each half-reaction has the same number of electrons (lowest common multiple).

This is the same as we did for the redox equation in the previous example.

In this case, the lowest common multiple is 6; therefore:

$$6 \times (Fe^{2+}(aq) \rightarrow Fe^{3+}(aq) + 1\ e^-)$$

$$1 \times (6\ e^- + 14\ H^+(aq) + Cr_2O_7^{2-}(aq) \rightarrow 2\ Cr^{3+}(aq) + 7\ H_2O(l))$$

This gives:

$$6\ Fe^{2+}(aq) \rightarrow 6\ Fe^{3+}(aq) + 6\ e^-$$

$$6\ e^- + 14\ H^+(aq) + Cr_2O_7^{2-}(aq) \rightarrow 2\ Cr^{3+}(aq) + 7\ H_2O(l)$$

<u>Rule 7</u>

Add the two half-reactions and cancel.

This is the same step as we did in the previous section.

$$6\ Fe^{2+}(aq) + \cancel{6\ e^-} + 14\ H^+(aq) + Cr_2O_7^{2-}(aq) \rightarrow 2\ Cr^{3+}(aq) + 7\ H_2O(l) + 6\ Fe^{3+}(aq) + \cancel{6\ e^-}$$

$$6\ Fe^{2+}(aq) + 14\ H^+(aq) + Cr_2O_7^{2-}(aq) \rightarrow 2\ Cr^{3+}(aq) + 7\ H_2O(l) + 6\ Fe^{3+}(aq)$$

Ideally, the equation is balanced at this point.

<u>Rule 8</u>

Check to make sure everything is balanced.

Do not forget to check the charges. You should do this step every time you balance an equation.

For this equation:

Reactant side	Species	Product side
6	Fe	6
14	H	14
2	Cr	2
7	O	7
+24	Charge	+24

Make sure you understand how this table was constructed and are able to do so yourself.

If something does not match, the two most likely causes are that you slipped up when adding up the charges on each side, or under rule 6, you accidently skipped one of the species while multiplying through. These should be easy for you to check. If neither of these is in error, you will be able to save time by simply starting over.

Now we will do a base example beginning with:

$$Bi(OH)_3(s) + SnO_2^{2-}(aq) \rightarrow Bi(s) + SnO_3^{2-}(aq) + H_2O(l)$$

This is a base example because there is a base ($Bi(OH)_3$) present.

For a base example, rules 3 and 4 need to be modified. All other rules remain the same.

EXAMPLE 3

<u>Rule 1</u>

Separate the equation into two half-reactions, one for oxidation and one for reduction.

$$Bi(OH)_3(s) \rightarrow Bi(s)$$

$$SnO_2^{2-}(aq) \rightarrow SnO_3^{2-}(aq)$$

These choices were made because Bi changed from +3 to 0, and Sn changed from +2 to +4. (For practice, you should check these for yourself.)

<u>Rule 2</u>

Balance all atoms in each half-reaction except H and O.
   Already done:

$$Bi(OH)_3(s) \rightarrow Bi(s)$$

$$SnO_2^{2-}(aq) \rightarrow SnO_3^{2-}(aq)$$

<u>Rule 3</u>

Since we are in a base, we balance the oxygen atoms by adding 2 OH^-(aq) for each oxygen needed on the side needed, plus 1 H_2O(l) per O needed to the opposite side.

$$3\ H_2O(l) + Bi(OH)_3(s) \rightarrow Bi(s) + 6\ OH^-(aq)$$

$$2\ OH^-(aq) + SnO_2^{2-}(aq) \rightarrow SnO_3^{2-}(aq) + H_2O(l)$$

As before, all atoms except H should be balanced at this point.

<u>Rule 4</u>

Since we are in a base, we balance the hydrogen atoms by adding 1 H_2O(l) to the side needing hydrogen, plus 1 OH^-(aq) per H to the opposite side.

$$3\ OH^-(aq) + 3\ H_2O(l) + Bi(OH)_3(s) \rightarrow Bi(s) + 6\ OH^-(aq) + 3\ H_2O(l)$$

$$2\ OH^-(aq) + SnO_2^{2-}(aq) \rightarrow SnO_3^{2-}(aq) + H_2O(l)$$

Notice rule 3 covered the O and in this case, the H, which is a common occurrence. At this point, all atoms should be balanced.

<u>Rule 5</u>

Balance the charges by adding electrons.

$$3\ e^- + 3\ OH^-(aq) + 3\ H_2O(l) + Bi(OH)_3(s) \rightarrow Bi(s) + 6\ OH^-(aq) + 3\ H_2O(l)$$

$$2\ OH^-(aq) + SnO_2^{2-}(aq) \rightarrow SnO_3^{2-}(aq) + H_2O(l) + 2\ e^-$$

This step is the same as in the preceding example.

<u>Rule 6</u>

Multiply each half-reaction so that each half-reaction has the same number of electrons (lowest common multiple).

$$2 \times (3\ e^- + 3\ OH^-(aq) + 3\ H_2O(l) + Bi(OH)_3(s) \rightarrow Bi(s) + 6\ OH^-(aq) + 3\ H_2O(l))$$

$$3 \times (2\ OH^-(aq) + SnO_2^{2-}(aq) \rightarrow SnO_3^{2-}(aq) + H_2O(l) + 2\ e^-)$$

$$6\ e^- + 6\ OH^-(aq) + 6\ H_2O(l) + 2\ Bi(OH)_3(s) \rightarrow 2\ Bi(s) + 12\ OH^-(aq) + 6\ H_2O(l)$$

$$6\ OH^-(aq) + 3\ SnO_2^{2-}(aq) \rightarrow 3\ SnO_3^{2-}(aq) + 3\ H_2O(aq) + 6\ e^-$$

<u>Rule 7</u>

Add the two half-reactions and cancel.

$$\cancel{6\,OH^-(aq)} + 3\,SnO_2^{2-}(aq) + \cancel{6\,e^-} + \cancel{6\,OH^-(aq)} + 6\,H_2O(l) + 2\,Bi(OH)_3(s) \rightarrow$$

$$2\,Bi(s) + \cancel{12\,OH^-(aq)} + 6\,H_2O(l) + 3\,SnO_3^{2-}(aq) + 3\,H_2O(l) + \cancel{6\,e^-}$$

Notice that when you canceled, more than the electrons canceled. Also, if you wished, you could have consolidated similar species when you added the two half reactions.

$$3\,SnO_2^{2-}(aq) + 2\,Bi(OH)_3(s) \rightarrow 2\,Bi(s) + 3\,SnO_3^{2-}(aq) + 3\,H_2O(l)$$

<u>Rule 8</u>

Check to make sure everything is balanced.
   This applies to all reactions under any and all circumstances.

Reactant side	Species	Product side
3	Sn	3
12	O	12
2	Bi	2
6	H	6
–6	Charge	–6

From this table, we can see that the equation is balanced.

Note that some people suggest that you can balance a basic equation by solving it as an acid and then converting this to a base. This procedure will work for people who are experts in balancing by the acid rules; however, people who are still learning how to follow the acid rules tend to get extremely confused, so staying with separate rules saves time and grades.

EXERCISE
31·2

Balancing redox equations

1. Without looking, what are the rules for balancing redox equations. List the rules in order.

2. Balance the following equation:
   $$MnO_4^-(aq) + C_2O_4^{2-}(aq) + H^+(aq) \rightarrow Mn^{2+}(aq) + CO_2(g) + H_2O(l)$$

3. Balance the following equation:
   $$Pt(s) + H^+(aq) + NO_3^-(aq) + Cl^-(aq) \rightarrow PtCl_6^{2-}(aq) + NO_2(g) + H_2O(l)$$

4. Balance the following equation:
   $$OH^-(aq) + C_2H_5OH(aq) + Cr_2O_7^{2-}(aq) \rightarrow CO_2(g) + H_2O + CrO_2^-(aq)$$

Electrochemical cells

It is possible to represent a redox equation as an electrochemical cell. In such a cell, the oxidation and reduction reactions are physically separated. An example of an electrochemical cell outside of the laboratory is the lead-acid automobile battery. Other electrochemical cells may appear more complicated; however, they may all be simplified to a simple electrochemical cell as follows:

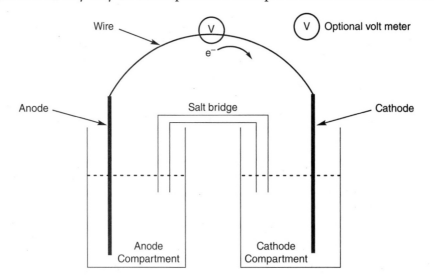

By convention, the cathode is placed on the right as shown here. The whole diagram may be reversed and still be correct; however, it is easier to remember if you always picture an electrochemical cell as shown here (cathode on right).

The anode, wire, and cathode must all conduct electrons, which usually means that they must be metals (however, there are special circumstances where other material may be used, such as a lead electrode coated with $PbO_2(s)$ as in a lead-acid automobile battery). The anode-wire-cathode-salt bridge complete an electrical circuit. The salt bridge contains an electrolyte solution containing ions that do not react, but with charges available to keep the charges in each compartment balanced. The system will not operate if the circuit is not complete (closed).

The terms *anode* and *cathode* refer to oxidation and reduction respectively. The **anode** is the electrode where oxidation occurs, and the **cathode** is the electrode where reduction occurs. The anode and cathode may or may not be involved in the redox reaction.

The anode compartment contains everything necessary for the oxidation half-reaction, while the cathode compartment contains everything necessary for the reduction half-reaction.

The voltmeter is present to measure the voltage of the cell. As stated in the diagram, it is optional. It is also possible to replace the voltmeter with an amp meter to measure the electrical current or to use a different meter to determine some other property.

EXAMPLE 4

Using the following balanced redox equation:

$$Zn(s) + Cu^{2+}(aq) \rightarrow Zn^{2+}(aq) + Cu(s)$$

describe how this may be expressed in the electrochemical cell shown previously.
This reaction involves the oxidation of Zn and the reduction of Cu. The half-reactions are:

$$Zn(s) \rightarrow Zn^{2+}(aq) + 2\ e^- \qquad \text{Oxidation (anode)}$$

$$Cu^{2+}(aq) + 2\ e^- \rightarrow Cu(s) \qquad \text{Reduction (cathode)}$$

According to these half-reactions, the anode is Zn(s), and the cathode is Cu(s). The oxidation half-reaction takes place in the anode compartment, and the reduction half-reaction takes place in the cathode compartment.

The salt bridge could contain any of several possible strong electrolytes such as $KNO_3(aq)$ or NaCl(aq).

The electrons leave the oxidation half-reaction and travel through the wire, to be gained by the cathode half-reaction.

The oxidation half-reaction produces Zn^{2+}, so the anode compartment gains a positive charge. To keep the charges balanced, some negative ions need to enter the anode compartment to keep the solution neutral. These ions are anions from the salt bridge. The reduction half-reaction removes positive ions from the cathode compartment, and cations from the salt bridge enter to replace the lost charge.

Instead of constructing a diagram such as the preceding one, it helps to have an abbreviated means of describing a cell.

EXAMPLE 5

The abbreviation begins with anode, goes through the anode compartment, then the salt bridge, then into the cathode compartment, and ends at the cathode. Sounds complicated, however, it simply travels from left to right in the diagram. If you drew the reverse cell (which is still correct) you need to remember to go from right to left.

Beginning with the anode, we write

$$Zn(s)$$

The (s) is optional in this abbreviation.

Next there is a phase change from the solid anode to the half-reaction solution. A phase change is indicated by a vertical line, |.

$$Zn(s)\,|$$

Next come the ions in the anode compartment:

$$Zn(s)\,|\,Zn^{2+}(aq)$$

The ion concentration may be added as, for example, $Zn(s)\,|\,Zn^{2+}(1.0\ M)$

The next thing we come to is the salt bridge, which is indicated by two vertical lines, ||. The formula for the electrolyte in the salt bridge may be written between the two vertical lines.

$$Zn(s)\,|\,Zn^{2+}(aq)\,||$$

Now, we need to deal with the ions in the cathode compartment:

$$Zn(s)\,|\,Zn^{2+}(aq)\,||\,Cu^{2+}(aq)$$

Next, we have a phase change from the liquid solution to the solid cathode:

$$Zn(s)\,|\,Zn^{2+}(aq)\,||\,Cu^{2+}(aq)\,|$$

Finally, we have the cathode:

$$Zn(s)\,|\,Zn^{2+}(aq)\,||\,Cu^{2+}(aq)\,|\,Cu(s)$$

Complications occur if a half-reaction does not involve a solid conductor. For example, what should we do in the case of the following half-reaction?

$$Fe^{3+}(aq) + 1\ e^{-} \rightarrow Fe^{2+}(aq)$$

The way this would be done is by using an electrode that does not react, an inert electrode. Platinum is a good selection. In this case, the anode side of the cell could be written as:

$$Pt(s)\,|\,Fe^{3+}(aq),\,Fe^{2+}(aq)\,||$$

This would simply be reversed for the cathode side.

Another complication occurs when the reaction involves something like $I_2(s)$ or $H_2(g)$, neither of which is water soluble nor an electrical conductor. In the case of $I_2(s)$, some is placed in the bottom of the compartment, and an inert electrode is lowered until it is in contact with the solid. For an anode:

$$Pt(s), I_2(s) \,|$$

This is simply reversed for the cathode.

In the case of a gas like $H_2(g)$, the gas is introduced through a tube with an inert electrode extending through the tube into contact with the solution. For an anode, this could be written as:

$$Pt(s), H_2(g) \,|$$

This is simply reversed for the cathode.

Finally, for reactions taking place in aqueous solution, $H_2O(l)$ is normally left out. However, it should be included if the water is actually being oxidized or reduced.

There are two broad categories of electrochemical cells. The category of that the cell depends upon whether the cell reaction is spontaneous or nonspontaneous. A **galvanic cell** or **voltaic cell** is an electrochemical cell where a spontaneous chemical reaction generates electricity. An **electrolysis cell** is an electrochemical cell where electricity is used to force a nonspontaneous chemical reaction to occur.

Both types of cells look identical except for the signs on the electrodes. For a spontaneous cell, the electrons travel through the wire in the spontaneous direction (from negative to positive). If you look at any battery, you will see these two signs at one end or the other or one side to the other. Sometimes only one is specified with the sign of the other being understood. For a nonspontaneous cell, the electrons travel through the wire in the nonspontaneous direction (from positive to negative). Thus, the only difference between these two types of cells is the signs on the electrodes.

EXERCISE
31·3

Electrochemical cells

1. Write an abbreviated diagram to represent the following redox reaction in an electrochemical cell.

$$Ni^{2+}(aq) + 2\, Ag(s) \rightarrow Ni(s) + 2\, Ag^+(aq)$$

2. Write an abbreviated diagram to represent the following redox reaction in an electrochemical cell.

$$2\, Cl^-(aq) + MnO_2(s) + 4\, H^+(aq) \rightarrow Mn^{2+}(aq) + Cl_2(g) + 2\, H_2O(l)$$

Standard reduction potentials

Let's begin with the following two half-reaction:

$$Zn(s) \rightarrow Zn^{2+}(aq) + 2\, e^- \qquad \text{Oxidation}$$

$$Cu^{2+}(aq) + 2\, e^- \rightarrow Cu(s) \qquad \text{Reduction}$$

It is sometimes helpful to know just what voltage the voltmeter would read if a cell were constructed using these two half-reactions. To predict this value, you will need to utilize the standard reduction potentials for each of these equations. Such values appear in tables such as the one at the end of this chapter, so all you need to do is locate these half-reactions. Note the reaction in the table is reversed in the zinc half-reaction. The two equations appear in the table as:

$$Zn^{2+}(aq) + 2\ e^- \rightarrow Zn(s) \qquad -0.7628\ V$$

$$Cu^{2+}(aq) + 2\ e^- \rightarrow Cu(s) \qquad +0.337\ V$$

To match the zinc half-reaction, the reaction from the table needs to be reversed, which also reverses the sign on the $E°$:

$$Zn(s) \rightarrow Zn^{2+}(aq) + 2\ e^- \qquad +0.7628\ V$$

$$Cu^{2+}(aq) + 2\ e^- \rightarrow Cu(s) \qquad +0.337\ V$$

Finally, to finish our determination of the cell potential we add the two half-reactions along with their cell potentials

$$Cu^{2+}(aq) + Zn(s) \rightarrow Zn^{2+}(aq) + Cu(s) \qquad +1.100\ V$$

The table lists all half-reaction potentials for reduction half-reactions because a simple reversal of the equation along with its sign gives the oxidation half-reaction and its potential.

> Older references tabulate standard oxidation potentials, which means that all the half-reactions are the reverse of the ones in the table at the end of this chapter, with the potentials having the opposite signs. Several years ago, it was decided that standard reduction potentials would be the preferred form.

These are standard potentials because the potentials were measured under standard conditions. For these half-reactions, this means that any gases are at 1 atm and any molarities are 1 M.

It would be possible to construct this table by running every conceivable combination of two half-reactions. However, it was deemed simpler to choose one reference half-reaction and compare everything to it. The choice involved $H_2(g)$ and $H^+(aq)$, and its standard potential was defined to have a potential as exactly 0. Therefore:

$$H_2(g) \rightarrow 2\ H^+(aq) + 2\ e^- \qquad E° = 0.0000\ V$$

$$2\ H^+(aq) + 2\ e^- \rightarrow H_2(g) \qquad E° = 0.0000\ V$$

This is the standard hydrogen electrode symbolized as a SHE.

Note that the procedure used here will always work when determining the standard potential. However, there are other methods. If your instructor insists that you use one of the other methods, erase this one from your mind and follow your instructor's method. Trying to use both methods will lead to confusion and will lead to incorrect answers.

Note that we will learn what to do if the cells are not standard in a later section.

Note that even cells that appear to be standard may show slight variations due to other factors.

EXAMPLE 6

Determine the cell potential for the following reaction:

$$2\,Ag^+(aq) + Zn(s) \rightarrow Zn^{2+}(aq) + 2\,Ag^+(aq)$$

First, we need to locate the half-reactions in the table at the end of this chapter. (If you are not sure what the half-reactions are, break down the redox equation into its half-reactions first.

$$Zn^{2+}(aq) + 2\,e^- \rightarrow Zn(s) \qquad -0.7628\,V$$

$$Ag^+(aq) + 1\,e^- \rightarrow Ag(s) \qquad +0.7994\,V$$

Reversing the zinc half-reaction:

$$Zn(s) \rightarrow Zn^{2+}(aq) + 2\,e^- \qquad +0.7628\,V$$

$$Ag^+(aq) + 1\,e^- \rightarrow Ag(s) \qquad +0.7994\,V$$

Before we can add the two half-reactions, we need to multiply the silver half-reaction by 2:

$$2\,Ag^+(aq) + 2\,e^- \rightarrow 2\,Ag(s) \qquad +0.7994\,V$$

It is very important to remember that you *do not multiply the voltage*. The only thing you can do with the cell potential is to change the sign.
Adding gives:

$$2\,Ag^+(aq) + Zn(s) \rightarrow Zn^{2+}(aq) + 2\,Ag(s) \qquad +1.5622\,V$$

Cell potentials may be either positive or negative. If the cell potential is positive, the redox reaction is spontaneous. If the cell potential is negative, the redox reaction is nonspontaneous. (A 0 potential means the reaction is in equilibrium.)

EXERCISE
31·4

Standard reduction potentials

Determine the cell potential for each of the following reactions:

1. $6\,Fe^{2+}(aq) + 14\,H^+(aq) + Cr_2O_7^{2-}(aq) \rightarrow 2\,Cr^{3+}(aq) + 7\,H_2O(l) + 6\,Fe^{3+}(aq)$

2. $2\,Cl^-(aq) + MnO_2(s) + 4\,H^+(aq) \rightarrow Mn^{2+}(aq) + Cl_2(g) + 2\,H_2O(l)$

Electrochemistry and thermodynamics

Of primary interest here is how much energy a spontaneous electrochemical cell can produce, or how much energy a nonspontaneous electrochemical cell requires to operate.

The cell potential may be used to calculate the Gibb's free energy through the use of one of the following two equations:

$$\Delta G° = -nFE°_{cell} \qquad \text{For standard cells}$$

$$\Delta G = -nFE_{cell} \qquad \text{For nonstandard cells}$$

We will cover nonstandard cell in the next section.

In these equations:

$\Delta G°$ and ΔG are the standard and nonstandard Gibbs free energies, respectively.

$E°_{cell}$ and E_{cell} are the standard and nonstandard cell potentials, respectively. It is possible to use the cell potentials for a half-reaction instead of the entire reaction.

n is the number of electrons in the reaction (the "lowest common multiple number")

F is the Faraday constant (either 96,485 coul mol^{-1} or 96,485 J V^{-1} mol^{-1} depending upon the units required.

> Note that *coul* is the abbreviation for *coulomb*, which is the number of electrons present. This means that Avogadro's number of electrons = 1 mol e$^-$ = 96,485 coulombs.

For problems in this section, normally 96,485 J V^{-1} mol^{-1} is used for F in the $\Delta G° = -nFE°_{cell}$ equations.

Because of the sign change we see:

$\Delta G < 0$ so $E > 0$, the process is spontaneous.

$\Delta G > 0$ so $E < 0$, the process is nonspontaneous.

$\Delta G = 0$ so $E = 0$, the process is at equilibrium.

From the thermodynamics chapter we have:

$$\Delta G° = -RT \ln K$$

This may be expanded to:

$$\Delta G° = -RT \ln K = -nFE°_{cell}$$

This leads to:

$$E°_{cell} = \frac{RT}{nF} \ln K$$

The temperature is normally 298.15 K unless stated otherwise.

Use 8.3141 J mol^{-1} V^{-1} = R for the gas constant.

EXAMPLE 7

Given the following balanced redox equation:

$$Cu^{2+}(aq) + Zn(s) \rightarrow Zn^{2+}(aq) + Cu(s) \qquad +1.100\,V$$

what is the value of $\Delta G°$?

This requires us to use the equation: $\Delta G° = -nFE°_{cell}$

We need to assign values to each term in this equation:

$$\Delta G° = ? \qquad n = 2 \qquad F = 96,485\,J\,V^{-1}\,mol^{-1} \qquad E°_{cell} = +1.100\,V$$

Note that if you are uncertain about where we got the value of n, break down the redox equation to the half-reactions, and check the numbers of electrons involved.

Note that n is an exact number and unitless.
Entering these values into the equation:

$$\Delta G° = -nFE°_{cell} = -(2)\left(\frac{96,485 \text{ J}}{\cancel{V} \text{ mol}}\right)(+1.100 \cancel{V}) = -2.123 \times 10^5 \text{ J mol}^{-1}$$

EXAMPLE 8

Given the following balanced redox equation:

$$Cu^{2+}(aq) + Zn(s) \rightarrow Zn^{2+}(aq) + Cu(s) \qquad +1.100 \text{ V}$$

what is the value of K?

This requires us to use the equation: $E°_{cell} = \dfrac{RT}{nF} \ln K$

We need to assign values to each term in this equation:

$$K = ? \qquad n = 2 \text{ mol e}^- \qquad F = 96485 \text{ J V}^{-1} \text{ mol}^{-1} \qquad E°_{cell} = +1.100 \text{ V}$$

$$R = 8.3141 \text{ J mol}^{-1} \text{ V}^{-1} \qquad T = 298.15 \text{ K}$$

We need to rearrange the equation and enter the known values:

$$\ln K = \frac{nFE°_{cell}}{RT} = \frac{(2)\left(\dfrac{96,485 \cancel{J}}{\cancel{V} \text{ mol}}\right)(+1.100 \cancel{V})}{\left(\dfrac{8.3141 \cancel{J}}{\text{mol } \cancel{K}}\right)(298.15 \cancel{K})} = 85.632 \text{ (unrounded)}$$

$$K = e^{85.632} = 1.55 \times 10^{37} \text{ (rounded to correct significant figures)}$$

EXERCISE
31·5

Electrochemistry and thermodynamics

1. Given the following balanced redox equation:

   $$2 Ag^+(aq) + Zn(s) \rightarrow Zn^{2+}(aq) + 2 Ag(s) \qquad +1.5622 \text{ V}$$

   what is the value of $\Delta G°$?

2. Given the following balanced redox equation:

   $$Cu^{2+}(aq) + Zn(s) \rightarrow Zn^{2+}(aq) + Cu(s) \qquad \Delta G° = -2.123 \times 10^5 \text{ J mol}^{-1}$$

   what is the value of $E°$?

3. Given the following balanced redox equation:

$$2 \, Ag^+(aq) + Zn(s) \rightarrow Zn^{2+}(aq) + 2 \, Ag(s) \qquad +1.5622 \, V$$

what is the value of K?

Nonstandard cells

For a cell to be nonstandard, at least one variable must not be standard. Normally, this means one or more concentrations is not 1 M.

To determine a nonstandard cell potential, E, an adjustment must be applied to the standard cell potential, $E°$. The equation for doing this is the Nernst equation:

$$E = E° - \frac{RT}{nF} \ln Q$$

All the terms in this equation have the same meanings as we saw previously. As a reminder, Q is the reaction quotient.

EXAMPLE 9

We will again use the following redox equation in this example:

$$2 \, Ag^+(aq) + Zn(s) \rightarrow Zn^{2+}(aq) + 2 \, Ag(s) \qquad +1.5622 \, V$$

To make the cell nonstandard, we will set $[Ag^+] = 0.500$ M and $[Zn^{2+}] = 0.750$ M.

For a nonstandard cell, we will need to use the Nernst equation:

$$E = E° - \frac{RT}{nF} \ln Q$$

Assigning values:

$E = ?$ $\qquad n = 2 \, mol \, e^-$ $\qquad F = 96{,}485 \, J \, V^{-1} \, mol^{-1}$ $\qquad E°_{cell} = +1.5622 \, V$

$$T = 298.15 \, K \qquad R = 8.3141 \, J \, mol^{-1} \, V^{-1}$$

$$Q = \frac{[Zn^{2+}]}{[Ag^+]^2} \qquad \text{with } [Ag^+] = 0.500 \text{ M and } [Zn^{2+}] = 0.750 \text{ M}$$

Entering the appropriate values into the Nernst equation:

$$E = E° - \frac{RT}{nF} \ln Q = E° - \frac{RT}{nF} \ln \frac{[Zn^{2+}]}{[Ag^+]^2} = 1.5622 \, V - \frac{\left(\frac{8.3141 \, J}{mol \, K}\right)(298.15 \, K)}{(2)\left(\frac{96485 \, J}{V \, mol}\right)} \ln \frac{[0.750]}{[0.500]^2} = 1.55 \, V$$

Nonstandard cells

1. We will again use the following redox equation in this example:

$$2 Ag^+(aq) + Zn(s) \rightarrow Zn^{2+}(aq) + 2 Ag(s) \qquad +1.5622\,V$$

   In this reaction, $[Ag^+] = 0.00500$ M and $[Zn^{2+}] = 0.750$ M.
   What is E?

2. We will again use the following redox equation in this example:

$$Cu^{2+}(aq) + Zn(s) \rightarrow Zn^{2+}(aq) + Cu(s) \qquad +1.100\,V$$

   In this reaction, $[Cu^{2+}] = 1.500$ M and $[Zn^{2+}] = 4.00$ M.
   What is E?

3. We will again use the following redox equation in this example:

$$VO_2^+(aq) + 2 H^+(aq) + 1\,e^- \rightarrow VO^{2+}(aq) + H_2O(l) \qquad +0.9994$$

   In this reaction, $E = 0.6264$ V, $[VO_2^+] = 0.10$ M and $[VO^{2+}] = 0.20$ M.
   What is $[H^+]$?

Electrolysis and calculations

Many important industrial chemicals are produced using electrolysis cells. These include NaOH, Cl_2, Mg, and Al. At one time 2% of the total electrical output of the United States was used just to produce $Cl_2(g)$.

Nearly everything we have done in this chapter works equally well for any type of cell. Indeed, the only real difference between the other cells we have seen is the $E°$ for electrolysis cells is negative (which makes $\Delta G°$ positive = nonspontaneous).

Instead of repeating the earlier calculations for electrolysis cells, we will use electrolysis cells as examples of stoichiometry calculations.

Stoichiometry calculations require moles; therefore, for these calculations we will normally use $F = 96{,}485$ coul mol^{-1}.

We will also need to consider another electrical term. The SI unit for electric current is the ampere, amp. An ampere is a coul s^{-1}.

We will cover more in the following examples.

EXAMPLE 10

If liquid titanium(IV) chloride (acidified with HCl) is electrolyzed by a current of 2.000 amp for 4.000 h, how many grams of titanium will be produced?

$$TiCl_4(l) \rightarrow Ti(s) + 2 Cl_2(g) \qquad \text{(not necessary)}$$

$$Ti^{4+}(l) + 4\,e^- \rightarrow Ti(s) \qquad \text{(necessary)}$$

$$(4.000\ \text{h})\left(\frac{3600\ \text{s}}{1\ \text{h}}\right)\left(\frac{2.000\ \text{coul}}{\text{s}}\right)\left(\frac{1\ \text{mol e}^-}{96485\ \text{coul}}\right)\left(\frac{1\ \text{mol Ti}}{4\ \text{mol e}^-}\right)\left(\frac{47.87\ \text{g Ti}}{1\ \text{mol Ti}}\right)=3.572\ \text{g Ti}$$

This is basically a unit conversion calculation based upon the titanium half-reaction.

EXAMPLE 11

If an aqueous solution of NaCl is electrolyzed by a current of 0.020 amp, how long will it take to liberate 0.030 mol of H_2? The appropriate half-reaction is:

$$2\ e^- + 2\ H_2O(l) \rightarrow H_2(g) + 2\ OH^-(aq)$$

$$(0.030\ \text{mol H}_2)\left(\frac{2\ \text{mol e}^-}{1\ \text{mol H}_2}\right)\left(\frac{96,485\ \text{coul}}{\text{mol e}^-}\right)\left(\frac{1\ \text{s}}{0.020\ \text{coul}}\right)=2.9\times10^5\ \text{s}$$

EXAMPLE 12

A solution containing a platinum salt was electrolyzed. A current of 0.00300 amp was passed for 1.00 h, and 5.46 mg of Pt was plated onto the cathode. What was the oxidation number of the platinum in the compound?

A stoichiometric ratio is necessary:

$$\frac{?\ \text{mol e}^-}{\text{mol Pt}}$$

This ratio is based on the following half-reaction:

$$Pt^{n+} + n\ e^- \rightarrow Pt$$

(since the Pt formed on the cathode)

$$\left(\frac{0.00300\ \text{coul}}{\text{s}}\right)\left(\frac{1\ \text{mg}}{0.001\ \text{g}}\right)\left(\frac{1.00\ \text{h}}{5.46\ \text{mg Pt}}\right)\left(\frac{3600\ \text{s}}{\text{h}}\right)\left(\frac{1\ \text{mol e}^-}{96485\ \text{coul}}\right)\left(\frac{195.1\ \text{g Pt}}{1\ \text{mol Pt}}\right)=\frac{4.00\ \text{mol e}^-}{\text{mol Pt}}$$

Therefore: Pt^{4+}

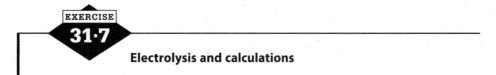

EXERCISE

31·7

Electrolysis and calculations

1. The half-reaction for the production of $Cl_2(g)$ from the electrolysis of a sodium chloride solution is:

   $$2\ Cl^-(aq) \rightarrow Cl_2(g) + 2\ e^-$$

   How many grams of $Cl_2(g)$ may be formed from the electrolysis of a NaCl solution by a current of 35.00 amp during an 8.000 h workday?

2. The half-reaction for the productions of $Cl_2(g)$ from the electrolysis of a sodium chloride solution is:

$$2\,Cl^-(aq) \rightarrow Cl_2(g) + 2\,e^-$$

The normal daily production of $Cl_2(g)$ in a small production plant is 1.00×10^5 g. How long (in seconds) will it take to produce this much $Cl_2(g)$ with a current of 1.000×10^4 amp?

3. A solution containing an iron salt was electrolyzed. A current of 30.00 amp was passed for 346 s, and 2.00 g of Fe was plated onto the cathode. Was the oxidation number of the iron in the compound +2 or +3?

Some selected standard reduction potentials, $E°$:

	$E°$ (volts)
$F_2(g) + 2\,e^- \rightarrow 2\,F^-(aq)$	+2.87
$BrO_4^-(aq) + 2\,H^+(aq) + 2\,e^- \rightarrow BrO_3^-(aq) + H_2O(l)$	+1.74
$PbO_2(s) + 4\,H^+(aq) + SO_4^{2-}(aq) + 2\,e^- \rightarrow PbSO_4(s) + 2\,H_2O(l)$	+1.70
$IO_4^-(aq) + 2\,H^+(aq) + 2\,e^- \rightarrow IO_3^-(aq) + H_2O(l)$	+1.65
$MnO_4^-(aq) + 8\,H^+(aq) + 5\,e^- \rightarrow Mn^{2+}(aq) + 4\,H_2O(l)$	+1.491
$PbO_2(s) + 4\,H^+(aq) + 2\,e^- \rightarrow Pb^{2+}(aq) + 2\,H_2O(l)$	+1.455
$Ce^{4+}(aq) + 1\,e^- \rightarrow Ce^{3+}(aq)$	+1.4430
$Cl_2(g) + 2\,e^- \rightarrow 2\,Cl^-(aq)$	+1.359
$Cr_2O_7^{2-}(aq) + 14\,H^+(aq) + 6\,e^- \rightarrow 2\,Cr^{3+}(aq) + 7\,H_2O(l)$	+1.33
$MnO_2(s) + 4\,H^+(aq) + 2\,e^- \rightarrow Mn^{2+}(aq) + 2\,H_2O(l)$	+1.23
$O_2(g) + 4\,H^+(aq) + 4\,e^- \rightarrow 2\,H_2O(l)$	+1.229
$Br_2(l) + 2\,e^- \rightarrow 2\,Br^-(aq)$	+1.087
$VO_2^+(aq) + 2\,H^+(aq) + 1\,e^- \rightarrow VO^{2+}(aq) + H_2O(l)$	+0.9994
$AuCl_4^-(aq) + 3\,e^- \rightarrow Au(s) + 4\,Cl^-$	+0.99
$2\,Hg^{2+}(aq) + 2\,e^- \rightarrow Hg_2^{2+}(aq)$	+0.907
$N_2O_4(g) + 2\,e^- \rightarrow 2\,NO_2^-(aq)$	+0.88
$HNO_2(aq) + 7\,H^+(aq) + 6\,e^- \rightarrow NH_4^+(aq) + 2\,H_2O(l)$	+0.86
$Hg^{2+}(aq) + 2\,e^- \rightarrow Hg(l)$	+0.850
$Ag^+(aq) + 1\,e^- \rightarrow Ag(s)$	+0.7994
$Hg_2^{2+}(aq) + 2\,e^- \rightarrow 2\,Hg(l)$	+0.792
$Fe^{3+}(aq) + 1\,e^- \rightarrow Fe^{2+}(aq)$	+0.771
$ClO_2^-(aq) + H_2O(l) + 2\,e^- \rightarrow ClO^-(aq) + 2\,OH^-(aq)$	+0.66
$BrO_3^-(aq) + 3\,H_2O(l) + 6\,e^- \rightarrow Br^-(aq) + 6\,OH^-(aq)$	+0.61
$ClO_4^-(aq) + 4\,H_2O(l) + 8\,e^- \rightarrow Cl^-(aq) + 8\,OH^-(aq)$	+0.56
$I_2(s) + 2\,e^- \rightarrow 2\,I^-(aq)$	+0.5355
$Cu^+(aq) + 1\,e^- \rightarrow Cu(s)$	+0.521
$2\,BrO^-(aq) + 2\,H_2O(l) + 2\,e^- \rightarrow Br_2(l) + 4\,OH^-(aq)$	+0.45
$Cu^{2+}(aq) + 2\,e^- \rightarrow Cu(s)$	+0.337
$HAsO_2(aq) + 3\,H^+(aq) + 3\,e^- \rightarrow As(s) + 2\,H_2O(l)$	+0.2475
$AgCl(s) + 1\,e^- \rightarrow Ag(s) + Cl^-(aq)$	+0.224
$Cu^{2+}(aq) + 1\,e^- \rightarrow Cu^+(aq)$	+0.153
$Sn^{4+}(aq) + 2\,e^- \rightarrow Sn^{2+}(aq)$	+0.15
$2\,H^+(aq) + 2\,e^- \rightarrow H_2(g)$	+0.0000
$Fe^{3+}(aq) + 3\,e^- \rightarrow Fe(s)$	−0.036

$Pb^{2+}(aq) + 2\ e^- \rightarrow Pb(s)$	-0.1263
$Sn^{2+}(aq) + 2\ e^- \rightarrow Sn(s)$	-0.1364
$PbSO_4(s) + 2\ e^- \rightarrow Pb(s) + SO_4^{2-}(aq)$	-0.356
$Tl^+(aq) + 1\ e^- \rightarrow Tl(s)$	-0.3363
$Cd^{2+}(aq) + 2\ e^- \rightarrow Cd(s)$	-0.4026
$Cr^{3+}(aq) + 1\ e^- \rightarrow Cr^{2+}(aq)$	-0.41
$Fe^{2+}(aq) + 2\ e^- \rightarrow Fe(s)$	-0.4402
$[Au(CN)_2]^-(aq) + 1\ e^- \rightarrow Au(s) + 2\ CN^-(aq)$	-0.50
$2\ SO_3^{2-}(aq) + 3\ H_2O(l) + 4\ e^- \rightarrow S_2O_3^{2-}(aq) + 6\ OH^-(aq)$	-0.58
$Cr^{3+}(aq) + 3\ e^- \rightarrow Cr(s)$	-0.74
$Zn^{2+}(aq) + 2\ e^- \rightarrow Zn(s)$	-0.7628
$2\ H_2O(l) + 2\ e^- \rightarrow H_2(g) + 2\ OH^-(aq)$	-0.8277
$Cr^{2+}(aq) + 2\ e^- \rightarrow Cr(s)$	-0.91
$Al^{3+}(aq) + 3\ e^- \rightarrow Al(s)$	-1.66
$Mg^{2+}(aq) + 2\ e^- \rightarrow Mg(s)$	-2.37
$Nb_2O_5(s) + 10\ H^+(aq) + 10\ e^- \rightarrow 2\ Nb(s) + 5\ H_2O(l)$	-2.7109
$Ca^{2+}(aq) + 2\ e^- \rightarrow Ca(s)$	-2.87
$K^+(aq) + 1\ e^- \rightarrow K(s)$	-2.924
$Li^+(aq) + 1\ e^- \rightarrow Li(s)$	-3.045

Nuclear chemistry

Radioactivity was discovered by accident in the late 1800s. The discovery was made when a sample of uranium ore accidently exposed a photographic plate.

Unlike nearly everything discussed previously in this book, nuclear chemistry deals with what happens in the nucleus of atoms and normally ignores anything outside the nucleus.

Terminology

Previously, we examined the structure of atoms and found that atoms consist of an extremely small nucleus surrounded by a much larger cloud of electrons. The nucleus contains the protons and neutrons, with the number of protons being the atomic number, and the number of protons plus the number of neutrons being the mass number (*not* the atomic mass). An atom with a specific number of protons and a specific number of neutrons is one isotope of an element. All elements have more than one known isotope, and of these known isotopes, at least some are radioactive. Some isotopes are natural, and some are artificial. The elements with all isotopes being artificial include all elements beyond uranium on the periodic table plus all isotopes of technetium and promethium. (Note that since the synthesis of some of the artificial isotopes, minute quantities of some of these isotopes have been discovered on Earth.)

The electrons are involved in bonding and chemical reactions, while the nucleus is not involved in these or any other general chemistry of the elements. Even the most violent of chemical reactions leave the nuclei unchanged.

Some nuclei are stable, and some are unstable. Unstable nuclei are radioactive, and such unstable nuclei are labeled radionuclides or radioisotopes. The breakdown of a radioisotope is termed *nuclear decay*. Nuclear decay processes are among the best examples of reactions following first-order kinetics.

During nuclear decay, an unstable isotope is transformed into a more stable isotope and emits radiation. This "more stable isotope" may be stable or unstable. If it is unstable, it will decay into an even more stable isotope. This radioactive decay series may extend for 14 steps or more until a stable isotope is reached.

Types of radiation

Soon after the discovery of radioactivity, there was a flurry of research attempting to discover what this newly discovered phenomenon was. This early research soon showed that there were three distinct types of radiation (more types were discovered later).

Initially, these types of radiation were labeled α, β, and γ (the first three letters of the Greek alphabet). Further research determined that α radiation consisted of the nuclei of helium atoms (two protons + two neutrons), β decay consisted of electrons, and γ radiation was electromagnetic radiation (light rays). Both α and β radiation involve particles and are normally accompanied by one or more γ rays.

Comparing these types of radiation to each other, we find the γ radiation is the most penetrating, being stopped only by thick walls of lead or concrete, as opposed to α radiation, which is stopped by a piece of paper. A 1-inch piece of wood will stop β radiation. To your body, the potentially most damaging is α radiation; γ radiation is potentially the least damaging. "Potentially" is qualified because the means of exposure makes a difference. In the case of the potentially most damaging α radiation, its inability to penetrate a piece of paper (or the outermost layer of your skin, which consists of a layer of dead skin cells) protects you from this damage. However, if you have a source of α radiation inside your body, you no longer have skin to protect you and can expect more damage. Radiation sources inside your body may be something ingested or dust particles in your lungs.

Other types of radiation have been discovered since the discovery of the "big three." All of these are less common than the big three.

The following three are slightly less common than the big three.

Electron capture is a type of radioactive decay where the nucleus "captures" an electron from the electron cloud and emits an X-ray, and possibly one or more γ rays. This is the only form of radioactive decay where something enters the nucleus.

Positron emission is a type of radioactive decay where the nucleus emits a positron. A positron is an electron with a positive charge instead of a negative charge. As the opposite of an electron, a positron is a form of antimatter. As a form of antimatter, if it comes in contact with normal matter, the two mutually annihilate each other, releasing two γ rays.

Spontaneous fission is a type of radioactive decay where a nucleus splits into two or more smaller nuclei. This breakup is relatively random; therefore, a prediction of the exact products cannot be done. During this process, one or more γ rays are emitted.

Other forms of radiation have been observed; however, these are extremely rare. These rare forms include proton emission, neutron emission, and the emission of other particles. Being so rare, these will not be your answers to questions other than "What are the rarest forms of radiation?"

We will have more to say about these types of radiation in the section on balancing nuclear equations.

EXERCISE

32·1

Types of radiation

1. What were the first three types of radiation discovered?

2. What is the most penetrating form of radiation?

3. What form of radiation is also a form of antimatter?

Nuclear stability

Why are some nuclei stable while others are unstable?

There are two basic reasons, the mass number and the neutron–proton ratio.

The highest mass number with a stable isotope is 209. All higher mass numbers are radioactive and tend to emit α radiation. Since α particles are the most massive form of radiation emitted, this is the fastest way to lower the mass number to less than 209. For nuclei with mass numbers much higher than 209, spontaneous fission becomes more probable.

The neutron–proton ratio applies to all nuclei. For light nuclei the ideal ratio is 1:1; however, the ideal gradually increases to about 1.5 for the heaviest stable nuclei. Since the ideal value changes with increasing mass number, one must be careful of what ratio to use. To get an idea of what the ideal is in different parts of the periodic table, look at the atomic *masses* of the elements in that region of the periodic table, the atomic masses are mostly based upon the average of the stable nuclei. While there are some exceptions, a mass number differing by more than 2 from the atomic mass tends to be radioactive (neutron–proton ratio too high or too low).

If the neutron–proton ratio is higher than ideal, β decay is preferred. If the neutron–proton ratio is lower than ideal, electron capture or positron emission are preferred. Note that these predictions are preferences, not a guarantee; thus, even though one type is preferred, it is unlikely that a particular radioisotope will exclusively undergo only one type of radioactive decay.

The numbers of protons and neutrons are important in another way, which depends only upon whether the numbers are even or odd. Nuclei containing both even numbers of protons and neutrons are more stable than other types. Nuclei with an even-odd combination of protons and neutrons are about equally stable. Finally, nuclei with both odd number of protons and neutrons are the least stable. Indeed, there are only four known stable odd-odd nuclei: hydrogen-2, lithium-6, boron-10, and nitrogen-14.

Related to the even-odd discussion of nuclear stability, are the "magic numbers." The **magic numbers** are either proton or neutron numbers that are especially stable. The magic numbers are: 2, 8, 20, 28, 50, 82, and 126. Looking at the protons and neutrons in a nucleus, if either is magic, the nucleus is more stable than otherwise, and if both are magic, the nucleus is even more stable. Some consequences of this are the double magic nuclei helium-4 (2 protons and 2 neutrons) and lead-208 (82 protons and 126 neutrons). The double magic stability of helium-4 explains the expulsion of α particles instead of some other particle. Another consequence is that the element with the greatest number of stable isotopes (10), is tin, with stable isotopes ranging from tin-112 to tin-124 (a range much greater than for any other element). It is no coincidence that tin has a magic number of protons (50).

The most stable isotope is iron-56. The reason for this is discussed later.

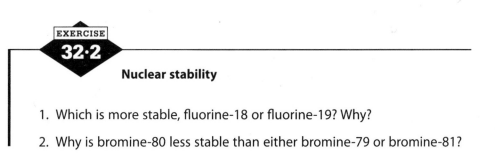

EXERCISE
32·2

Nuclear stability

1. Which is more stable, fluorine-18 or fluorine-19? Why?

2. Why is bromine-80 less stable than either bromine-79 or bromine-81?

3. Why is calcium-48 less stable than calcium-40?

Balancing nuclear equations

Unlike normal chemical equations, nuclear equations require you to keep track of the atomic numbers and mass numbers of the atoms involved. This means that instead of using Kr in an equation, we must use $^{84}_{36}\text{Kr}$. Do not make the mistake of using the atomic mass instead of the mass number. Mass numbers are *always* integers (exact numbers), since they are counted not calculated. In addition, the charges on ions are largely ignored. If there is more than one of a particle, there will be a coefficient.

In addition, some particles resulting from the radioactive decay of radioisotopes appear quite often, and unfortunately, not always in the same form. These are:

Alpha particle $= {}^{4}_{2}\alpha$ or ${}^{4}_{2}\text{He}$ Beta particle $= {}^{0}_{-1}\beta$ or ${}^{0}_{-1}e$ Gamma rays $= {}^{0}_{0}\gamma$ or γ

Proton $= {}^{1}_{1}p$ Electron $= {}^{0}_{-1}\beta$ or ${}^{0}_{-1}e$ Neutron $= {}^{1}_{0}n$

Positron $= {}^{0}_{+1}\beta$ or ${}^{0}_{+1}e$

The option of γ is allowed because the mass number and atomic number are 0.

EXAMPLE 1

Here is an example of a balanced nuclear equation for us to analyze:

$$^{238}_{92}\text{U} \rightarrow {}^{234}_{90}\text{Th} + {}^{4}_{2}\alpha + {}^{0}_{0}\gamma$$

or

$$^{238}_{92}\text{U} \rightarrow {}^{234}_{90}\text{Th} + {}^{4}_{2}\text{He} + \gamma$$

There may be more than one ${}^{0}_{0}\gamma$ or γ. (Note that due to there being 0 and 0, γ rays are sometimes ignored.)

Now back to the equation. There are two things to balance, the mass numbers and the atomic numbers. It does not matter which you do first. We will start with the mass numbers: on the reactant side we have 238 and on the product side will have 234 + 4 + 0 = 238, so the mass numbers are balanced. Now, we move on to the atomic numbers. On the reactant side we have 92 and on the product side we have 90 + 2 + 0 = 92, so they are balanced also.

EXAMPLE 2

Here is an example of a balanced nuclear equation for us to analyze:

$$^{40}_{19}\text{K} \rightarrow {}^{40}_{20}\text{Ca} + {}^{0}_{-1}\beta + {}^{0}_{0}\gamma$$

Mass numbers: 40 = 40 + 0 + 0, so balanced.
Atomic numbers: 19 = 20 − 1 + 0, so balanced.

EXAMPLE 3

Here is an example of a balanced nuclear equation for us to analyze:

$$^{235}_{92}\text{U} + {}^{1}_{0}n \rightarrow {}^{91}_{36}\text{Kr} + {}^{142}_{56}Ba + 3\,{}^{1}_{0}n + {}^{0}_{0}\gamma$$

Mass numbers: 235 + 1 = 91 + 142 + 3(1) + 0, so balanced.
Atomic numbers: 92 + 0 = 36 + 56 + 3(0) + 0, so balanced.

Now try the modification in the next example.

EXAMPLE 4

Here is an example of a balanced nuclear equation for us to analyze:

$$^{239}_{94}\text{Pu} + {}^{1}_{0}n \rightarrow {}^{102}_{40}\text{Zr} + ? + 4\,{}^{1}_{0}n + {}^{0}_{0}\gamma$$

Mass numbers: 239 + 1 = 102 + ? + 4(1) + 0, so ? = 134
Atomic numbers: 94 + 0 = 40 + ? + 4(0) + 0, so ? = 54
Therefore, the missing isotope is $^{134}_{54}\text{Xe}$.

EXERCISE

32·3

Balancing nuclear equations

1. Complete the following nuclear equation by supplying the identity of the missing isotope.

$$^{238}_{92}\text{U} + {}^{1}_{0}n \rightarrow ? + {}^{0}_{0}\gamma$$

2. Complete the following nuclear equation by supplying the identity of the missing isotope.

$$^{14}_{6}\text{C} \rightarrow ? + {}^{0}_{-1}\beta + {}^{0}_{0}\gamma$$

Kinetics of nuclear decay

Nuclear decay obeys the rules of first-order kinetics. As a reminder, these are the equations associated with first-order kinetics:

$$Rate = k\,[\text{A}]^1 \qquad \ln\frac{[\text{A}]_0}{[\text{A}]_t} = kt \qquad t_{1/2} = \frac{\ln 2}{k} = \frac{0.693}{k}$$

As noted in the kinetics chapter, these equations sometimes appear in different forms.
Also, as a reminder, these are the meanings of the terms:

$Rate$ = how fast is the reaction going k = rate constant   A is the substance reacting
$[\text{A}]_t$ the amount of A after some time has passed $[\text{A}]_0$ the original amount of A
t is the time passed $t_{1/2}$ is the half-life

In the kinetics chapter A was expressed as molarity; however, when dealing with nuclear decay A may be represented not only by molarity, but also by anything proportional to molarity. Time and half-life may have any time unit. The rate constant has units of time^{-1}.

EXAMPLE 5

The half-life of carbon-14 is 5,730 years. What is the value of the rate constant for carbon-14?

We need to use the half-life equation: $t_{1/2} = \dfrac{\ln 2}{k} = \dfrac{0.693}{k}$

This needs to be rearranged to $k = \dfrac{\ln 2}{t_{1/2}} = \dfrac{0.693147}{5,730\ y} = 1.210 \times 10^{-4}\ y^{-1}$

One use of naturally occurring radioisotopes is in determining the age of a sample. This general method works for up to about 10 times the half-life of the radioisotope being studied. Historic samples usually employ carbon dating, while geological samples may use U-Pb dating, Rb-Sr dating, or K-Ar dating.

In the case of carbon-14 dating, this isotope is continually being generated in Earth's upper atmosphere, which maintains an approximately constant level. Living organisms absorb carbon-14 from the atmosphere by one means or another and maintain a characteristic level. When the organism dies, the carbon-14 is no longer replenished, so the level begins to decrease based upon the half-life of carbon-14 (5,730 years).

Dating using other radioisotopes works similar to carbon-14 dating. These methods also work up to about 10 times the half-life of the radioisotope.

EXAMPLE 6

A sample of wood from an Egyptian tomb has a decay rate that is 61.5 percent of the decay rate of carbon-14 in contemporary samples. If the rate constant is $1.213 \times 10^{-4}\ y^{-1}$, what is the age of the wood?

The equation we need is $\ln \dfrac{[A]_0}{[A]_t} = kt$

From the data in the problem, we have $[A]_0 = 100\%$, $[A]_t = 61.5\%$, $k = 1.213 \times 10^{-4}\ y^{-1}$, and $t = ?$
Entering the information into the equation gives:

$$\ln \frac{[A]_0}{[A]_t} = kt = \ln \frac{[100]_0}{[61.5]_t} = (1.213 \times 10^{-4}\ y^{-1})\,t$$

Calculating:

$$\ln \frac{[100]_0}{[61.5]_t} = 0.4861 = (1.213 \times 10^{-4}\ y^{-1})\,t$$

$$t = \frac{0.4861}{1.213 \times 10^{-4}\,y^{-1}} = 4.00 \times 10^3\ y$$

Kinetics of nuclear decay

1. The half-life of hydrogen-3 (tritium) is 13.6 y. What is the value of the rate constant?

2. The half-life of iodine-131, used in hospitals for the diagnosis of thyroid problems, is 8.04. What is the value of the rate constant?

3. The half-life of technetium-99, used in hospitals for the diagnosis of several medical problems, is 6.0 h. What is the value of the rate constant?

4. Atmospheric testing of nuclear weapons released many substances into the atmosphere. One of those substances was strontium-90 with a half-life of 29.1. The last atmospheric nuclear test was exploded on September 16, 1980. What percentage of this amount still remains exactly 45 y after this date?

5. Cobalt-60 is used in hospitals for radiation treatments. The half-life of cobalt-60 is 5.271 y. The hospital's cobalt-60 needs to be replaced when 20.00% of the cobalt-60 has decayed. If the hospital replaced their last cobalt-60 unit on February 1, 2023, when will this unit need to be replaced?

6. One of the isotopes used to date geological samples is potassium-40. The half-life of potassium-40 is 1.251×10^9 y. Analysis of a rock sample from northern Minnesota shows that 8.415% of the potassium-40 remains in the sample. What is the age of the rock?

7. Historical samples are often dated using carbon-14. A museum obtains an artifact claimed to be ancient Egyptian from the third dynasty (2686 to 2613 BCE). Carbon-14 analysis shows that 69.6% of the carbon-14 remains in the sample. The half-life of carbon-14 is 5,730 years. Is the artifact genuine or a fraud? Why?

Nuclear energy

Nuclear energy is one application of Einstein's famous equation $E = mc^2$. This equation relates energy to mass times the speed of light squared. In other words, matter and energy are two aspects of the same thing. This equation leads to the ability to convert mass to energy or to convert energy to mass. Both of these processes have been observed. We will examine two uses of this equation in this section. The first deals with nuclear reactions and the second deals with the energy holding the nuclei of atoms together.

All nuclear reactions involve a release of energy. Part of this energy is the kinetic energy ($E = \frac{1}{2} mv^2$) of the product particles and the remainder is the energy ($E = h\nu$) of the γ rays produced. The total amount of energy may be found by determining the total mass of the products and subtracting to total mass of the reactants. This gives a mass differences, which is the m in $E = mc^2$.

> In problems such as these, use the isotopic masses instead of the atomic masses of the elements, as atomic masses are the average of different isotopic masses.

EXAMPLE 7

Determine the energy change occurring when exactly 1 mol of uranium-235 undergoes fission according to the following equation:

$$^{235}_{92}U + ^{1}_{0}n \rightarrow ^{91}_{36}Kr + ^{142}_{56}Ba + 3\,^{1}_{0}n + ^{0}_{0}\gamma$$

Accurate masses of the substances involved are:

Mass of neutron = 1.0086649 amu   Mass of krypton-91 = 90.9234 amu

Mass of uranium-235 = 235.0439 amu   Mass of barium-142 = 141.9165 amu

The values given are in amu, which means they are for individual neutrons or atoms. In order to deal with moles of these particles, technically we need to convert amu per particle to grams per mole. To do this, the amu-to-grams conversion is $1.000\ g = 6.022 \times 10^{23}$ amu, and the particle to mole conversion is that $6.022 \times 10^{23} = 1.000$ mol. For example, for the neutron:

$$\text{Mole neutrons} = (1.0086649\ \cancel{amu})\left(\frac{1.000\ g}{6.022 \times 10^{23}\ \cancel{amu}}\right)\left(\frac{6.022 \times 10^{23}}{1.000\ mol}\right) = 1.0086649\ g\ mol^{-1}$$

If you find it necessary, you can do the other conversions; otherwise:

Mass of neutron = 1.0086649 g mol^{-1} Mass of krypton-91 = 90.9234 g mol^{-1}

Mass of uranium-235 = 235.0439 g mol^{-1} Mass of barium-142 = 141.9165 g mol^{-1}

The alternative would be to postpone the amu-to-gram conversion until the mass change has been calculated, so only one such conversion is necessary.

Next, we need to calculate the masses of the reactants and products:

Mass of reactants = 235.0439 g mol^{-1} + 1.0086649 g mol^{-1} = 236.0526 g mol^{-1}

Mass of products = 90.9234 g mol^{-1} + 141.9165 g mol^{-1} + 3(1.0086649 g mol^{-1})

= 235.8659 g mol^{-1}

The change in mass is products − reactants = 235.8659 g mol^{-1} − 236.0526 g mol^{-1}

= −0.1867 g mol^{-1}

The negative sign means that energy will be released.
Entering this information into the Einstein relation:

$$E = mc^2 = \left(-0.1867\,\frac{g}{mol}\right)\left(2.9979 \times 10^8\,\frac{m}{s}\right)^2$$

We now need to add a conversion to energy units (joules) and a gram to kilogram conversion:

$$E = mc^2 = \left(-0.1867\,\frac{\cancel{g}}{mol}\right)\left(2.9979 \times 10^8\,\frac{m}{s}\right)^2 \left(\frac{J}{\frac{kg\,m^2}{s^2}}\right)\left(\frac{1\ kg}{1{,}000\ \cancel{g}}\right) = -1.678 \times 10^{13}\ J\ mol^{-1}$$

This extremely large number explains the power of an atomic bomb.

Fission reactions such this one are random processes; the nucleus breaks at random, not always to the same product nuclei.

If you compare the mass of an atom to the sum of the masses of all the protons, neutrons, and electrons, the actual mass of the atom is less than the sum of the parts for all atoms except hydrogen-1. This mass difference, the nuclear binding energy, is the mass that has been converted to the energy required to hold the nucleus together. Again, this mass difference is the m in $E = mc^2$.

When dealing with problems of this type, very accurate masses are necessary, otherwise the very small differences may be missed. The reason for this is that even though the mass difference is very small, it is to be multiplied by the speed of light squared. The very accurate masses are:

$$\text{Mass of proton} = 1.0072765 \text{ amu}$$
$$\text{Mass of neutron} = 1.0086649 \text{ amu}$$
$$\text{Mass of electron} = 0.0005488 \text{ amu}$$

It is also necessary to have very accurate nuclear masses.

EXAMPLE 8

Given the following information, how much energy is released when exactly 1 mol of iron-56 atoms are formed from protons, neutrons, and electrons?

Mass of iron-56 atom = 55.934942 amu	Mass of proton = 1.0072765 amu
Mass of neutron = 1.0086649 amu	Mass of electron = 0.0005488 amu

As shown in the preceding example, the amu masses of individual particles are numerically the same as the molar masses of the same particles in grams; therefore, without redoing the conversions, we will replace the amu in the table with grams mole^{-1}.

We shall begin by determining the number of each constituent of an iron-56 atom. The atomic number iron is 26, so there are 26 protons and 26 electrons present. Since the mass number of this isotope, $56 - 26 = 30$ neutrons.

The mass of a mole of each of the constituents of iron-56 atoms are:

$$26 \text{ p} = 26(1.0072765 \text{ g mol}^{-1}) = 26.1891890 \text{ g mol}^{-1}$$
$$26 \text{ e} = 26(0.0005488 \text{ g mol}^{-1}) = 0.0142688 \text{ g mol}^{-1}$$
$$30 \text{ p} = 30(1.0086649 \text{ g mol}^{-1}) = 30.2599470 \text{ g mol}^{-1}$$
$$\text{Total} = 56.4634048 \text{ g mol}^{-1}$$

Now, subtract the molar mass given from the molar mass just calculated:
$$56.4634048 \text{ g mol}^{-1} - 55.934942 \text{ g mol}^{-1} = 0.528463 \text{ g mol}^{-1}$$

This value is known as the mass defect. All atoms except hydrogen-1 have a mass defect. The mass defect is the mass that has been converted to energy and is the m in $E = mc^2$. We can enter this value into the Einstein relationship and add the appropriate conversions:

$$E = mc^2 = \left(0.528463 \frac{g}{mol} \right) \left(2.9979 \times 10^8 \frac{m}{s} \right)^2 \left(\frac{J}{\frac{kg \cdot m^2}{s^2}} \right) \left(\frac{1 \text{ kg}}{1{,}000 \text{ g}} \right) = 4.750 \times 10^{13} \text{ J mol}^{-1}$$

The energy just calculated is the nuclear binding energy. The **nuclear binding energy** is the energy holding the nucleus together. This amount of energy is required to pull a mole of iron-56 atoms apart into protons, electrons, and neutrons, or this amount of energy is the energy released when a mole of iron-56 is assembled from the component parts.

The negative of this value is the energy "released when exactly one mole of iron-56 atoms is formed from protons, neutrons, and electrons," the answer to the original question.

Nuclear binding energies are often expressed as joules per nucleon, where the number of nucleons is the number of protons plus neutrons in the nucleus (mass number). Plotted on a graph, the nuclear binding energies per nucleon begin low and then increase up to iron-56 and then decrease gradually.

EXERCISE
32·5

Nuclear energy

You may use the following information while doing these problems:

Mass of neutron = 1.0086649 amu   Mass of electron = 0.0005488 amu
Mass of proton = 1.0072765 amu   Mass of uranium-233 atom = 233.039627 amu
Mass of strontium-94 = 93.91537 amu   Mass of xenon-137 = 136.91156 amu
Mass of strontium-94 = 93.91537 amu   Mass of xenon-134 = 133.905395 amu
Mass of zirconium-103 = 102.9266 amu   Mass of plutonium-239 = 239.052156 amu

1. Determine the energy change occurring when exactly 1 mol of uranium-233 undergoes fission according to the following equation:

$$^{233}_{92}U + ^{1}_{0}n \rightarrow ^{94}_{38}Sr + ^{137}_{54}Xe + 3\,^{1}_{0}n + ^{0}_{0}\gamma$$

2. Determine the energy change occurring when exactly 1 mol of plutonium-239 undergoes fission according to the following equation:

$$^{239}_{94}Pu + ^{1}_{0}n \rightarrow ^{103}_{40}Zr + ^{134}_{54}Xe + 3\,^{1}_{0}n + ^{0}_{0}\gamma$$

3. Given the preceding information, how much energy is released when exactly 1 mol of uranium-233 atoms are formed from protons, neutrons, and electrons?

Constants and equations

Basics

T = temperature n = moles m = mass P = pressure
V = volume D = density v = velocity M = molar mass
KE = kinetic energy t = time

electron charge = -1.602×10^{-19} coulombs
1 electron volt per atom = 96.5 kJ mol^{-1}
Avogadro's number = 6.0221429×10^{23} mol^{-1}
K = °C + 273 $D = m/V$
mass of proton = 1.007276466583 amu
mass of neutron = 1.00866491600 amu
mass of electron = $5.4857990946 \times 10^{-4}$ amu
density of water (4°C) = 1.000 g cm^{-3}
1 g = 6.0221429×10^{23} amu
1 joule = 1 j = 1 kg m^{-2} s^{-2}

Gases

r = rate of effusion
STP = 0.000°C and 1.000 atm
$PV = nRT$

$$P_A = P_{total} \times X_A, \text{ where } X_A = \frac{\text{moles A}}{\text{total moles}}$$

$$P_{total} = P_A + P_B + P_C + \ldots$$

$$\frac{P_1 V_1}{n_1 T_1} = \frac{P_2 V_2}{n_2 T_2}$$

$$u_{rms} = \sqrt{\frac{3\,kT}{m}} = \sqrt{\frac{3\,RT}{M}}$$

$$r_1/r_2 = \sqrt{M_2/M_1}$$

1 atm = 760 mm Hg = 760 torr
gas constant, R = 8.3144621 J mol^{-1} K^{-1} = 0.082057361 L atm mol^{-1} K^{-1}

Thermodynamics

$S°$ = standard entropy $H°$ = standard enthalpy
$G°$ = standard free energy q = heat
c = specific heat capacity C_p = molar heat capacity at constant pressure

specific heat of water = 4.18 J g^{-1} °C^{-1}

$\Delta S° = \Sigma\, S°$ products $- \Sigma\, S°$ reactants

$\Delta H° = \Sigma\, \Delta H_f°$ products $- \Sigma\, \Delta H_f°$ reactants

$\Delta G° = \Sigma\, \Delta G_f°$ products $- \Sigma\, \Delta G_f°$ reactants

$$\Delta G° = \Delta H° - T\Delta S°$$
$$= -RT \ln K = -2.303\, RT \log K$$
$$= -nFE°$$
$$\Delta G = \Delta G° + RT \ln Q = \Delta G° + 2.303\, RT \log Q$$
$$q = mc\,\Delta T$$
$$C_p = \Delta H/\Delta T$$

Light and electrons

E = energy $\quad$ v = frequency $\quad$ λ = wavelength

p = momentum $\quad$ v = velocity $\quad$ n = principal quantum number

m = mass

$$E = hv = \frac{hc}{\lambda}$$

$c = \lambda v$

speed of light, $c = 2.99792458 \times 10^8$ m s^{-1} $\quad$ Planck's constant, $h = 6.62606957$ J s

Solutions

molarity, M = moles solute per liter solution

Kinetics

$$\frac{1}{[A]_t} - \frac{1}{[A]_0} = kt$$

$$\ln\frac{[A]_0}{[A]_t} = kt$$

$$[A]_0 - [A]_t = kt$$

E_a = activation energy

k = rate constant

first-order: $\quad t_{1/2} = \dfrac{\ln 2}{k} = \dfrac{0.693}{k}$

second-order: $\quad t_{1/2} = \dfrac{1}{k[A]_0}$

zero-order: $\quad t_{1/2} = \dfrac{[A]_0}{2k}$

Electrochemistry

I = current (amperes) $\qquad\qquad$ q = charge (coulombs)

$E°$ = standard reduction potential $\qquad$ K = equilibrium constant

Faraday's constant, F = 96,485.3365 coulomb mol^{-1} = 96485.3365 J volt^{-1} mol^{-1}

SI units and conversions

SI base units

	Unit	Abbreviation
length	meter	m
mass	kilogram	kg
time	second	s
temperature	Kelvin	K
amount of a substance	mole	mol
electric current	ampere	A
luminous intensity	candela	cd

SI prefixes

exa	E	10^{18}	1 000 000 000 000 000 000
peta	P	10^{15}	1 000 000 000 000 000
tera	T	10^{12}	1 000 000 000 000
giga	G	10^{9}	1 000 000 000
mega	M	10^{6}	1 000 000
kilo	k	10^{3}	1 000
hecto	h	10^{2}	100
deka	da	10^{1}	10
deci	d	10^{-1}	0.1
centi	c	10^{-2}	0.01
milli	m	10^{-3}	0.001
micro	μ	10^{-6}	0.000 001
nano	n	10^{-9}	0.000 000 001
pico	p	10^{-12}	0.000 000 000 001
femto	f	10^{-15}	0.000 000 000 000 001
atto	a	10^{-18}	0.000 000 000 000 000 001

Some English–SI conversions

1 oz	=	28.349 g		1 in.	=	2.54 cm (exactly)
1 lb	=	453.59 g		1 ft	=	30.48 cm (exactly)
1 kg	=	2.2046 lbs		1 mi	=	1.6093 km

Some properties of water

density of water $(4°C) = 1.000$ g cm^{-3}
specific heat of water $= 4.18$ J g^{-1} °C^{-1}

	Solvent	Boiling Point (°C)	K_b (°C/m)	Melting Point (°C)	K_f (°C/m)
Water	H_2O	100.00	0.512	0.00	1.858

Vapor pressure of water

T (°C)	P (torr)
−10.00	2.149
−5.00	3.163
0.00	4.579
5.00	6.543
10.00	9.209
15.00	12.788
20.00	17.535
25.00	23.756
30.00	31.824
35.00	41.175
40.00	55.324
45.00	71.88
50.00	92.51
55.00	118.04
60.00	149.38
65.00	187.54
70.00	233.7
75.00	289.1
80.00	355.1
85.00	433.6
90.00	525.76
95.00	633.90
100.00	760.00
105.00	906.07
110.00	1,074.56

Thermodynamic quantities

Specific heats of selected materials

	J/g °C		J/g °C
Al(s)	0.900	Fe(s)	0.445
Au(s)	0.129	H_2O(l)	4.184
C_2H_5OH(l)	2.45	Hg(l)	0.139
Cu(s)	0.384		

Properties of some solvents

	Solvent	Boiling Point (°C)	K_b (°C/m)	Melting Point (°C)	K_f (°C/m)
acetic acid	$HC_2H_3O_2$	117.9	2.98	16.60	3.85
benzene	C_6H_6	80.1	2.58	5.45	5.09
camphor	$C_{10}H_{16}O$	sublimes	—	179.5	40
carbon tetrachloride	CCl_4	76.8	5.02	−22.3	29.8
chloroform	$CHCl_3$	61.2	3.63	−63.5	4.7
cyclohexane	C_6H_{12}	80.74	2.79	6.55	20.0
ethyl alcohol	C_2H_5OH	78.4	1.17	−115.3	1.99
water	H_2O	100.00	0.512	0.00	1.858

Average bond energies in kJ/mole

Br:Br	192.5	C:::N	891	F:N	272	N:O	176
Br:C	276	C:O	351	F:O	185	N::O	607
Br:Cl	218	C::O*	781	F:P	485	N:P	209
Br:F	237	C:::O	1,072	F:S	285	O:O	142
Br:H	366.1	C:P	263	F:Si	540.	O::O	498.7
Br:I	180.	C:S	255	H:H	436.4	O:P	350.
Br:N	243	C::S	477	H:I	298.3	O::P	502
Br:P	270.	C:Si	360.	H:N	393	O:S	347
Br:S	215	Cl:Cl	242.7	H:O	464	O::S	469
Br:Si	290.	Cl:F	253	H:P	326	O:Si	370.
C:C	347	Cl:H	431.9	H:S	340.	P:P	215
C::C	615	Cl:I	210.	H:Si	395	P::P	489
C:::C	812	Cl:N	200.	I:I	151.0	P:S	230.
C:Cl	331	Cl:O	205	I:O	200.	P:Si	215
C:F	439	Cl:P	330.	I:P	215	S:S	215
C:H	414	Cl:S	250.	I:Si	215	S::S	352
C:I	240.	Cl:Si	359	N:N	159	S:Si	225
C:N	293	F:F	150.6	N::N	418	Si:Si	230.
C::N	615	F:H	568.2	N:::N	941.4		

*C::O 799 in CO_2

Thermodynamic properties of some elements and compounds

	ΔH_f° kJ mol^{-1}	S° J mol^{-1} K^{-1}	ΔG_f° kJ mol^{-1}
Ag(s)	0	42.7	0
Ag$^+$(aq)	105.9	73.9	77.1
AgCl(s)	−109.7	96.11	−127
Ag$_2$CrO$_4$(s)	−712	214	−622
AgN$_3$(s)	378.5	99.2	310.3
Ag$_2$O(s)	−31.05	121.3	−11.20
Al(s)	0	28.32	0
Al$_2$O$_3$(s)	−1,669.8	50.99	−1,576.4
Al$_2$(SO$_4$)$_3$(s)	−3,441	239	−3,100.
B$_2$O$_3$(s)	−1,272	53.8	−1,193
BaCO$_3$(s)	−1,218.8	112.1	−1,138.9
BaO(s)	−558.2	70.3	−528.4
BaSO$_4$(s)	−1,464.4	132.2	−1,353.1
BaCl$_2$·2H$_2$O(s)	−1,460	202.9	−1,296
Ba(OH)$_2$·8H$_2$O(s)	−3,350	400.	−1,366
Br$_2$(g)	−30.91	245.38	−3.13
Br$_2$(l)	0	152.3	0
C(s) (graphite)	0	5.686	0
C(s) (diamond)	1.896	2.439	2.866

	ΔH_f° kJ mol^{-1}	S° J mol^{-1} K^{-1}	ΔG_f° kJ mol^{-1}
$CH_4(g)$	−74.9	186.19	−50.8
$C_2H_2(g)$	54.19	48.00	50.0
$C_2H_4(g)$	52.47	219.22	68.36
$C_2H_6(g)$	−84.7	229.49	−32.89
$C_4H_{10}(g)$	−125	310.0	−16
$C_6H_{12}(l)$ (cyclohexane)	−123.1	298.2	31.76
$C_8H_{18}(l)$ n-octane	−250.3	361.2	−358
$CH_3OH(g)$	−201.2	237.6	−161.9
$CH_3OH(l)$	−238.6	126.8	−166.23
$C_2H_5OH(l)$	−277.63	161.04	−174.8
$C_2H_5OH(g)$	−235.1	282.6	−168.6
$C_6H_{12}O_6(s)$	−1,273.3	212.1	−910.56
$CO(g)$	−110.5	197.5	−137.2
$CO_2(g)$	−393.5	213.7	−394.4
$COCl_2(g)$	−223.0	289.2	−210.5
$Ca(s)$	0	41.63	0
$CaCO_3(s)$	−288.45	22.2	−269.78
$CaCl_2(s)$	−796	104.6	−750.2
$CaH_2(s)$	−45.1	10	−35.8
$CaI_2(s)$	−533.5	142	−528.9
$Ca(OH)_2(s)$	−986.6	76.1	−896.76
$CaO(s)$	−635.1	38.2	−603.5
$CaSO_4(s)$	−1,434	107	−1,320.3
$CaSO_4 \cdot 2H_2O(s)$	?	194.0	?
$Cl_2(g)$	0	222.96	0
$Cr_2O_3(s)$	−1,140	81	−1058
$Cu(s)$	0	33.1	0
$CuO(s)$	−157.3	42.6	−129.7
$CuS(s)$	−48.5	66.5	−49.0
$Cu_2S(s)$	−79.5	120.9	−86.2
$FeCl_3(s)$	−499.5	142	−334
$FeO(s)$	−271.9	60.75	−255.2
$Fe_2O_3(s)$	−822.16	89.96	−740.98
$H_2(g)$	0	130.6	0
$HBr(g)$	−36.23	198.49	−53.22
$HCl(aq)$	−167.46	55.06	−131.17
$HCl(g)$	−95.3	186.8	−92.3
$HI(g)$	26	206	1.30
$HF(g)$	−268.61	173.51	−270.70
$HC_2H_3O_2(l)$	−487.0	159.8	−392.4
$H_2O(g)$	−241.8	188.7	−228.6
$H_2O(l)$	−285.8	69.94	−237.2
$H_3PO_4(aq)$	−1,288.3	158	−1,142.7
$H_3PO_4(s)$	−1,279	110.	−1,119
$H_2S(g)$	−20.2	205.6	−33
$H_2SO_4(aq)$	−907.51	17	−741.99
$H_2SO_4(l)$	−813.989	156.90	−690.059
$KCl(s)$	−436.7	82.59	−409.2
$KClO_3(s)$	−397.7	143.1	−296.3

	ΔH_f° kJ mol^{-1}	S° J mol^{-1} K^{-1}	ΔG_f° kJ mol^{-1}
KI(s)	−327.9	106.3	−325
KNO$_3$(s)	−492.7	133	−393
MgCO$_3$(s)	−1,112.9	65.69	−1,029.3
MgO(s)	−601.8	26.78	−569.6
Mn(s)	0	32.0	0
MnO(s)	−385.2	59.7	−362.9
MnO$_2$(s)	−520.9	53.1	−466.1
N$_2$(g)	0	191.5	0
NH$_3$(g)	−46.1	192.2	−16.7
NH$_3$(aq)	−80.8	110.	−26.7
NH$_4$Cl(s)	−315	94.6	−203.9
N$_2$O(g)	81.6	220.0	51.84
NO(g)	90.37	210.62	86.71
NO$_2$(g)	33.84	240.45	51.84
NOCl(g)	51.71	261.6	66.07
N$_2$O$_4$(g)	9.66	304.3	98.28
NaBr(s)	−86.30	20.75	−83.409
NaCl(s)	−411.0	72.38	−384.0
NaI(s)	−288	91.2	−286.1
Na$_2$O$_2$(s)	−504.6	94.98	−447.69
NaOH(s)	−426.8	64.454	−379.53
Na$_2$CO$_3$(s)	−1,131	136	−1,044.49
NaHCO$_3$(s)	−947.7	155	−851.0
O(g)	249.2	160.95	231.7
O$_2$(g)	0	205	0
O$_3$(g)	143	238.82	163
P$_4$O$_{10}$(s)	−3,110.	229	−2,698
Pb(s)	0	68.85	0
PbI$_2$(s)	−175.5	175	−173.6
Pb(NO$_3$)$_2$(s)	−452		
PbO(s)	−217.3	68.70	−187.9
S(s) [S$_8$(s)]	0	31.88	0
SO$_2$(g)	−296.9	248.5	−300.4
SO$_3$(g)	−396	256.22	−370.4
SO$_4^{2-}$(aq)	−907.51	17	−741.99
SiBr$_4$(l)	−95.1	66.4	−106.1
SiO$_2$(s)	−914.4	41.3	−856.0
Sr^{2+}(aq)	−545.51	−39	−557.3
SrSO$_4$(s)	−1,445	122	−1,334
TiCl$_4$(g)	−763.2	354.9	−726.8
TiO$_2$(s) (rutile)	−944.7	50.29	−889.4
ZnO(s)	−83.24	10.43	−76.08
ZnS(s)	−203	57.7	−198

Equilibrium constants

K_a and K_b for selected acids and bases

Acid		K_{a1}	K_{a2}	K_{a3}	pK_{a1}	pK_{a2}	pK_{a3}
acetic acid	$HC_2H_3O_2$	1.74×10^{-5}			4.759		
arsenic acid	H_3AsO_4	6.0×10^{-3}	1.0×10^{-7}	3.0×10^{-12}	2.22	6.98	11.53
benzoic acid	C_6H_5COOH	6.5×10^{-5}			4.19		
carbonic acid	H_2CO_3	4.5×10^{-7}	4.8×10^{-11}		6.35	10.32	
chlorous acid	$HClO_2$	1.1×10^{-2}			1.97		
chromic acid	H_2CrO_4	1.1×10^{-1}	3.2×10^{-7}		0.98	6.50	
cyanic acid	$HOCN$	3.5×10^{-4}			3.46		
formic acid	$HCOOH$	1.8×10^{-4}			3.75		
hydrazoic acid	HN_3	1.9×10^{-5}			4.72		
hydrocyanic acid	HCN	6.2×10^{-10}			9.21		
hydrofluoric acid	HF	6.8×10^{-4}			3.17		
hydroselenic acid	H_2Se	1.3×10^{-4}	1.0×10^{-11}		3.89	11.00	
hydrosulfuric acid	H_2S	1.0×10^{-7}	1.3×10^{-13}		6.99	12.89	
hydrotelluric acid	H_2Te	2.3×10^{-3}	1×10^{-11}		2.64	11.0	
hypobromous acid	$HBrO$	2.5×10^{-9}			8.60		
hypochlorous acid	$HOCl$	5.0×10^{-8}			7.30		
hypoiodous acid	HIO	2.3×10^{-11}			10.64		
hyponitrous acid	$H_2N_2O_2$	6.2×10^{-8}	2.9×10^{-12}		7.21	11.54	
iodic acid	HIO_3	1.6×10^{-1}			0.79		
nitrous acid	HNO_2	5.1×10^{-4}			3.29		
oxalic acid	$H_2C_2O_4$	5.6×10^{-2}	5.4×10^{-5}		1.25	4.27	
periodic acid	HIO_4	2.8×10^{-2}			1.55		
phenol	C_6H_5OH	1.0×10^{-10}			10.00		
phosphoric acid	H_3PO_4	7.6×10^{-3}	6.2×10^{-8}	4.2×10^{-13}	2.12	7.21	12.38
propanoic acid	$HC_3H_5O_2$	1.35×10^{-5}			4.870		
selenious acid	H_2SeO_3	2.4×10^{-3}	4.8×10^{-9}		2.62	8.32	
sulfuric acid	H_2SO_4	—	1.2×10^{-2}		—	1.94	
sulfurous acid	H_2SO_3	1.7×10^{-2}	6.2×10^{-8}		1.76	7.20	
tellurous acid	H_2TeO_3	2.7×10^{-3}	1.8×10^{-8}		2.57	7.74	
thiocyanic acid	$HSCN$	1.4×10^{-1}			0.85		
thiosulfuric acid	$H_2S_2O_3$	2.5×10^{-1}	1.9×10^{-2}		0.60	1.72	

Base		K_b	pK_b
ammonia	NH_3	1.76×10^{-5}	4.755
aniline	$C_6H_5NH_2$	4.2×10^{-10}	9.38
dimethylamine	$(CH_3)_2NH$	1.1×10^{-3}	2.97
hydrazine	N_2H_4	9.8×10^{-7}	6.01
hydroxylamine	NH_2OH	9.6×10^{-9}	8.02
methylamine	CH_3NH_2	5.25×10^{-3}	2.280
pyridine	C_5H_5N	1.5×10^{-9}	8.82

Formation constants for selected complex ions

	K_f		K_f		K_f
$[AgCl_2]^-$	2.0×10^5	$[Co(NH_3)_6]^{2+}$	2.5×10^4	$[Fe(CN)_6]^{4-}$	1.0×10^{24}
$[Ag(NH_3)_2]^+$	1.7×10^7	$[Co(NH_3)_6]^{3+}$	1.6×10^{35}	$[HgCl_4]^{2-}$	1.7×10^{16}
$[Cd(NH_3)_4]^{2+}$	3.6×10^6	$[Cu(NH_3)_2]^+$	7.2×10^{10}	$[Ni(NH_3)_6]^{2+}$	5.5×10^8
$[Co(CN)_6]^{3-}$	1.0×10^{64}	$[Cu(NH_3)_4]^{2+}$	1.1×10^{12}	$[Zn(NH_3)_4]^{2+}$	5.0×10^8
$[Co(CN)_6]^{4-}$	1.0×10^{19}	$[Fe(CN)_6]^{3-}$	1.0×10^{31}		

Solubility product constants of selected compounds

	K_{sp}		K_{sp}		K_{sp}
$Al(OH)_3$	2.7×10^{-20}	GeS	3.0×10^{-35}	$Pt(OH)_2$	1.0×10^{-35}
Sb_2S_3	1.6×10^{-93}	$Au(OH)_3$	5.5×10^{-46}	PtS	8.0×10^{-17}
As_2S_3	3.7×10^{-27}	$In(OH)_2$	5.0×10^{-34}	$Pu(OH)_3$	2.0×10^{-20}
As_2S_5	3.9×10^{-34}	In_2S_3	1.0×10^{-88}	$Pu(OH)_4$	1.0×10^{-52}
$BaSO_4$	1.1×10^{-10}	IrS_2	1.0×10^{-75}	PoS	5.0×10^{-29}
$Bi(OH)_3$	3.2×10^{-32}	$Fe(OH)_2$	7.9×10^{-16}	$AgC_2H_3O_2$	4×10^{-3}
Bi_2S_3	1.0×10^{-97}	$Fe(OH)_3$	3.2×10^{-38}	AgBr	5.3×10^{-13}
$Cd(OH)_2$	2.2×10^{-14}	$Fe_3(PO_4)_2$	1.0×10^{-36}	Ag_2CO_3	8.2×10^{-12}
CdS	7.9×10^{-27}	$FePO_4$	4.0×10^{-27}	AgCl	1.78×10^{-10}
$Ca_3(AsO_4)_2$	6.8×10^{-19}	FeS	5.0×10^{-18}	Ag_2CrO_4	1.2×10^{-12}
$CaCO_3$	8.7×10^{-9}	$La(OH)_3$	2.0×10^{-19}	AgOCN	2.3×10^{-7}
CaF_2	4.0×10^{-11}	La_2S_3	2.0×10^{-13}	AgCN	2.2×10^{-16}
$Ca(OH)_2$	5.5×10^{-6}	$PbCl_2$	1.6×10^{-5}	$Ag_2Cr_2O_7$	1×10^{-10}
$Ca_3(PO_4)_2$	1.2×10^{-26}	$Pb(OH)_2$	1.1×10^{-20}	AgI	8.3×10^{-17}
$CaSO_3$	1.3×10^{-8}	$Pb(OH)_4$	6.5×10^{-71}	$AgNO_2$	1.6×10^{-4}
$CaSO_4$	9.1×10^{-6}	$Pb(IO_3)_2$	2.6×10^{-13}	$Ag_2C_2O_4$	3.5×10^{-11}
$Ce(OH)_3$	1.5×10^{-20}	PbI_2	1.1×10^{-9}	Ag_3PO_4	1.3×10^{-20}
$Cr(OH)_2$	1.0×10^{-17}	$PbSO_4$	1.6×10^{-8}	Ag_2SO_4	1.6×10^{-5}
$Cr(OH)_3$	6.3×10^{-31}	PbS	2.5×10^{-27}	Ag_2S	6.3×10^{-50}
$CrPO_4$	1.0×10^{-17}	PbS_2O_3	4.0×10^{-7}	Tl_2S	5.0×10^{-21}
$Co(OH)_2$	2.0×10^{-15}	$Mg(OH)_2$	6.0×10^{-10}	$Th(OH)_4$	3.2×10^{-45}
$Co(OH)_3$	4.0×10^{-45}	$Mn(OH)_2$	1.9×10^{-13}	$Sn(OH)_2$	6.3×10^{-27}

	K_{sp}		K_{sp}		K_{sp}
CoS	4.0×10^{-21}	$Mn(OH)_3$	1.0×10^{-36}	$Sn(OH)_4$	1.0×10^{-57}
$Cu_3(AsO_4)_2$	7.6×10^{-36}	MnS	2.5×10^{-13}	SnS	1.0×10^{-25}
CuBr	5.25×10^{-9}	Hg_2Br_2	5.8×10^{-23}	SnS_2	5.3×10^{-15}
$CuBr_2$	3.2×10^{-1}	Hg_2Cl_2	3.5×10^{-18}	$W(OH)_4$	1.0×10^{-50}
CuCl	1.2×10^{-6}	$Hg_2C_2O_4$	1.0×10^{-13}	$Zn(OH)_2$	7.1×10^{-18}
$Cu(OH)_2$	2.2×10^{-20}	Hg_2S	1.0×10^{-47}	ZnS	1.6×10^{-24}
CuI	1.1×10^{-12}	HgS	1.6×10^{-52}	$Zr_3(PO_4)_4$	1.00×10^{-132}
Cu_2S	6.3×10^{-36}	$Mo(OH)_4$	1.0×10^{-50}		
CuS	2.5×10^{-48}	$Ni(OH)_2$	2.0×10^{-15}		
$Ga(OH)_3$	1.5×10^{-36}	NiS	3.2×10^{-19}		

Oxidation-reduction properties

Activity series

$Li(s) \rightarrow Li^+(aq) + 1\ e^-$
$K(s) \rightarrow K^+(aq) + 1\ e^-$
$Ba(s) \rightarrow Ba^{2+}(aq) + 2\ e^-$
$Sr(s) \rightarrow Sr^{2+}(aq) + 2\ e^-$
$Ca(s) \rightarrow Ca^{2+}(aq) + 2\ e^-$
$Na(s) \rightarrow Na^+(aq) + 1\ e^-$
$Mg(s) \rightarrow Mg^{2+}(aq) + 2\ e^-$
$Al(s) \rightarrow Al^{3+}(aq) + 3\ e^-$
$Mn(s) \rightarrow Mn^{2+}(aq) + 2\ e^-$
$Zn(s) \rightarrow Zn^{2+}(aq) + 2\ e^-$
$Cr(s) \rightarrow Cr^{2+}(aq) + 2\ e^-$
$Fe(s) \rightarrow Fe^{2+}(aq) + 2\ e^-$
$Cd(s) \rightarrow Cd^{2+}(aq) + 2\ e^-$
$Co(s) \rightarrow Co^{2+}(aq) + 2\ e^-$
$Ni(s) \rightarrow Ni^{2+}(aq) + 2\ e^-$
$Sn(s) \rightarrow Sn^{2+}(aq) + 2\ e^-$
$Pb(s) \rightarrow Pb^{2+}(aq) + 2\ e^-$
$H_2(g) \rightarrow 2\ H^+(aq) + 2\ e^-$
$Cu(s) \rightarrow Cu^{2+}(aq) + 2\ e^-$
$Hg(s) \rightarrow Hg^{2+}(aq) + 2\ e^-$
$Ag(s) \rightarrow Ag^+(aq) + 1\ e^-$
$Pd(s) \rightarrow Pd^{2+}(aq) + 2\ e^-$
$Pt(s) \rightarrow Pt^{2+}(aq) + 2\ e^-$
$Au(s) \rightarrow Au^{3+}(aq) + 3\ e^-$

Some selected standard reduction potentials, $E°$

	$E°$ (volts)
$F_2(g) + 2\ e^- \rightarrow 2\ F^-(aq)$	+2.87
$BrO_4^-(aq) + 2\ H^+(aq) + 2\ e^- \rightarrow BrO_3^-(aq) + H_2O(l)$	+1.74
$PbO_2(s) + 4\ H^+(aq) + SO_4^{2-}(aq) + 2\ e^- \rightarrow PbSO_4(s) + 2\ H_2O(l)$	+1.70
$IO_4^-(aq) + 2\ H^+(aq) + 2\ e^- \rightarrow IO_3^-(aq) + H_2O(l)$	+1.65
$MnO_4^-(aq) + 8\ H^+(aq) + 5\ e^- \rightarrow Mn^{2+}(aq) + 4\ H_2O(l)$	+1.491
$PbO_2(s) + 4\ H^+(aq) + 2\ e^- \rightarrow Pb^{2+}(aq) + 2\ H_2O(l)$	+1.455
$Ce^{4+}(aq) + 1\ e^- \rightarrow Ce^{3+}(aq)$	+1.4430

	$E°$ (volts)
$Cl_2(g) + 2\,e^- \rightarrow 2\,Cl^-(aq)$	+1.359
$Cr_2O_7^{2-}(aq) + 14\,H^+(aq) + 6\,e^- \rightarrow 2\,Cr^{3+}(aq) + 7\,H_2O(l)$	+1.33
$MnO_2(s) + 4\,H^+(aq) + 2\,e^- \rightarrow Mn^{2+}(aq) + 2\,H_2O(l)$	+1.23
$O_2(g) + 4\,H^+(aq) + 4\,e^- \rightarrow 2\,H_2O(l)$	+1.229
$Br_2(l) + 2\,e^- \rightarrow 2\,Br^-(aq)$	+1.087
$VO_2^+(aq) + 2\,H^+(aq) + 1\,e^- \rightarrow VO^{2+}(aq) + H_2O(l)$	+0.9994
$AuCl_4^-(aq) + 3\,e^- \rightarrow Au(s) + 4\,Cl^-$	+0.99
$2\,Hg^{2+}(aq) + 2\,e^- \rightarrow Hg_2^{2+}(aq)$	+0.907
$N_2O_4(g) + 2\,e^- \rightarrow 2\,NO_2^-(aq)$	+0.88
$HNO_2(aq) + 7\,H^+(aq) + 6\,e^- \rightarrow NH_4^+(aq) + 2\,H_2O(l)$	+0.86
$Hg^{2+}(aq) + 2\,e^- \rightarrow Hg(l)$	+0.850
$Ag^+(aq) + 1\,e^- \rightarrow Ag(s)$	+0.7994
$Hg_2^{2+}(aq) + 2\,e^- \rightarrow 2\,Hg(l)$	+0.792
$Fe^{3+}(aq) + 1\,e^- \rightarrow Fe^{2+}(aq)$	+0.771
$ClO_2^-(aq) + H_2O(l) + 2\,e^- \rightarrow ClO^-(aq) + 2\,OH^-(aq)$	+0.66
$BrO_3^-(aq) + 3\,H_2O(l) + 6\,e^- \rightarrow Br^-(aq) + 6\,OH^-(aq)$	+0.61
$ClO_4^-(aq) + 4\,H_2O(l) + 8\,e^- \rightarrow Cl^-(aq) + 8\,OH^-(aq)$	+0.56
$I_2(s) + 2\,e^- \rightarrow 2\,I^-(aq)$	+0.5355
$Cu^+(aq) + 1\,e^- \rightarrow Cu(s)$	+0.521
$2\,BrO^-(aq) + 2\,H_2O(l) + 2\,e^- \rightarrow Br_2(l) + 4\,OH^-(aq)$	+0.45
$Cu^{2+}(aq) + 2\,e^- \rightarrow Cu(s)$	+0.337
$HAsO_2(aq) + 3\,H^+(aq) + 3\,e^- \rightarrow As(s) + 2\,H_2O(l)$	+0.2475
$AgCl(s) + 1\,e^- \rightarrow Ag(s) + Cl^-(aq)$	+0.224
$Cu^{2+}(aq) + 1\,e^- \rightarrow Cu^+(aq)$	+0.153
$Sn^{4+}(aq) + 2\,e^- \rightarrow Sn^{2+}(aq)$	+0.15
$2\,H^+(aq) + 2\,e^- \rightarrow H_2(g)$	+0.0000
$Fe^{3+}(aq) + 3\,e^- \rightarrow Fe(s)$	−0.036
$Pb^{2+}(aq) + 2\,e^- \rightarrow Pb(s)$	−0.1263
$Sn^{2+}(aq) + 2\,e^- \rightarrow Sn(s)$	−0.1364
$PbSO_4(s) + 2\,e^- \rightarrow Pb(s) + SO_4^{2-}(aq)$	−0.356
$Tl^+(aq) + 1\,e^- \rightarrow Tl(s)$	−0.3363
$Cd^{2+}(aq) + 2\,e^- \rightarrow Cd(s)$	−0.4026
$Cr^{3+}(aq) + 1\,e^- \rightarrow Cr^{2+}(aq)$	−0.41
$Fe^{2+}(aq) + 2\,e^- \rightarrow Fe(s)$	−0.4402
$[Au(CN)_2]^-(aq) + 1\,e^- \rightarrow Au(s) + 2\,CN^-(aq)$	−0.50
$2\,SO_3^{2-}(aq) + 3\,H_2O(l) + 4\,e^- \rightarrow S_2O_3^{2-}(aq) + 6\,OH^-(aq)$	−0.58
$Cr^{3+}(aq) + 3\,e^- \rightarrow Cr(s)$	−0.74
$Zn^{2+}(aq) + 2\,e^- \rightarrow Zn(s)$	−0.7628
$2\,H_2O(l) + 2\,e^- \rightarrow H_2(g) + 2\,OH^-(aq)$	−0.8277
$Cr^{2+}(aq) + 2\,e^- \rightarrow Cr(s)$	−0.91
$Al^{3+}(aq) + 3\,e^- \rightarrow Al(s)$	−1.66
$Mg^{2+}(aq) + 2\,e^- \rightarrow Mg(s)$	−2.37
$Nb_2O_5(s) + 10\,H^+(aq) + 10\,e^- \rightarrow 2\,Nb(s) + 5\,H_2O(l)$	−2.7109
$Ca^{2+}(aq) + 2\,e^- \rightarrow Ca(s)$	−2.87
$K^+(aq) + 1\,e^- \rightarrow K(s)$	−2.924
$Li^+(aq) + 1\,e^- \rightarrow Li(s)$	−3.045

Nomenclature

The nomenclature is divided into three parts; in Appendix H you will find a related discussion of oxidation numbers.

Nomenclature I

In this section, the following elements will be used extensively.

H	hydrogen	H_2O	water	Sn	tin	He	helium
N	nitrogen	NH_3	ammonia	Pb	lead	Ne	neon
O	oxygen	CH_4	methane	P	phosphorus	Ar	argon
F	fluorine	C	carbon	As	arsenic	Kr	krypton
Cl	chlorine	Si	silicon	Sb	antimony	Xe	xenon
Br	bromine	Ge	germanium	Bi	bismuth	Rn	radon
I	iodine	B	boron	Se	selenium		
S	sulfur	Al	aluminum	Te	tellurium		

The first step in learning nomenclature is to learn the names of the individual elements, polyatomic ions, and the special names. The second step is to take the names which you have learned and combine them through a set of rules. To be able to name compounds will require both of these steps. Many people make the mistake of doing only the first step. This step is straightforward memorization, so flash cards or some other rote memory method should be applied on a daily basis. The second step requires practice, because there are so many possible combinations that it is not possible to make flash cards for all of them (there are thousands of possibilities).

By learning a few of the new names on your own plus a little review, you should be able to do the first step. While you should concentrate on the new names, you will need to continue looking over the older names. Short study sessions, such as five minutes before or after each class, work best.

The second step requires learning and applying the nomenclature rules for binary and ternary compounds. This appendix examines only binary compounds of the nonmetals. These rules also include the metalloids and a few ternary compounds that contain *only* nonmetals and/or metalloids.

NOTE: The metalloids may also be named using the rules discussed for metals.

Special names

Some binary compounds, such as water, have special names. For these few compounds the method used in the first nomenclature step is sufficient. These names are fixed; there are no other rules to apply. Along these lines, do not try to name one of these compounds by the following rules. Thus, H_2O is water, and no other name is correct.

Binary compounds

All other binary compounds follow one rule—their names end in *-ide*. (There are a few situations where a ternary compound may have this ending also.) Other rules in addition to this one are applied depending on the type and number of atoms present.

If only nonmetals are present, then a multiplying prefix may be needed. (Hydrogen is the exception to this, there are no prefixes used for hydrogen in binary compounds.) Multiplying prefixes tell how many of each type of atom is present. The prefixes are:

1	=	mono-	5	=	penta-	8	=	octa-
2	=	di-	6	=	hexa-	9	=	nona-
3	=	tri-	7	=	hepta-	10	=	deca-
4	=	tetra-						

There are other prefixes, but these are the most common ones. The prefix *mono-* is seldom used, the lack of any prefix is used to indicate the fact that there is only one atom present. The only common compound where *mono-* is used is CO, carbon monoxide.

Since prefixes are not used to designate the number of hydrogen atoms present in a compound, it is not always easy to predict the correct formula. In general, in the absence of other information the number of hydrogen atoms present in a binary compound may be predicted from the position of the element on the periodic table. Beginning with the far right of the table (column VIIIA or column 18) and moving to the left, the number of hydrogens expected are 0, 1, 2, 3, and 4. Thus:

$$He \qquad HF \qquad H_2O \qquad NH_3 \qquad CH_4$$

When putting together the name of a compound, normally the element to the right, and/or higher on the periodic table goes last. Exceptions include the noble gases (VIIIA), which are never last, and hydrogen, which is often last. Whichever element goes last in the name goes last in the written formula too. The ending of the elemental name that is at the end is changed to *-ide*. For example:

hydrogen	= hydride	carbon	= carbide
nitrogen	= nitride	phosphorus	= phosphide
oxygen	= oxide	sulfur	= sulfide
fluorine	= fluoride	chlorine	= chloride
bromine	= bromide	iodine	= iodide

In compounds containing only nonmetals the name *hydride* is seldom used. These same names will be used when these elements are in binary compounds containing metals.

Again, this change to an *-ide* suffix is used *only* for the element appearing last in the formula or name. Do not change the ending of the other element.

Binary acids

As with many rules there are exceptions. Most of these are covered under the special names; the others are certain compounds of hydrogen. In nomenclature, as elsewhere, hydrogen is a common exception. One reason why hydrogen is an exception for binary compounds is because some of its compounds are acids. The binary acids (hydrogen plus certain nonmetals) also have two-word names. In their names, binary acids have a *hydro-* prefix and an *-ic acid* ending. For example, HCl may be named *hydro*chlor*ic acid*. There are only a few binary acids; the following list are the common ones:

Binary Compound		Binary Acid
hydrogen fluoride	HF	hydrofluoric acid
hydrogen chloride	HCl	hydrochloric acid
hydrogen bromide	HBr	hydrobromic acid
hydrogen iodide	HI	hydroiodic acid
hydrogen sulfide	H_2S	hydrosulfuric acid

As with other acids (see later) the formulas are normally written with the H in front. Nonacidic hydrogen compounds are usually written with some other element first in the formula (except for H_2O).

Technically the binary acids should be named as acids only when they are dissolved in water. In other situations they should be named as ordinary binary compounds (for example, HCl is hydrogen chloride). Many times either name will be acceptable unless information is given that indicates which name to use. This information may be given in a problem, or it may be included with the formula. To indicate in the formula that one of the previous compounds must be named as an acid, the designation *(aq)* is used (for example, HCl*(aq)*). The *aq* in parentheses is the abbreviation for aqueous solution—a solution where something is dissolved in water. If anything else is included in parentheses after the name (for example, HCl*(g)*), it should be named as a binary compound.

General

There are some situations where other nomenclature systems are used. One major alternate system is for many compounds of carbon—this is the naming of organic compounds. Other systems are used for many binary hydrogen compounds. If you see another name that does not seem to follow these rules, just use it.

The presence or lack of prefixes determines the formulas for other binary compounds. *Mono-* or no prefix indicates that only one atom is present, while the number of the other atoms is given directly as a prefix.

carbon dioxide	CO_2	(di = 2)
chlorine trifluoride	ClF_3	(tri = 3)
dinitrogen pentoxide	N_2O_5	(di = 2 and pent(a) = 5)

These rules may be extended to some ternary compounds. This occurs only when there are no metals or ions present. You should be able to recognize such names should you see one.

The following are some examples of binary compounds. Look them over and make sure you understand *why* the names and formulas are written as they are. These are just a few examples of the hundreds of binary compounds, so do not expect this list to be complete. (Yes, these compounds all exist.)

AsF_3	arsenic trifluoride	GeO_2	germanium dioxide
$TeCl_4$	tellurium tetrachloride	Sb_2S_3	diantimony trisulfide
XeO_3	xenon trioxide	KrF_2	krypton difluoride
$SiBr_4$	silicon tetrabromide	SeF_6	selenium hexafluoride
PCl_5	phosphorus pentachloride	BI_3	boron triiodide
CO	carbon monoxide	NH_3	ammonia
PN	phosphorus nitride	H_2S	hydrogen sulfide
CO_2	carbon dioxide		or hydrosulfuric acid
IF_7	iodine heptafluoride	$H_2S(g)$	hydrogen sulfide
P_4O_{10}	tetraphosphorus decaoxide	$H_2S(aq)$	hydrosulfuric acid
CH_4	methane	P_4O_8	tetraphosphorus octaoxide
N_2O	dinitrogen oxide	H_2O	water
NO	nitrogen oxide	As_4S_{10}	tetraarsenic decasulfide
N_2O_3	dinitrogen trioxide	$HBr(g)$	hydrogen bromide
NO_2	nitrogen dioxide	P_4S_9	tetraphosphorus nonasulfide
N_2O_4	dinitrogen tetraoxide	OF_2	oxygen difluoride
N_2O_5	dinitrogen pentaoxide	O_2F_2	dioxygen difluoride
S_2F_{10}	disulfur decafluoride	$SeBr_4$	selenium tetrabromide
S_7O_2	heptasulfur dioxide	TeO_3	tellurium trioxide
I_4O_9	tetraiodine nonaoxide	$XeOF_4$	xenon oxide tetrafluoride
$HI(g)$	hydrogen iodide	$HF(l)$	hydrogen fluoride
$HF(aq)$	hydrofluoric acid	HCl	hydrogen chloride
Cl_2O_7	dichlorine heptoxide		or hydrochloric acid

Nomenclature II

In this section, the following elements and ions will be used extensively.

Li	lithium	Be	beryllium	PO_4^{3-}	phosphate ion
Na	sodium	OH^-	hydroxide ion	HPO_4^{2-}	hydrogen phosphate ion
K	potassium	NH_4^+	ammonium ion	$H_2PO_4^-$	dihydrogen phosphate ion
Mg	magnesium	Hg_2^{2+}	mercury(I) ion	SO_3^{2-}	sulfite ion
Ca	calcium	NO_3^-	nitrate ion	NO_2^-	nitrite ion
Sr	strontium	CO_3^{2-}	carbonate ion	$C_2O_4^{2-}$	oxalate ion
Ba	barium	SO_4^{2-}	sulfate ion	$S_2O_3^{2-}$	thiosulfate ion
Ra	radium	MnO_4^-	permanganate ion	AsO_4^{3-}	arsenate ion
Zn	zinc	CN^-	cyanide ion	AsO_3^{3-}	arsenite ion
Cd	cadmium	OCN^-	cyanate ion	$C_2H_3O_2^-$	acetate ion
Hg	mercury	SCN^-	thiocyanate ion	O_2^{2-}	peroxide ion
Rb	rubidium	CrO_4^{2-}	chromate ion	O_2^-	superoxide ion
Cs	cesium	$Cr_2O_7^-$	dichromate ion		
HCO_3^-	bicarbonate ion	or	hydrogen carbonate ion		
HSO_4^-	bisulfate ion	or	hydrogen sulfate ion		

Four types of compounds will be examined in this section. The first type are the binary compounds containing a metal and a nonmetal. The second and third types are ternary compounds; these are either "simple" ternary compounds or ternary acids. The final category includes the hydrates.

Binary compounds of metals and nonmetals

In most cases a compound containing a metal and a nonmetal or polyatomic ion is ionic. For the ionic compounds covered in this workbook the metal will always be present as a cation (positively charged ion). The charges on these ions may be predicted from the position of the element on the periodic table. This is fairly simple for the main group (representative) elements. To determine the charge of a main group metal, use the number at the top of the column on the periodic table. Thus, the **metals** in columns IA, IIA, IIIA, IVA, and VA have charges of +1, +2, +3, +4, and +5 respectively. To this may be added the fact that elements in column IIB are usually +2. There are a few exceptions to this, especially on the right-hand side of the periodic table. These metals are exceptions: tin and lead, which are often +2, and bismuth, which is often +3. (Later you will see how to tell which charge to choose and how to indicate the value in a name.)

You should practice predicting the charges of the metals by looking at a periodic table. Remember, these charges do not apply to these metals in their elemental form (where the metal has no charge). That is, the aluminum in a piece of aluminum foil does not have a charge—just the aluminum in a compound such as aluminum chloride is +3.

The metal in a binary compound must be accompanied by a nonmetal. The nonmetal will be present as an anion (negatively charged ion). The prediction of the charge on a nonmetal may also be obtained from its position on the periodic table. However, the counting is started for the opposite side. Thus **nonmetals** in columns VIIIA, VIIA, VIA, VA, and IVA have charges of 0, −1, −2, −3, and −4, respectively. The only other nonmetal is hydrogen, which has a charge of −1 when it is combined with a metal. Nonmetals may have other charges, but not in the simple binary compounds with metals covered in this course.

In all compounds the total of all positive and negative charges *must* be 0. To obtain a net charge of zero it will be necessary to combine the appropriate numbers of cations and anions. Thus a +2 ion could be combined with either a −2 ion or two −1 ions. In some cases, the canceling may be a little more involved; for example, it takes three +2 ions to cancel two −3 ions. In these cases, just keep track of the total charges, so three +2 ions = +6, and two −3 ions = −6, thus the compound has +6 − 6 = 0. In this case 6 is the lowest common multiple (LCM) of 2 and 3. The LCM method will work for any binary (or ternary) compound.

Some texts use the crisscross rule instead of the LCM method for predicting formulas. To use this method, use the charge of the anion as the number of cations that are present and use the charge of the cation to predict the number of anions. This rule works directly in most cases; however, some of the predictions that it gives must be simplified. Examples of these two methods follow.

In all ionic compounds the name or symbol of the cation *must* go first. The name or symbol of the anion *must* go last. The name of the anion has an *-ide* suffix, the same as they have in binary compounds containing only nonmetals. The major difference when naming binary compounds containing metals is that *no* multiplying prefixes are used.

Example 1: Give the name and formula of a compound containing aluminum and oxygen.

Based on their positions on the periodic table, aluminum (Al) should have a charge of +3, and oxygen (O) should have a charge of −2.

$$Al^{3+} \quad O^{2-}$$

By the LCM method: The LCM of 3 and 2 is 6, and it takes two 3's and three 2's to equal 6. Thus:

$$Al_2O_3$$

By the crisscross rule: Al needs a 2 (from O^{2-}) and the O needs a 3 (from Al^{3+}). Thus:

$$Al_2O_3$$

Notice, that even though the compound contains +3 and −2 ions, the charges are not written in the formulas.

The name, in this case, will be aluminum oxide.

Example 2: Give the formula for a compound containing Pb^{4+} and O^{2-}.

Since the charges on the ions are already given it is not necessary to use the periodic table. In a situation such as this make sure you do not inadvertently alter the charges, you must use the ones given, whether they make sense or not.

By the LCM method: The LCM of 4 and 2 is 4, and it takes one 4 and two 2's.

$$PbO_2$$

By the crisscross method: Pb needs 2 and O needs 4.

$$Pb_2O_4$$

This is the problem with using the crisscross method. When you get a formula, such as this one, where the subscripts may be reduced by a common divisor, the subscripts *must* be reduced. In this case both subscripts may be divided by 2, so you should do so:

$$PbO_2$$

Only after this additional step do you get the same (correct) answer as the LCM method.

Ternary compounds of the main group elements

The ternary compounds that are covered in this workbook normally contain a polyatomic ion. As with the names of the elements the names of these ions must be learned directly. In addition to their names and elemental composition, it is also necessary to learn the charges on these ions. The reason why polyatomic ions are treated as distinct entities is that their behavior is as such. Thus both in their names and in chemical formulas containing these ions they are always kept together. Also, when naming these ions, or any ions, as individuals the word *ion* is part of their name. (The word *ion* never appears in the name of a compound even if ions are present.)

For this workbook only two polyatomic cations will be given. These cations are the ammonium ion (NH_4^+) and the mercury(I) ion (Hg_2^{2+}). These two ions are treated like normal metal cations covered in the binary compound section of this handout. For example, even though NH_4Cl is technically a ternary compound, it is named like a binary (ammonium chloride).

Most of the polyatomic ions covered in this workbook are anions. Thus in compounds they will be combined with a cation. As before, the net charge *must* be zero. The charges for the cations are predicted in the same manner as for binary compounds. The same LCM or crisscross rules are still applicable.

Example 3: Give the name and formula for a compound containing aluminum ions and sulfate ions.

$$Al_2(SO_4)_3$$

aluminum sulfate

Notice that parentheses are used around the sulfate ion to maintain its identity as a single entity of which three are needed for the compound.

The compound is understood to contain ions; however, neither the charges nor the word ion, appear anywhere in either the formula or the name.

Ternary acids

The remaining category of ternary compounds are the ternary acids. These compounds contain hydrogen (as do all acids) and a polyatomic ion. The naming is based upon the name of the category of the anion. The categories are based upon the anion suffix. The possible suffixes are -ide, -ite, or -ate.

The -ide suffix is not very common for polyatomic ions, and even when it is present, the resultant combination of the ion with hydrogen is not necessary an acid. Thus combining hydrogen with a hydroxide ion gives HOH, which is more commonly written as H_2O, and is called water. Hydrogen compounds with the hydroxide ion, peroxide ion or the superoxide ion are not normally treated as acids. In the cases where the compounds are treated as acids, the hydrogen will usually be listed first in the formula followed by the formula for the polyatomic ion. When naming acids arising from polyatomic ions with an -ide suffix, the binary acids rule is applied (hydro- prefix and -ic acid ending).

Example 4: HCN = hydrogen plus cyanide = hydrocyanic acid

To name the acid resulting when hydrogen is combined with a polyatomic ion with an -ite suffix, simply replace the -ite with -ous acid. Thus, the nitrite ion (NO_2^-) gives nitrous acid (HNO_2). In a few cases, more letters are added to make the name easier to pronounce. For example, the sulfite ion (SO_3^{2-}) ion gives sulfurous acid (H_2SO_3).

Polyatomic anions with an -ate suffix are named by changing the -ate to -ic acid. Using this rule means that the nitrate ion (NO_3^-) yields nitric acid (HNO_3). Again, sulfur and phosphorus use more letters to make them easier to pronounce. So the phosphate ion (PO_4^{3-}) ion changes to phosphoric acid (H_3PO_4).

Do not forget, acids are compounds; therefore, the net charge *must* be zero. The hydrogen is treated as +1 in all acids. Thus the charge on the anion always tells you how many hydrogens are needed.

Hydrates

The final category of compound to be discussed in this nomenclature part is the hydrates. These are examples of a type of compound, which are made by combining two or more different compounds with each of the different compounds retaining its own identity. An example is $MgSO_4 \cdot 7H_2O$. These compounds form because certain other compounds, water in this case, may become very strongly associated with a particular ion. For nomenclature, only those containing water will be considered. The naming is simple—just name the nonwater portion as a normal compound followed by hydrate. If there is more than one water present, use a multiplying prefix on the hydrate. Thus this example would be magnesium sulfate heptahydrate. Such compounds are considered a single unit, just like any simpler compound.

Examples

The following are some examples of the type of compounds discussed. Look over them and make sure you understand *why* the names and formulas are written the way that they are. As in the first nomenclature part, these are just a few of the possible compounds, and, yes, they all could exist.

NaCl	sodium chloride	AlF_3	aluminum fluoride
KNO_3	potassium nitrate	$CaSO_4$	calcium sulfate
$MgCO_3$	magnesium carbonate	$Zn(MnO_4)_2$	zinc permanganate
$CdSO_3$	cadmium sulfite	$Hg_2(CN)_2$	mercury(I) cyanide
Li_2O_2	lithium peroxide	$Ba(O_2)_2$	barium superoxide
$Sr(OCN)_2$	strontium cyanate	RbSCN	rubidium thiocyanate
$Cs_2S_2O_3$	cesium thiosulfate	$Be_3(PO_3)_2$	beryllium phosphite
$K_2C_2O_4$	potassium oxalate	$Ra_3(PO_4)_2$	radium phosphate
$Al(NO_2)_3$	aluminum nitrite	$RaHPO_4$	radium hydrogen phosphate
NH_4BiO_3	ammonium bismuthate	$Ra(H_2PO_4)_2$	radium dihydrogen phosphate
NaOH	sodium hydroxide	K_2CrO_4	potassium chromate
$AlPO_4$	aluminum phosphate	$LiHSO_4$	lithium bisulfate
HOH	water		or lithium hydrogen sulfate
H_2CO_3	carbonic acid	$Hg_2C_2O_4$	mercury(I) oxalate
H_2SO_4	sulfuric acid	$(NH_4)_2Cr_2O_7$	ammonium dichromate
$HMnO_4$	permanganic acid	$Al(C_2H_3O_2)_3$	aluminum acetate
HCN	hydrocyanic acid	$NaHCO_3$	sodium bicarbonate
H_2O_2	hydrogen peroxide		or sodium hydrogen carbonate
$H_2S_2O_3$	thiosulfuric acid	$Na_2SO_4 \cdot 10H_2O$	sodium sulfate decahydrate
$HC_2H_3O_2$	acetic acid	$CaSO_4 \cdot 2H_2O$	calcium sulfate dihydrate

Nomenclature III

In this section, the following elements and ions will be used extensively.

Cu	copper	Ti	titanium	OF^-	hypofluorite ion	ClO_3^-	chlorate ion
Ag	silver	V	vanadium	ClO^-	hypochlorite ion	BrO_3^-	bromate ion
Au	gold	Cr	chromium	BrO^-	hypobromite ion	IO_3^-	iodate ion
Fe	iron	Mo	molybdenum	IO^-	hypoiodite ion	ClO_4^-	perchlorate ion
Co	cobalt	W	tungsten	ClO_2^-	chlorite ion	BrO_4^-	perbromate ion
Ni	nickel	Th	thorium	BrO_2^-	bromite ion	IO_4^-	periodate ion
Pt	platinum	U	uranium	IO_2^-	iodite ion		
Mn	manganese	Pu	plutonium				

This is the final nomenclature section. In order to cover this last stage of nomenclature, you must first understand the rules for oxidation numbers. If you have not already gone over how to derive oxidation numbers for the various elements, do so before working through this handout. Rules for assigning oxidation numbers can be found in the next appendix, Appendix H.

Many metals, particularly the transition metals, may exist as more than one type of ion (different charges). When such a metal is present in a compound, the name of the compound *must* contain some indication of the cationic charge. The most widely used method is to attach the charge of the metal to the name of the metal using Stock notation. Stock notation consists of placing the charge of the metal in parentheses in Roman numerals immediately after the name of the metal. Thus if iron is present with a +3 charge, this would be indicated in the name as iron(III).

Note that the parentheses are required and that there is no space between the metal name and the parentheses (the name and the parentheses are treated as one word).

In this workbook, *any metal* that may have more than one oxidation state, *must* be named using Stock notation. There are other methods, but this is the only acceptable for this workbook. The metals that fall into this category are most of the transition metal (other than Zn and Cd), and those metals adjacent to lead on the periodic table. This method may be used for the metalloids, but they are usually treated as nonmetals in nomenclature. This method should also be used for some of the inner transition metals; however, you will not be expected to predict which ions may form.

On the periodic table the elements that require Stock notation and you are required to know are:

Transition elements										p-Block	
Ti	V	Cr	Mn	Fe	Co	Ni	Cu				
		Mo					Ag*			Sn	Sb
		W				Pt	Au	Hg		Pb	Bi

Inner transition elements
Th* U Pu

*In most compounds Ag has a +1 oxidation number; for this reason, the (I) designation is not always included in the name of silver(I) compounds. Similarly, Th is nearly always +4, and so the (IV) is often neglected in naming its compounds.

Other than the addition of Stock notation to the name of the metal in a compound, there is no difference in how these compounds are named. Thus the rules discussed in nomenclature number II should be reviewed at this time.

Example 1   Name the compound FeO
From its position on the periodic table O is normally -2, then the Fe must be $+2$ for the overall charge to be 0 (the same total as for all compounds).
Since the iron is present as Fe^{2+}, it is called iron(II) (treated as only one word).
By previous rules the oxygen should be called oxide.
Thus the name of the compound is iron(II) oxide.

Example 2   What are the formulas for copper(I) oxide and copper(II) oxide?
The oxide in both cases is O^{2-}.
copper(I) = Cu^+ copper(II) = Cu^{2+}
For the total charge to be 0, the formulas must be:
copper(I) oxide = Cu_2O copper(II) oxide = CuO

Example 3   Give the formula for tungsten(VI) oxide.
The oxide must be O^{2-}.
The tungsten is W^{6+}.
By the LCM method, the lowest multiple of 2 and 6 is 6:
One 6 and three 2's are required giving:
WO_3
By the crisscross method
2 (from O) W and 6 (from W) O's are needed:
W_2O_6

This must now be reduced (divide by 2):

WO_3

Do not make the common mistake of using the Stock notation like a multiplying prefix; they are designed to indicate different things. By coincidence, there are a few cases where they will give the same result, but these situations are rare.

The following are some examples of the type of compound discussed in this section, plus a few other new names. As with the previous appendixes, look them over and make sure you understand *why* the names and formulas are written the way they are. Again, these are just a few of the possible compounds, and, yes, they could exist.

AgF_2	silver(II) fluoride	$AuCl_3$	gold(III) chloride
FeO	iron(II) oxide	Fe_2O_3	iron(III) oxide
$CuSCN$	copper(I) thiocyanate	$CoSO_4 \cdot 7H_2O$	cobalt(II) sulfate heptahydrate
BiF_5	bismuth(V) fluoride	$Ni(OF)_2$	nickel(II) hypofluorite
$V(ClO_4)_3$	vanadium(III) perchlorate	MoO_3	molybdenum(VI) oxide
$CrAsO_4$	chromium(III) arsenate	$Mn(BrO_3)_2$	manganese(II) bromate
WS_2	tungsten(IV) sulfide	PtS	platinum(II) sulfide
TiO_2	titanium(IV) oxide	$Th(NO_3)_4$	thorium(IV) nitrate
UF_6	uranium(VI) fluoride	$HgCr_2O_7$	mercury(II) dichromate
$Pu(C_2O_4)_2$	plutonium(IV) oxalate	SnS_2O_3	tin(II) thiosulfate
$Pb(ClO)_4$	lead(IV) hypochlorite	Na_3AsO_3	sodium arsenite
Bi_2O_3	bismuth(III) oxide	$Co(ClO_3)_2$	cobalt(II) chlorate
$KClO_2$	potassium chlorite	$HBrO$	hypobromous acid
$RbIO$	rubidium hypoiodite	$HBrO_2$	bromous acid
$RbIO_2$	rubidium iodite	$HBrO_3$	bromic acid
$RbIO_3$	rubidium iodate	$HBrO_4$	perbromic acid
$RbIO_4$	rubidium periodate	$Fe_2(Cr_2O_7)_3$	iron(III) dichromate
Hg_2Cl_2	mercury(I) chloride	$Ti(C_2H_3O_2)_3$	titanium(III) acetate
Sb_2O_3	antimony(III) oxide	CrC_2O_4	chromium(II) oxalate
VF_5	vanadium(V) fluoride	WO_3	tungsten(VI) oxide
$MoCl_4$	molybdenum(IV) chloride	$Fe(MnO_4)_3$	iron(III) permanganate
$MnAsO_4$	manganese(III) arsenate	$Ni_2(Cr_2O_7)_3$	nickel(III) dichromate
Co_2O_3	cobalt(III) oxide	$CuSO_4 \cdot 5H_2O$	copper(II) sulfate pentahydrate
$PtCl_4$	platinum(IV) chloride	$AuCN$	gold(I) cyanide
$SbCl_5$	antimony(V) chloride	$Sn_3(AsO_3)_4$	tin(IV) arsenite
$AgNO_3$	silver(I) nitrate	$PbSO_4$	lead(II) sulfate

$Fe(NH_4)_2(SO_4)_2 \cdot 6H_2O$	iron(II) ammonium sulfate hexahydrate
$(NH_4)_2Mg(CO_3)_2 \cdot 4H_2O$	ammonium magnesium carbonate tetrahydrate
$Al(BrO_3)_3 \cdot 9H_2O$	aluminum bromate nonahydrate
$BeC_2O_4 \cdot 3H_2O$	beryllium oxalate trihydrate
$Cr(C_2H_3O_2)_3 \cdot H_2O$	chromium(III) acetate hydrate (or monohydrate)
$Co_3(AsO_4)_2 \cdot 8H_2O$	cobalt(II) arsenate octahydrate
$CuCr_2O_7 \cdot 2H_2O$	copper(II) dichromate dihydrate
$Na_2CO_3 \cdot 10H_2O$	sodium carbonate decahydrate

Oxidation numbers

The following rules may be used to predict the oxidation number for each of the atoms/ions in a compound or ion. Oxidation numbers are theoretical values of assigned charges for atoms, which treats the atoms as if they were ions.

General rules

(*Always* applicable—no exceptions)

1. For any uncombined element, the oxidation number is zero.

   Examples: Fe, Xe, O_2, H_2 (0 for all of these)

2. For any monatomic ion, the oxidation number is equal to the charge on the ion.

   Examples: Fe^{2+}, F^-, O^{2-}, H^+ (+2, –1, –2, and +1 respectively)

3. For any compound, the sum of the oxidation numbers must be zero.

   Examples: $NaCl = Na^+ + Cl^- = +1 + (-1) = 0$
   $MgF_2 = Mg^{2+} + 2\ F^- = +2 + 2(-1) = 0$
   $Fe_2(Cr_2O_7)_3 = 2\ Fe^{3+} + 6\ Cr^{6+} + 21\ O^{2-} = 2(+3) + 6(+6) + 21(-2) = 0$

4. For any ion, the sum of the oxidation numbers must equal the charge on the ion.

   Examples: $NO_3^{1-} = N^{5+} + 3\ O^{2-} = +5 + 3(-2) = -1$
   $C_2H_3O_2^{1-} = 2C^0 + 3H^+ + 2O^{2-} = 2(0) + 3(+1) + 2(-2) = -1$

Special rules

These rules apply only to certain columns on the periodic table. They often must be used in conjunction with General Rules 3 and 4.

Note: Rules 1–8 deal with the main group elements. Different periodic tables label these in different ways, so two types of numbers are given. The currently accepted method is to use the numbers 1–2 and 13–18 for the representative elements. Older tables used IA-VIIIA (or IB–VIIIB). Each of the rules will use both of these methods with the newer method of labeling given in parentheses.

Note: These are the most common/stable oxidation numbers. There are exceptions in addition to the ones that follow.

Note: These values refer to these elements in compounds. General Rule 1 still applies to the elements in the elemental state.

1. Column IA(1)   +1(except H with a metal, then H =−1)

2. Column IIA(2)   +2

3. Column IIIA(13)   +3(normally, however, +1 possible near the bottom of the table)

4. Column IVA(14)   +4 to −4

5. Column VA(15)   +5 to −3

6. Column VIA(16)   +6 to −2(oxygen is −2 except when combined with F, or in O_2^{2-} or O_2^-)

7. Column VIIA(17)   +7 to −1(fluorine is −1 in compounds)

8. Column VIIIA(18)   +8 to 0(usually only 0)

9. Transition metals IIIB−IIB (3−12)

If the B designation is used, the values may range from +2 to the group number. If the other designation is used, the values may range from +2 to the group number for columns 3−8 and to the group number minus 10 for groups 11 and 12 (treat columns 9 and 10 as if they were 7 and 6 respectively).

Exceptions: Hg_2^{2+} (Hg =+1) and Au^{3+} (Au =+3)

When an element may have a positive or negative oxidation number, it will normally be negative if it is to the upper right on the periodic table, with respect to the other elements in the compound or ion. When it is negative, under these circumstances, it will probably have the most negative of the possible values noted in the previous rules.

Note that in Special Rules 4−8, the range of possible values is always eight. With these elements, the more probable oxidation states may be determined by counting from highest to lowest by twos.

In naming compounds containing a metal with a variable oxidation number (most of the transition metals and those around lead on the periodic table), Stock numbers should be used in the name.

Older nomenclature systems used different names to designate variations in the oxidation number (for example: ferrous and ferric). We will not do so in this workbook.

Examples:

Cr   possible values: 0, +2, +3, +4, +5, +6
Cl   possible values $\underline{-1}$, 0, $\underline{+1}$, +2, $\underline{+3}$, +4, $\underline{+5}$, +6, $\underline{+7}$
                          (underlined preferred)
F   possible values -1, 0
C   possible values -4, -3, -2, -1, 0, +1, +2, +3, +4
Pb   possible values   0, +2, +4   (even though Pb and C are in the same column, Pb only has positive values since it is a metal.)

NaH	Na^+	H^-	
BaO_2	Ba^{2+}	O^-	(barium peroxide)
Cl_2O_7	Cl^{7+}	O^{2-}	
CuS	Cu^{2+}	S^{2-}	
Hg_2Cl_2	Hg^+	Cl^-	
$HgCl_2$	Hg^{2+}	Cl^-	
Pb_3O_4	$Pb^{8/3+}$	O^{2-}	(actually $2\,Pb^{2+} + 1\,Pb^{4+}$)
H_2SO_4	H^+	S^{6+}	O^{2-}
MoO_3	Mo^{6+}	O^{2-}	

Periodic table

1A	2A	3B	4B	5B	6B	7B	8B	8B	8B	1B	2B	3A	4A	5A	6A	7A	8A
1 H 1.00794																	2 He 4.002602
3 Li 6.941	4 Be 9.0121831											5 B 10.811	6 C 12.0107	7 N 14.0067	8 O 15.9994	9 F 18.9984032	10 Ne 20.1797
11 Na 22.989769	12 Mg 24.3050	3	4	5	6	7	8	9	10	11	12	13 Al 26.9815385	14 Si 28.0855	15 P 30.973762	16 S 32.065	17 Cl 35.453	18 Ar 39.948
19 K 39.0983	20 Ca 40.078	21 Sc 44.955908	22 Ti 47.867	23 V 50.9415	24 Cr 51.9961	25 Mn 54.938044	26 Fe 55.845	27 Co 58.933194	28 Ni 58.6934	29 Cu 63.546	30 Zn 65.38	31 Ga 69.723	32 Ge 72.63	33 As 74.92160	34 Se 78.971	35 Br 79.904	36 Kr 83.798
37 Rb 85.4678	38 Sr 87.62	39 Y 88.90584	40 Zr 91.224	41 Nb 92.90637	42 Mo 95.95	43 Tc (98)	44 Ru 101.07	45 Rh 102.90550	46 Pd 106.42	47 Ag 107.8682	48 Cd 112.414	49 In 114.818	50 Sn 118.710	51 Sb 121.760	52 Te 127.60	53 I 126.90447	54 Xe 131.293
55 Cs 132.905452	56 Ba 137.327	57* La 138.9047	72 Hf 178.49	73 Ta 180.9479	74 W 183.84	75 Re 186.207	76 Os 190.23	77 Ir 192.217	78 Pt 195.084	79 Au 196.966569	80 Hg 200.592	81 Tl 204.38	82 Pb 207.2	83 Bi 208.9804	84 Po (209)	85 At (210)	86 Rn (222)
87 Fr (223)	88 Ra (226)	89** Ac (227)	104 Rf (268)	105 Db (270)	106 Sg (269)	107 Bh (270)	108 Hs (277)	109 Mt (278)	110 Ds (281)	111 Rg (281)	112 Cn (285)	113 Nh (286)	114 Fl (289)	115 Mc (289)	116 Lv (293)	117 Ts (294)	118 Og (294)

* Lanthanides

58 Ce 140.116	59 Pr 140.90766	60 Nd 144.242	61 Pm (145)	62 Sm 150.36	63 Eu 151.964	64 Gd 157.25	65 Tb 158.92535	66 Dy 162.500	67 Ho 164.93033	68 Er 167.259	69 Tm 168.93422	70 Yb 173.045	71 Lu 174.9668

** Actinides

90 Th 232.0377	91 Pa 231.03588	92 U 238.02891	93 Np (237)	94 Pu (244)	95 Am (243)	96 Cm (247)	97 Bk (247)	98 Cf (251)	99 Es (252)	100 Fm (257)	101 Md (258)	102 No (259)	103 Lr (262)

Glossary

Absolute zero   (The lowest possible temperature) is 0 K and is the point at which all molecular motion ceases.

Activation energy   The minimum amount of energy necessary for a reaction to occur.

Activity series   Lists metals and hydrogen in order of decreasing ease of oxidation.

Actual yield   The amount actually formed in a reaction.

Alpha particle   A helium nucleus with two protons and two neutrons.

Amorphous solids   Solids that lack extensive ordering of the particles.

Amu   One-twelfth the mass of a carbon atom that contains 6 protons and 6 neutrons (C-12).

Angular momentum quantum number (l)   The quantum number that describes the shape of the orbital.

Anion   A negatively charged ion.

Anode   The electrode at which oxidation is taking place.

Anode compartment   Contains the electrolyte solution in which the anode is immersed.

Aqueous solution   A solution in which water is the solvent.

Atomic number (Z)   The number of protons in the nucleus of an element.

Atomic orbital   The region of space in which it is most likely to find a specific electron in an atom.

Aufbau principle   States that the electrons enter an atom in order of increasing energy.

Avogadro's law   For a gas (P and T constant), the volume is directly proportional to the moles.

Avogadro's number   The number of particles (atoms or molecules or ions) in a mole and is numerically equal to 6.022×10^{23} particles.

Barometer   An instrument for measuring atmospheric pressure.

Base dissociation constant (K_b)   The equilibrium constant associated with the dissociation of a weak base in water.

Bronsted-Lowry acids   Proton (H^+) donors.

Bronsted-Lowry bases   Proton (H^+) acceptors.

Basic   Describes a solution whose $[OH^-] > [H^+]$.

Beta particle   An electron emitted through radioactive decay.

Binary compounds   Compounds that consist of only two elements.

Body-centered cubic unit cell   Has particles located at the corners of a cube and in the center of the cube.

Boiling   The process of going from the liquid state to the gaseous state; vaporization.

Boiling point (b.p.)   The temperature where the vapor pressure of a liquid equals the external pressure.

Bond energy   The energy required (endothermic) to break a bond between two atoms.

Boyle's law   For a gas (n and T constant), the volume is inversely proportional to the pressure.

Buffer capacity   For a solution, it is the maximum amount of acid or base that may be added to a buffer before the solution ceases to be a buffer.

Buffer   A solution that resists changes in pH.

calorie   (not capitalized) The amount of energy needed to raise the temperature of 1 g of water 1°C.

Calorie   (capitalized) The nutritional calorie used to express the energy content of food. It is equal to 1,000 calories.

Calorimetry   The laboratory technique used to measure the heat released or absorbed during a chemical or physical change.

Capillary action   The spontaneous rising of a liquid through a narrow tube against the force of gravity.

Catalyst   A substance that alters the rate of a reaction without being consumed by the reaction.

Cathode   The electrode in an electrochemical cell where reduction takes place.

Cathode compartment   Contains the electrolyte solution in which the cathode is immersed.

Cation   A positively charged ion.

Charles's law   For a gas (n and P constant), the volume is directly proportional to the temperature.

Chemical equilibrium   Reached when two exactly opposite reactions are occurring at the same place, at the same time, and with the same rates of reaction (equal rates in opposite directions).

Chemistry   The study of matter.

Colligative property   A characteristic of a solution that depends only upon the concentration of the solute particles and not on the identity of the solute.

Combination reactions   Involve two or more substances reacting to form one product.

Combined gas law   The combined gas equation relates the pressure, temperature, moles, and volume of a gas.

Combustion reactions   Redox reactions in which the chemical species rapidly combines with oxygen and usually emits heat and light.

Common ion   A species added to an equilibrium from an outside source.

Common ion effect   The alteration of an equilibrium by the addition of a common ion.

Complex   Composed of a central atom, normally a metal, surrounded by atoms or groups of atoms called ligands.

Compound   A pure substance that has a fixed proportion of two or more elements.

Concentrated   A qualitative way of describing a solution that has a relatively large amount of solute in comparison to the solvent.

Concentration   A measure of the amount of solute dissolved in the solvent.

Conjugate acid-base pair   An acid–base pair that differs by only a single H^+.

Coordinate covalent bonds   Covalent bonds in which one of the atoms furnishes both electrons for the bond.

Covalent bond   A bond where one or more electron pairs are shared between two atoms.

Critical point   The highest temperature and pressure for which it is possible to distinguish between the liquid and gas phases.

Crystal lattice   A three-dimensional structure that describes crystalline solids.

Crystalline solids   Display a very regular ordering of the particles (atoms, molecules, or ions) in a three-dimensional structure called the crystal lattice.

Dalton's law   The total pressure of a mixture equals the sum of the partial pressures of the gases in the mixture, or $P_{\text{Total}} = P_A + P_B + P_C + \ldots$.

Decomposition reactions   One substance reacts to form two or more products.

Dilute   A qualitative term that refers to a solution that has a relatively small amount of solute in comparison to the amount of solvent.

Dipole-dipole force   The intermolecular force between polar molecules.

Displacement reactions   Involves two elements changing places. In general, one element enters a compound to push (displace) another element.

Double displacement (replacement) or metathesis reaction   A chemical reaction where at least one insoluble or unionized product is formed from the mixing of two solutions.

Ductility   is the ability of a piece of metal to be drawn to a wire.

Effective nuclear charge   The overall attraction that an electron experiences. This is less than the actual nuclear charge because other electrons interfere (shield) the nuclear charge from the electron being considered.

Electrochemical cell   Voltaic cells and electrolysis cells use indirect electron transfer to produce electricity by a redox reaction, or they use electricity to produce a desired redox reaction.

Electrochemistry   The study of chemical reactions involving electricity.

Electrode   That part of the electrochemical cell that transfers the electrons that are involved in an electrochemical cell.

Electrode compartment   The solutions in which the electrodes are immersed.

Electrolysis   A reaction in which electricity is used to decompose a compound in terms of an electrolytic cell.

Electrolyte   A substance that separates into ions in solution.

Electrolytic cells   Use electricity from an external source to force a desired redox reaction.

Electron affinity   The energy change when an electron is added to a gaseous atom in its ground state.

Electron capture   A type of radioactive decay where the nucleus "captures" an electron from the electron cloud and emits an X-ray, and possibly one or more γ rays.

Electron cloud   A volume of space in which the probability of finding a particular electron is high.

Electronegativity (EN)   A measure of the attractive force that an atom exerts on a bonding pair of electrons in a compound.

Electron configuration   A condensed way of representing the arrangement of electrons in an atom.

Empirical formula   The simplest formula of a compound.

Endothermic   A reaction that absorbs energy (heat) from the surroundings.

Endpoint   The experimental end of a titration.

Energy   The capacity or ability to do work.

Enthalpy   The enthalpy change, ΔH, is heat content of the system.

Entropy (S)   A thermodynamic quantity representing the system's energy that is unavailable for conversion to work. Some sources consider entropy to be a measure of the disorder of a system; however, this definition may lead to some erroneous conclusions.

Equilibrium constant   The quantity calculated when the equilibrium concentrations of the chemical species are substituted into the reaction quotient.

Equivalence point   The theoretical end of a titration.

Excited state   An energy state of an atom where one or more electrons is not as close to the nucleus as possible.

Exothermic   A reaction that releases energy (heat) to its surroundings.

Face-centered cubic unit cell   Has particles at the corners and one in the center of each face of the cube, but not in the center of the cube.

First law of thermodynamics   The total energy of the universe is constant.

Formation constant The equilibrium constant for the formation of a complex ion from a metal ion and ligands.

Frequency (v) The number of waves that pass a point per second.

Galvanic (voltaic) cell An electrochemical cell where a spontaneous chemical reaction generates electricity.

Gamma emission A radioactive decay process in which high-energy, short-wavelength photons that are similar to X-rays are given off.

Gas A state of matter that has neither definite shape nor volume.

Gibbs free energy (G) A thermodynamic function that combines the enthalpy, entropy, and temperature. ΔG is the best indicator of whether or not a reaction will be spontaneous.

Graham's law The speed of gas diffusion (mixing of gases due to their kinetic energy) or effusion (movement of a gas through a tiny opening) is inversely proportional to the square root of the gases' molecular mass.

Ground state The lowest energy state where all the electrons are as near the nucleus as possible.

Group A vertical column on the periodic table; family.

Half-life ($t_{1/2}$) The amount of time that it takes for a reactant concentration to decrease to one-half its initial concentration.

Halogens Elements in Group 17 (7A) on the periodic table.

Heat capacity The quantity of heat needed to change the temperature of the system by 1 K (or 1°C).

Henderson-Hasselbalch equation Can be used to calculate the pH or pOH of a buffer.

Hess's law Combines thermochemical equations to produce new thermochemical equations.

Hund's rule When dealing with a set of orbitals, the electrons enter the orbitals individually with the same spin (spins parallel).

hybrid orbitals Atomic orbitals formed as a result of the mixing of the atomic orbitals of the atoms involved in covalent bonding.

Hydrogen bonding Occurs when a hydrogen atom is bonded to a very electronegative atom. The only atoms sufficiently electronegative are atoms of fluorine, oxygen, and nitrogen.

Ideal gas A gas that obeys the five postulates of the kinetic molecular theory of gases.

Ideal gas equation The equation that relates the temperature, volume, pressure, and amount of a gas, and has the mathematical form of $PV = nRT$.

Indicator A substance that indicates the end of a reaction, usually through a change in color. The point in a titration where the indicator signals that the reaction is over is the endpoint.

Inert (inactive) electrode An electrode in an electrochemical cell that does not take part in the redox reaction.

Inner transition elements The elements in the two horizontal groups that have been pulled out of the main body of the periodic table where the f-orbitals are being filled.

Integrated rate law Ties the rate law to the time involved in a reaction.

Intermolecular forces Attractive or repulsive forces between molecules.

Ion-dipole force Attractive forces that occur between ions and polar molecules.

Ion-induced dipole forces Attractive forces that occur between an ion and a nonpolar molecule.

Ionic bond The result of the mutual attraction of cations and anions.

Total ionic equation The equation that shows the soluble reactants and products in the form of separated ions.

Ionic solid Have their lattices composed of ions held together by the attraction of opposite charges of the ions.

Ionization energy (IE) The energy required to remove an electron from a gaseous atom in its ground state.

Isotope An atom of an element with a specific number of neutrons.

Joule (J)   The SI unit of energy.

Kinetic energy   Energy of motion.

Kinetics   The study of reaction rates.

Law of conservation of matter   In ordinary chemical reactions matter is neither created nor destroyed.

Le Châtelier's principle   If a stress is applied to a system at equilibrium, the system will respond to counter the stress and go to a new equilibrium.

Lewis structure   A structural formula that represents an atom of an element and its valence electrons.

Limiting reactant   Controls (limits) a chemical reaction.

Liquid   A state of matter that has a definite volume but no definite shape.

Magnetic quantum number (m_l)   Describes the orientation of the orbital around the nucleus.

Malleability   The ability of a piece of metal to be flattened by using, for example, a hammer.

Mass number (A)   The sum of the protons and neutrons in an atom.

Mechanism   The sequence of steps that a reaction undergoes in going from reactants to products.

Melting point (m.p.)   The temperature at which a solid converts into the liquid state.

Metallic bonding   The electrons of the atoms are delocalized and are free to move throughout the entire solid.

Metallic solids   Solids that are held together by metallic bonds.

Metalloids   A group of elements that have properties of both metals and nonmetals.

Metals   Normally solids (mercury being an exception), shiny, and good conductors of heat and electricity. Chemically, metals tend to lose electrons in reactions.

Metathesis reaction   A double displacement, where two compounds are reacting. Two exchanges take place. The formation of a precipitate (solid) when two solutions are mixed is a common example.

Molality (m)   The moles of solute per kilogram of solvent.

Molar mass   The mass in grams of 1 mole of a substance.

Molarity (M) or sometimes [ ]   The moles of solute per liter (dm^3) of solution.

Mole (mol)   The amount of a substance that contains the same number of particles as atoms in exactly 12 g of carbon-12.

Molecular equation   An equation in which all reactants and products are shown in the undissociated form.

Molecular formula   The actual formula that shows which elements are in the molecule and the actual number of atoms of each element.

Molecular solids   Solids that are held together by van der Waals forces.

Molecule   A covalently bonded species.

Nernst equation   The equation that calculates the cell potential of a electrochemical cell that is not under standard conditions.

Net ionic equation   A total ionic equation with all the spectator ions removed.

Network covalent solids   Solids that are held together by covalent bonds.

Neutral   7.00 on the pH scale (where $[H^+] = [OH^-]$).

Neutralization reactions   Involves an acid reacting with a base to form a salt, and in most cases water. This is a special type of metathesis reaction.

Noble gases   Elements in Group 18 (8A) on the periodic table. They are very unreactive owing to their filled valence shell.

Nonelectrolyte   A substance that does not separate into ions in solution.

Nonmetals   These have properties that are generally the opposite of metals. Some are gases, are poor conductors of heat and electricity, are neither malleable nor ductile, and tend to gain or share electrons in their chemical reactions.

Nucleus   A dense core of protons and neutrons.

Octet rule Atoms lose, gain, or share electrons in order to achieve a filled valence shell, to complete their octet.

Orbital An orbital or wave function is a quantum mechanical mathematical description of the location of electrons. The electrons in a specific subshell are distributed among these volumes of space of equal energies.

Order The exponent in the rate equation that indicates what effect a change in concentration of that particular reactant species will have on the reaction rate.

Osmotic pressure The pressure generated through osmosis.

Oxidation The loss of electrons.

Oxidation numbers Theoretical values of assigned charges for atoms, which treats the atoms as if they were ions.

Pascal The SI unit of pressure ($kg\ m^{-1}\ s^{-2}$).

Percent yield (% yield) The actual yield divided by the theoretical yield, with the result multiplied by 100%.

Period A horizontal row on the periodic table with elements having consecutive atomic numbers.

Phase changes Changes of state.

Phase diagram Summarizes the phases present under a certain set of conditions.

Pi (π) bonds Bonds resulting from the overlap of atomic orbitals on both sides of a line connecting two atomic nuclei.

Polar covalent bonds Covalent bonds in which there is an unequal sharing of the bonding pair of electrons.

Polyprotic acids Acids that can donate more than one proton.

Potential energy Stored energy.

Positron An electron with a positive charge instead of a negative charge. As the opposite of an electron, a positron is a form of antimatter.

Precipitate An insoluble product that forms from a solution; the formation of a solid from ions in solution.

Precipitation reactions Reactions that involve the formation of an insoluble compound, a precipitate, from the mixing of two soluble compounds.

Pressure The force exerted per unit of surface area.

Principal quantum number (n) Describes the size of the orbital, its energy, and relative distance from the nucleus.

Quantized There could be only certain distinct energies associated with a state of the atom.

Quantum numbers Used to describe each electron within an atom corresponding to the orbital size, shape, and orientation in space.

Radioactivity The spontaneous decay of an unstable isotope to a more stable one.

Rate constant (k) A proportionality constant that appears in the rate law and relates the concentration of reactants to the speed of reaction.

Rate law Relates the speed of reaction to the concentration of reactants and has the form: Rate = $k[A]^m[B]^n$. . . . where k is the rate constant and m and n are the orders of reaction with respect to that specific reactant.

Reactants The starting materials in a chemical reaction, which get converted into different substances called products.

Reaction quotient (Q) The numerical value that results when nonequilibrium concentrations are inserted into the equilibrium expression. When the system reaches equilibrium, the reaction quotient becomes the equilibrium constant.

Redox reactions Chemical reactions in which electrons are lost and gained.

Reduction The gain of electrons.

Resonance A way of describing a molecular structure that cannot be represented by a single Lewis structure. Several different Lewis structures are used, each differing only by the position of the electrons.

Salt bridge Contains an electrolyte solution that has ions that do not react, but with charges available to keep the charges in each compartment balanced.

Saturated solution Contains the maximum amount of solute that will normally dissolve.

Second law of thermodynamics The second law of thermodynamics says that for any spontaneous process, the entropy change is positive.

Shells The electrons in an atom are found in various energy levels or shells that are located at different distances from the nucleus.

SI system (Système International) The system of units used in science, which is related to the metric system.

Sigma (σ) bonds Bonds that have the orbital overlap on a line drawn between the two nuclei. All single bonds between two atoms are σ bonds.

Simple cubic unit cell Has particles located at the corners of a simple cube.

Displacement (replacement) reactions Reactions in which atoms of an element replace the atoms of another element in a compound.

Solid A state of matter that has both a definite shape and a definite volume.

Solubility product constant (K_{sp}) The equilibrium constant associated with sparingly soluble salts and the product of the ionic concentrations, each one raised to the power of the coefficient in the balanced chemical equation.

Solute The component of the solution that is present in the smallest amount.

Solution A homogeneous mixture composed of solvent and one or more solutes.

Solvent That component of a solution that is present in largest amount.

Specific heat (heat capacity) (c) The quantity of heat needed to raise the temperature of 1 g of the substance by 1 K (or °C).

Spectator ions Ions that appear unchanged on both side of the reaction arrow.

Speed of light (c) The speed at which all electromagnetic radiation travels in a vacuum, 3.0×10^8 m/s.

Spin quantum number (m_s) Indicates the direction the electron is spinning.

Standard cell potential ($E°$) The potential (voltage) associated with an electrochemical cell at standard conditions.

Standard enthalpy of formation ($\Delta H_f°$) The change in enthalpy when 1 mole of a substance formed from its elements with all substances are in their standard states.

Standard molar entropies ($S°$) For elements and compounds, the entropies associated with 1 mole of a substance in its standard state.

Standard reduction potentials The voltage associated with a half-reaction shown in the form of reduction.

Stoichiometry Examines the mole relationships between substances.

Strong acid An acid that ionizes completely in solution.

Strong base A base that ionizes completely in solution.

Strong electrolytes Electrolytes that completely ionize or dissociate in solution.

Sublimation Going directly from the solid state to the gaseous state without ever having become a liquid.

Subshells Within the shells, the electrons are grouped in subshells of slightly different energies.

Supersaturated solution Contains more solute than a saturated solution.

Surface tension The amount of force that is required to break through the molecular layer at the surface of a liquid.

Surroundings The part of the universe that is not the system being studied.

System The part of the universe that we are studying.

Ternary compounds Those containing three (or more) elements.

Theoretical yield The amount calculated to form in a reaction.

Thermochemistry The part of thermodynamics dealing with the changes in heat that take place during chemical processes, for example, ΔH_{fusion}.

Thermodynamics   The study of energy and its transformations.

Titration   A laboratory procedure in which a solution of known concentration is used to determine the concentration of an unknown solution.

Transition elements   Groups 3–12 (3B–2B) on the periodic table contain the transition elements where the *d*-orbitals are being filled.

Triple point   A fixed point in a phase diagram when three phases are in equilibrium. The three phases are not necessarily solid, liquid, and gas.

Unit cell   The smallest repeating unit in a crystal lattice.

Valence bond theory (VBT)   Describes covalent bonding as the overlap of atomic orbitals to form a new set of orbitals, hybrid orbitals.

Valence electrons   The electrons in the outermost energy level (outermost shell).

Van't Hoff factor (*i*)   The number of ions generated by a strong electrolyte.

Vapor pressure   The pressure exerted by the gaseous molecules that are at equilibrium with a liquid in a closed container.

Viscosity   The resistance to flow of a liquid.

VSEPR theory   The VSEPR (valence-shell electron-pair repulsion) theory says that the electron pairs around a central atom will try to get as far as possible from each other in order to minimize the repulsive forces. This theory is used to predict molecular geometry.

Water dissociation constant (K_w)   The water dissociation constant is the equilibrium constant associated with the ionization of pure water or a solution.

Wave function   A mathematical description of the electron's behavior.

Wavelength (λ)   The distance between two identical points on a wave.

Weak acid   An acid that only partially ionizes in solution.

Weak base   A base that only partially ionizes in solution.

Weak electrolytes   Electrolytes that only partially ionize or dissociate in solution.

Answers to exercises

1   What is chemistry?

1.1   Terminology

1. Extensive   2. Intensive   3. Intensive   4. Intensive   5. Extensive

1.2   Precision and accuracy

1. Student A (the values are closer together)   2. Impossible to tell

1.3   Basic concepts

1. 80 lb (40% of 200 lb)   2. 90 kg   3. Both are the same.   4. Pan A
5. Question 4   6. Question 3   7. More heat is needed to boil more water.
8. Water has a boiling point regardless of quantity.

2   Chemistry and numbers

2.1   Units

1. Meter, m   2. Second, s   3. Ampere, A

2.2   Modifying units

1. Meter $\times$ meter $= m \times m = m^2$ 2. Meter $\times$ meter $\times$ meter $= m \times m \times m = m^3$

3. $\dfrac{\text{mass}}{\text{volume}} = \dfrac{kg}{m \times m \times m} = \dfrac{kg}{m^3}$

2.3   Types of numbers

1. Measured   2. Exact   3. Exact   4. Measured

2.4   Significant figures

1. 4   2. 3   3. 4

2.5   Significant figures in mathematical operations

1. 124.5 in.   2. $5.3 \times 10^2 \dfrac{kg}{m^3}$ 3. 88 cm^2 4. 2.5×10^4 mm^3 5. $0.030 \dfrac{g}{cm^3}$

3   What is matter?

3.1   Atomic theory

1. 11   2. No. It is an ion.   3. These are two different isotopes of the same element.

3.2 Locating the components of atoms

1. The electrons are not buried in the nucleus; therefore, their location brings them near the surface of an atom where it is easier to remove or add an electron.
2. Protons are located in the nucleus of the atom which is difficult to reach in order to change the number of protons.

3.3 Modern atoms

1. The protons are in the nucleus at center of the atom.   2. The electrons are orbiting the nucleus.

3.4 Atomic weight (mass)

1. 35.45 amu   2. 24.305 amu   3. 65.40 amu

3.5 Introduction to the periodic table

1. 157.25 amu   2. 81   3. The elements in the same column: F, Cl, I, At, and Ts
4. The element with the higher atomic number (I) has a lower atomic mass than the element with the lower atomic number (Te).   5. H

3.6 Chemical substances

1. Pure substance, compound   2. Mixture   3. Pure substance, element

4 Nomenclature

4.1 Special names

1. NH_3 2. Methane

4.2 Nonmetal-nonmetal

1. Chlorine trifluoride   2. Hydrogen sulfide   3. Dichlorine tetroxide   4. CO_2 5. XeF_4 6. P_4O_{10}

4.3 Metal-nonmetal

1. Sodium phosphide
2. K_2O 3. AlN   4. TiO_2 5. Chromium(II) chloride   6. Cobalt(III) oxide

4.4 Compound containing polyatomic ions

1. K_3PO_4 2. Calcium nitrate   3. $(NH_4)_2SO_4$ 4. $FeSO_3$ 5. Chromium(III) sulfate   6. $Ni(NO_3)_2$

4.5 Acids

1. Hydrobromic acid   2. HI(aq)   3. Hydrogen selenide   4. Phosphoric acid   5. $HMnO_4$
6. Dichromic acid

5 Chemical equations

5.1 Terminology

1. $Al_2(SO_4)_3(aq) + 3\ H_2(g)$ 2. $2\ Al(s) + 3\ H_2SO_4(aq)$

5.2 Balancing equations

1. _3_ $Ba(OH)_2(aq)$ + _2_ $H_3PO_4(aq) \rightarrow$ _1_ $Ba_3(PO_4)_2(s)$ + _6_ $H_2O(l)$
2. _3_ $Mg(s)$ + _2_ $FeCl_3(s) \rightarrow$ _2_ $Fe(s)$ + _3_ $MgCl_2(s)$
3. _2_ $Si_4H_{10}(l)$ + _13_ $O_2(g) \rightarrow$ _8_ $SiO_2(s)$ + _10_ $H_2O(g)$
4. _1_ $P_4(s)$ + _5_ $O_2(g) \rightarrow$ _1_ $P_4O_{10}(s)$
5. _1_ $(NH_4)_2Cr_2O_7(s) \rightarrow$ _1_ $N_2(g)$ + _1_ $Cr_2O_3(s)$ + _4_ $H_2O(g)$

6   Light and matter

6.1   Light

1. $6.00 \times 10^{16}\, s^{-1}$ 2. $9.23 \times 10^{-10}\, m$ 3. $2.15 \times 10^{-16}\, J$ 4. $3.98 \times 10^{-17}\, J$ 5. $1.14 \times 10^{-16}\, J$

6.2   Quantized energy and the Bohr atom

1. 2.044×10^{-18} joules   2. 1.64×10^{-18} joules

6.3   Wave behavior of matter

1. $5.27 \times 10^{-11}\, m$ 2. $3.43 \times 10^{-8}\, m$ 3. $5.03 \times 10^{-37}\, m$

7   Quantum chemistry

7.1   Quantum numbers

1. l may equal 0, 1, or 2.
2. m_l may equal −2, −1, 0, +1, or +2
3. m_s may equal −1/2 or +1/2
4. l may be 0 or 1. If $l = 0$, then $m_l = 0$, and if $m_l = 1$, then $m_l = -1$, 0, or +1
5. 8 electrons (Determine this by writing the sets of quantum numbers until no more are possible for $n = 2$.)
6. The sets are as follows:

n	l	m_l	m_s
2	1	−1	−1/2
2	1	−1	+1/2
2	1	0	−1/2
2	1	0	+1/2
2	1	+1	−1/2
2	1	+1	+1/2

The order you place these sets is not important.
(Note 2p, says that for every electron $n = 2$ and $l = 1$ (p))

7.

Electron	n	l	m_l	m_s
1	1	0	0	+1/2
2	1	0	0	−1/2
3	2	0	0	+1/2
4	2	0	0	−1/2
5	2	1	+1	+1/2

The order may be varied. For electron 5, m_l could also be 0 or −1, and m_s may be −1/2.

8. Each electron in the shell must have a different set of the four quantum numbers. There are only two sets possible for electrons in the first shell ($n = 1$). (Pauli exclusion principle)
9. A 2d subshell would have $n = 2$ and $l = 2$; however, if $n = 2$, l can only equal 0 or 1 never 2. (Pauli exclusion principle)
10. Lower energy subshells fill before higher energy subshells. (Aufbau principle)

8 Atomic structure

8.1 Atomic orbitals

1. $4s$, $4p$, $4d$, and $4f$
2. 9 (one $3s$, three $3p$, and five $3d$)
3. There are three correct answers. Any of the following lines are one of the correct answers.

$n = 2$	$l = 1$	$m_l = -1$
$n = 2$	$l = 1$	$m_l = 0$
$n = 2$	$l = 1$	$m_l = +1$

4. 2 (for any of the answers in question 3, there can be only the electrons with $m_s = +1/2$ and $m_s = -1/2$, and no others.)
5. 18 electrons (from question 4, 2 electrons per orbital)
6. 32 electrons (2 electrons in each orbital, with 4s having 1 orbital, 4p having 3 orbitals, 4d having 5 orbitals, and 4f having 7 orbitals OR $[1 + 3 + 5 + 7] \times 2$)

8.2 Many electron atoms

1. The Bohr model does not work for atoms other than hydrogen because of interactions between the electrons.
2. It is expected to be equal to the nuclear charge (+1) because there are no core electrons to shield the nucleus.
3. They should be similar even though aluminum has 8 more protons (+8) in its nucleus, because it also has 8 more electrons in its core to shield the increased positive charge.
4. 10 (two in the $1s$, two in the $2s$, and 6 in the $2p$)

9 Electronic structure of atoms

9.1 Electron configurations

1. (a) 2, (b) 14, (c) 27, (d) 64, (e) 97
2. (a) 2, (b) 18, (c) 44, (d) 36, (e) 54
3. 48
4. $1s^2 2s^2 2p^6 3s^2 3p^1$
5. $1s^2 2s^2 2p^6 3s^2 3p^6 4s^2 3d^8$
6. $1s^2 2s^2 2p^6 3s^2 3p^6 4s^2 3d^{10} 4p^6 5s^2 4d^{10} 5p^3$
7. Hund's rule only applies to partially filled subshells; therefore, only the $2p$ orbitals need to be considered.

n	l	m_l	m_s
2	1	-1	$+1/2$
2	1	0	$+1/2$
2	1	$+1$	$+1/2$

   (The m_l values may be in any order, and all the m_s values could be $-1/2$.)
8. All atoms with an odd number of electrons
9. 5
10. 7 (a half-filled f subshell; to get to the f subshell, all preceding subshells must be filled)
11. $1s^2$
12. $1s^2 2s^2 2p^6 3s^2 3p^6$
13. $1s^2 2s^2 2p^6 3s^2 3p^6 3d^3$
14. The electron configurations of all three end with $s^2 p^5$.
15. The electron configurations of all three end with $s^2 p^6$.
16. Ar
17. Ne
18. The electron configurations of these elements (the noble gases) are especially stable.
19. All end with $s^2 p^6$ (an octet).

10 Periodic properties

10.1 Atomic radii

1. The effective nuclear charge increases toward the right in any period. (Note, saying the trend is to increase to the right is not an explanation.)
2. F < Cl < Br < I < At
3. P < Si < Al < Mg < Na
4. Be < Al < Ca < Sb < Mo
5. Ru because there are exceptions in the transition elements.

10.2 Ionization energy

1. The effective nuclear charge increases toward the right in any period. (Note, saying the trend is to increase to the right is not an explanation.)
2. Po < Te < Se < S < O
3. B < C < O < F < Ne
4. Be < Al < Ca < Sb < Mo
5. N because it has a half-filled subshell.

10.3 Electron affinity

1. The effective nuclear charge increases toward the top in any column. (Note, saying the trend is to increase to the top is not an explanation.)
2. Bi < Sb < As < P < N
3. Li < B < C < O < F
4. Mo < Sb < Ca < Al
5. Ne because it has a filled subshell.

11 Types of chemical reactions

11.1 Combination reactions

1. $2 C(s) + 1 O_2(g) \rightarrow 2 CO(g)$
2. $1 P_4(s) + 10 Cl_2(g) \rightarrow 4 PCl_5(l)$

11.2 Decomposition reactions

1. $2 XeO_3(s) \rightarrow 2 Xe(g) + 3 O_2(g)$
2. $4 C_3H_5N_3O_9(l) \rightarrow 12 CO_2(g) + 6 N_2(g) + 10 H_2O(g) + 1 O_2(g)$

11.3 Displacement reactions

1. $1 Ca(s) + 2 HCl(g) \rightarrow 1 CaCl_2(s) + 1 H_2(g)$
2. $2 Al(s) + 3 H_2SO_4(aq) \rightarrow 1 Al_2(SO_4)_3(aq) + 3 H_2(g)$
3. $3 CoCl_2(aq) + 2 Al(s) \rightarrow 3 Co(s) + 2 AlCl_3(aq)$

11.4 Metathesis reactions

1. $3 SrCl_2(aq) + 1 Fe_2(SO_4)_3(aq) \rightarrow 3 SrSO_4(s) + 2 FeCl_3(aq)$
2. $3 Ba(NO_3)_2(aq) + 2 H_3PO_4(aq) \rightarrow 1 Ba_3(PO_4)_2(s) + 6 HNO_3(aq)$

11.5 Neutralization reactions

1. $2 HCl(aq) + 1 Ca(OH)_2(aq) \rightarrow 1 CaCl_2(aq) + 2 H_2O(l)$
2. $3 H_2SO_4(aq) + 2 Al(OH)_3(s) \rightarrow 1 Al_2(SO_4)_3(aq) + 6 H_2O(l)$
3. $1 H_3PO_4(aq) + 3 NH_3(aq) \rightarrow 1 (NH_4)_3PO_4(aq)$

11.6 Combustion reactions

1. $1 P_4(s) + 5 O_2(g) \rightarrow 1 P_4O_{10}(s)$
2. $2 C_4H_{10}(g) + 9 O_2(g) \rightarrow 8 CO_2(g) + 10 H_2O(g)$
3. $2 H_2O(l) + 2 F_2(g) \rightarrow 4 HF(g) + 1 O_2(g)$

Redox reactions

1. $3\ Cu(s) + 8\ HNO_3(aq) \rightarrow 3\ Cu(NO_3)_2(aq) + 2\ NO(g) + 4\ H_2O(l)$
2. $1\ K_2Cr_2O_7(aq) + 6\ FeCl_2(aq) + 14\ HCl(aq) \rightarrow 2\ CrCl_3(aq) + 6\ FeCl_3(aq) + 2\ KCl(aq) + 7\ H_2O(aq)$

12 Ionic bonds

12.1 Bond polarities and electronegativity

1. 1.54   2. 1.74   3. Ionic   4. Covalent   5. Covalent (actually polar covalent)

12.2 Lewis symbols of atoms

There are alternate ways of correctly drawing the Lewis symbols shown.

1. 4   2. Na·   3. ·Äl   4. ·N̈·   5. ·S̈:

12.3 Ionic bonding

1. K^+ 2. Ba^{2+} 3. $\left[:\ddot{P}: \right]^{3-}$ 4. $\left[:\ddot{O}: \right]^{2-}$

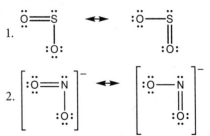
5.

13 Covalent bonds and Lewis structures

13.1 Covalent bonding

1. Yes, because both are nonmetals.
2. No, because one is a metal, and the other is a nonmetal.

13.2 Lewis structures

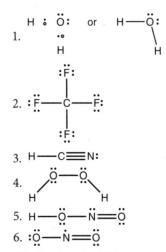

1.

2.

3. H—C≡N:

4.

5. H—Ö—N̈=Ö

6. :Ö—N̈=Ö

Notice that the N does not get an octet. There are only 17 valence electrons available, the Lewis structure may not include any other number of electrons. (The double bond could be on the other side.) N comes up short because it is less electronegative than O.

13.3 Resonance

1.

2.

3.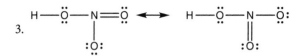

13.4   Exceptions to the octet rule

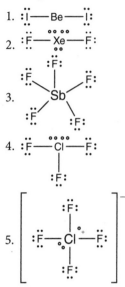

1. :Ï—Be—Ï:

2. :F̈—Xe—F̈:

3. (Sb structure with five F)

4. :F̈—Cl—F̈:

5. [ClF₅ structure]⁻

13.5   Bond energies

1. −549 kJ   2. −1,285 kJ   3. −107 kJ

14   Intermolecular forces

14.1   Types of intermolecular forces

1. C_8H_{18} 2. HCl   3. H_2S, HCN, $C_2H_2F_2$ 4. HF

14.2   Consequences of intermolecular forces

1. CH_4 2. NO   3. O_2 4. C_8H_{18} 5. HCl   6. HF   7. NaCl   8. $NaNO_2$
9. The structure (shape) of the molecule
10. It must be linear.   11. It cannot be linear.   (It is bent.)

15   Molecular structure (VSEPR)

15.1   Electron groups

1. 4 and 0   2. 4 and 1   3. 4 and 2   4. 5 and 1   5. 5 and 3

15.2   Molecular geometry and polarity

1. Tetrahedral and nonpolar   2. Linear and polar   3. Trigonal bipyramidal and nonpolar
4. Octahedral and nonpolar   5. Tetrahedral and polar

15.3   Lone pairs

1. Linear and nonpolar   2. Bent (angular) and polar   3. Bent (angular) and polar (The same is true around the other central atom [O].)   4. Square pyramid and polar   5. Trigonal planar and polar
6. The geometry around each O is bent (angular) and each is polar.   7. Tetrahedral and polar   8. Square planar and nonpolar   9. T-shaped and polar   10. Linear and nonpolar

16 Molecular structure (VBT)

16.1 Hybridization

1. sp^3 2. sp^3 3. sp^3 4. sp^2 5. sp^2 6. sp 7. sp^3d 8. sp^3d^2 9. sp^3 10. sp^2 11. sp^3d 12. sp^3
13. sp^3d^2 14. sp^3d
15. The XF_6^{2-} ion requires sp^3d^2 hybridization. Second period elements like C do not have d-orbitals available to hybridize. All elements in the third period or below, such as Si, do have d-orbitals to hybridize.
16. The C with the O atoms attached is sp^2 and the C without any O atoms attached is sp^3.
17. In H_2O hybridization of the orbitals around the O are sp^3, which ideally gives a bond angle of 109.5°. The two lone pairs on the O decreases this angle slightly. In H_2S, hybridization is not needed around the sulfur atom because it is larger than an O atom. Without hybridization the bond angle should be about 90°.

17 Molar relationships

NOTE: If your units and significant figures do not match those in the following answers, your answer is no more than partially correct.

17.1 Finding moles

1. 1.412×10^{-14} mol people (see Example 17.2) 2. 7.2×10^{18} grains of sand (see Example 17.1)
3. 0.1261 mol $KClO_3$ (see Example 17.3) 4. 488.0 g O_2 (see Example 17.4)
5. 182 mol C (see Example 17.5) 6. 9.18 mol CO_2 (see Example 17.6)

17.2 Mole conversions

1. 1.28×10^3 g CO_2 2. 9.64×10^{23} SO_2 molecules 3. 199 g O_2 4. 1.778 g $H_2C_2O_4$
5. 1.524×10^{22} H_2O molecules 6. 3.47 g $MnSO_4$ 7. 5.7852×10^{23} CO_2 molecules
8. 0.44998 g H_2O 9. 32.64 g $CaCO_3$ 10. 13.13 g CO 11. 3.04 g of H_2O 12. 6.06 g CO_2

18 Stoichiometry

18.1 Stoichiometry

1. 3.70 g O_2(g) 2. 48.62 g Cl_2(g) 3. 154 g HCl(g) 4. 97.43 g H_2SO_4 (aq) 5. 35.62 g $H_2C_2O_4$ (aq)

18.2 Limiting reactants

1. 41.14 g HCl(g) 2. 280 g CO(g) 3. 23.48 g $MnSO_4$(s) 4. 10.06 g $MnSO_4$(s) 5. 12.42 g H_2SO_4(l)

18.3 Actual and percent yield

1. 92.7% 2. 96.3% 3. 70.84% 4. 65.5% 5. 95.8%

18.4 Empirical formulas

1. B_2O_3 2. Fe_3O_4 3. Empirical: HO; actual H_2O_2 4. Empirical: CH_2O; actual $C_6H_{12}O_6$
5. $C_{10}H_5N_3O_6$

19 Molarity

19.1 Molarity

1. 0.300 M KNO_3 2. 0.200 L 3. 0.4500 mol HCl 4. 125.0 mL 5. 0.4800 M KCl

19.2 More molarity calculations

1. 45.96 g $KClO_3$ 2. 0.05551 M $C_6H_{12}O_6$ 3. 1.75 M $HC_2H_3O_2$ 4. 0.2077 M $MgSO_4$
5. 45.04 g $HC_2H_3O_2$ 6. 21.99 mL

19.3   Titrations

1. 0.3500 M KOH   2. 0.1214 M HCl   3. 0.06067 M H_3PO_4 4. 0.09000 M $Ca(OH)_2$
5. 0.01875 mol $Ca(OH)_2$ 6. 0.4250 g NaOH

20   Gases

20.1   Pressure

1. 76.5 kPa   2. 143.6 kPa

20.2   Temperature

1. 118 K   2. 25°C

20.3   Gas laws

1. 58.3 atm   2. 888 torr   3. 103°C   4. 1.04 atm

20.4   More gas laws

1. 1.4 atm N_2 and 0.35 for O_2 2. 146 g mol^{-1}

20.5   Gas stoichiometry

1. 70.78 L   2. 8.90 L   3. 21.5 L   4. 926.2 torr

20.6   Kinetic molecular theory

1. The second assumption because, for example, if the volume decreases, the pressure will increase due to the fact that the particles do not need to travel as far to push on the sides of the container and increase the pressure.
2. The fourth assumption because Graham's law assumes that at the same temperature, lighter particles travel faster than heavier molecules (lighter molecules effuse/diffuse faster than heavier ones). This is required to keep the kinetic energies the same.

20.7   Real gases

1. H_2O 2. O_3

21   Solids and liquids

21.1   Some properties of liquids

1. H_2O because its intermolecular forces (hydrogen bonds) are stronger than those in CCl_4 (London dispersion forces) or those in SO_2 (dipole-dipole forces).
2. $C_{15}H_{32}(l)$ because larger molecules do not move (flow) as easily as smaller molecules. (In addition, larger molecules tend to have greater London dispersion forces.)

21.2   Phase changes

1. NaCl(s) because ionic bonds are stronger than hydrogen bonds ($H_2O(l)$) or London dispersion forces ($N_2(s)$)
2. He(l), because, with fewer electrons, it has the weakest London dispersion forces of the three.

21.3   Vapor pressure

1. The pressure in a pressure cooker is higher than the pressure outside. A higher pressure increases the boiling point above that of outside.
2. At a higher temperature, according to KMT, the average kinetic energy of the particles is greater, which leads to more molecules having sufficient energy to escape the liquid (evaporate).
3. The rate of evaporation depends on the particles with sufficient kinetic energy being at or near the surface of the liquid. Spilling the water out of the glass spreads it out more, and more molecules will have an opportunity to evaporate.

4. The temperature will decrease, because as more of the particles with high kinetic energy evaporate, the average kinetic energy of the remaining particles is lower. A lower average kinetic energy equates to a lower temperature.

21.4 Phase diagrams

1. The triple point
2. The triple point is above "normal" pressure (around 1 atm); therefore, on warming, solid CO_2 sublimes instead of melting first. Since the CO_2 does not melt, there is no liquid CO_2 to make it wet.
3. The CO_2 would remain solid. The H_2O would melt.

21.5 Types of solids

1. $SiO_2(s)$ because it is a covalent solid. $CsI(s)$ is an ionic solid; however the presence of the large Cs^+ and I^- ions means that its melting point will be relatively low when compared to other ionic solids. $CH_3OH(s)$ is a molecular solid held together by van der Waals forces (specifically hydrogen bonds).
2. $NaCl(s)$ because ionic substances will conduct electricity if they are molten. The other two choices are molecular solids.
3. Ionic solid

21.6 Structures of solids

1. There is one CsCl formula unit. The Cs ions at the corners supply one Cs^+, while the I^- is the center of the unit cell. With one formula unit, this is a simple cubic unit cell. Do not mistake this for a body-centered unit cell, which would have two formula units. This is because all the lattice points are not identical.
2. There are two lattice points containing one P_4 each; therefore, the total number of atoms is 8.

22 Thermochemistry

22.1 Energy

1. 3.94×10^3 J
2. 83 J
3. −83 J

22.2 Heat

1. 9.14×10^4 J
2. 0.271 K
3. 51.4 g
4. 1.13 J g^{-1} °C^{-1}

22.3 Heats of reaction

1. 487 kJ
2. −3,120 kJ
3. −150 kJ

22.4 Calorimetry

1. +22.0 kJ mol^{-1} NaNO$_3$
2. -3.23×10^3 kJ mol^{-1} $C_6H_{12}O_6$
3. -5.150×10^3 kJ/mol $C_{10}H_8$

22.5 Hess's law

1. −918.3 kJ
2. −1,778.3 kJ
3. −81 kJ

22.6 Enthalpies of formation

1. $2 Na(s) + C(s) + 1.5 O_2(g) \rightarrow Na_2CO_3(s)$ $\Delta H_f° = -1{,}130.8$ kJ
2. $2 H_2(g) + 2 C(s) + O_2(g) \rightarrow HC_2H_3O_2(l)$ $\Delta H_f° = -487.0$ kJ
3. −871.7 kJ   4. −226 kJ

23 Solutions I

23.1 Aqueous solutions

1. Water is a bent molecule consisting of two very polar O–H bonds.
2. Water can form hydrogen bonds because it contains a hydrogen atom bonded to an oxygen atom.
3. CH_3F because the H is bonded to a C and not to an N, O, or F.

23.2 Electrolytes

1. No, because it does not conduct electricity. (It is a nonelectrolyte.)
2. Yes, because it does conduct electricity. (It is an electrolyte.)
3. Yes, because it does conduct electricity. (Vinegar [acetic acid] is a weak electrolyte.)

23.3 Solubility

1. SO_2 is polar and CH_3OH is polar and can for hydrogen bonds.
   Going through each of the three factors:
   1. It is necessary to overcome some of the hydrogen bonds in the CH_3OH.
   2. It is necessary to overcome the dipole-dipole forces in SO_2.
   3. Dipole-dipole forces form between the CH_3OH and the SO_2 molecules.

23.4 Acids and bases

1. Acid: HCl; base: NaOH; salt: NaCl
2. Acid: H_2SO_4; base: $Ba(OH)_2$; salt: $BaSO_4$
3. Acid: HBr; Base: NH_3; Salt: NH_4Br

23.5 Net ionic equations

1. $Ag^+(aq) + NO_3^-(aq) + Na^+(aq) + Cl^-(aq) \rightarrow AgCl(s) + Na^+(aq) + NO_3^-(aq)$
2. $CaCO_3(s) + 2 H^+(aq) \rightarrow Ca^{2+}(aq) + H_2CO_3(aq)$
3. $Ba^{2+}(aq) + 2 C_2H_3O_2^-(aq) + 2 H^+(aq) + SO_4^{2-}(aq) \rightarrow BaSO_4(s) + 2 HC_2H_3O_2(aq)$
   $Ba^{2+}(aq) + 2 C_2H_3O_2^-(aq) + 2 H^+(aq) + SO_4^{2-}(aq) \rightarrow BaSO_4(s) + 2 HC_2H_3O_2(aq)$
4. $Ca^{2+}(aq) + 2 Cl^-(aq) + 2 Na^+(aq) + C_2O_4^{2-}(aq) \rightarrow CaC_2O_4(s) + 2 Na^+(aq) + 2 Cl^-(aq)$
   $Ca^{2+}(aq) + C_2O_4^{2-}(aq) \rightarrow CaC_2O_4(s)$

24 Solutions II

24.1 Concentrations

1. 0.6001 M Cl^-
2. 0.200
3. 0.200

24.2 More concentrations

1. 1.300 m Na^+
2. 0.507152 m CH_3OH

24.3 Colligative properties

1. 90.9 torr
2. −1.86°C
3. 78.7°C
4. 18.0 atm

25 Kinetics

25.1 Introduction to kinetics

1. $\dfrac{\Delta[A]}{\Delta t} = \dfrac{[\text{Molarity}]}{\text{time}}$, any time unit works.

25.2 Reaction rates

1. $-\dfrac{\Delta[C_3H_8]}{\Delta t} = -\dfrac{1}{10}\dfrac{\Delta[F_2]}{\Delta t} = \dfrac{1}{3}\dfrac{\Delta[CF_4]}{\Delta t} = \dfrac{1}{8}\dfrac{\Delta[HF]}{\Delta t}$

25.3 Rate laws

1. $\text{Rate} = k[NO]^2[Cl_2]$, $k = 5.68 \text{ M}^{-2}\text{ s}^{-1}$ 2. $\text{Rate} = k[S_2O_8{}^{2-}][I^-]$, $6.0 \times 10^{-3}\text{ M}^{-1}\text{ s}^{-1}$

25.4 Integrated rate laws

1. 6.54×10^{-6} s   2. 0.075 M (If you got 0.499 M, you forgot to change hours to seconds, or vice versa.)

25.5 Half-life

1. 8.40×10^3 s   2. 1.47×10^4 s   3. 3.2×10^4 s

26 Introduction to equilibrium

26.1 Equilibrium reactions

1. $K_c = \dfrac{[H^+][CN^-]}{[HCN]}$

2. $K_p = \dfrac{P_{SO_3}^2}{P_{SO_2}^2\, P_{O_2}}$

3. $K_c = [Ag^+]^2[SO_4{}^{2-}]$

4. $K_c = \dfrac{[H^+]\,[C_2H_3O_2^-]}{[HC_2H_3O_2]}$

26.2 Equilibrium constants

1. 1.27   2. 1.6×10^7 3. 1.4×10^3

26.3 Multiple equilibria

1. 2.49×10^{-17} 2. 1.3×10^{-15}

26.4 Le Châtelier's principle

1. It shifts left. Adding a product shifts a reaction toward the reactants.
2. It shifts right. For endothermic process, heat may be treated as a reactant, and adding a reactant shifts the reaction toward the products.
3. No change. Adding (or removing) some solid makes no difference.
4. It will shift to the left. $K > Q\ (= 0.977)$, so the reaction will shift to the right.
5. It will shift to the right. $K < Q\ (= 1.69)$, so the reaction will shift to the left.

26.5 Equilibrium calculations

1. $P_{CO} = 4.93 \times 10^{-7}$ atm
2. $[HOI] = 0.500$ M $[H^+] = 3.39 \times 10^{-6}$ M $[OI^-] = 3.39 \times 10^{-6}$ M
3. $[Zn^{2+}] = 4.6 \times 10^{-7}$ M $[PO_4{}^{3-}] = 3.1 \times 10^{-7}$ M

27 Acids and bases

27.1 Types of acids and bases

1. HSO_4^- 2. $HC_2O_4^-$ 3. NH_2^- 4. OH^- 5. NH_3

27.2 Autoionization of water and pH

1. 2.46 2. 11.54 3. 1.870 4. 12.826

27.3 K_a and K_b calculations

1. 3.92 2. 2.51 3. 1.74 4. 12.17

27.5 Acid-base properties of salts

1. Neutral 2. Acidic 3. Basic 4. Acidic

28 Other equilibria

28.1 Solubility equilibria

1. $LiF(s) \rightleftharpoons Li^+(aq) + F^-(aq)$ $K_{sp} = [Li^+][F^-] = 3.8 \times 10^{-3}$
2. $PbSO_4(s) \rightleftharpoons Pb^{2+}(aq) + SO_4^{2-}(aq)$ $K_{sp} = [Pb^{2+}][SO_4^{2-}] = 1.6 \times 10^{-8}$
3. $Cr(OH)_3(s) \rightleftharpoons Cr^{3+}(aq) + 3\ OH^-(aq)$ $K_{sp} = [Cr^{3+}][OH^-]^3 = 6.3 \times 10^{-31}$
4. $Cu_3(AsO_4)_2(s) \rightleftharpoons 3\ Cu^{2+}(aq) + 2\ AsO_4^{3-}(aq)$ $K_{sp} = [Cu^{2+}]^3[AsO_4^{3-}]^2 = 7.6 \times 10^{-36}$
5. $MgNH_4PO_4(s) \rightleftharpoons Mg^{2+}(aq) + NH_4^+(aq) + PO_4^{3-}(aq)$ $K_{sp} = [Mg^{2+}][NH_4^+][PO_4^{3-}] = 2.5 \times 10^{-13}$

28.2 Complex ion equilibria

1. $Al^{3+}(aq) + 4\ OH^-aq) \rightleftharpoons [Al(OH)_4]^-(aq)$ $K_f = \dfrac{[Al(OH)_4^-]}{[Al^{3+}][OH^-]^4} = 1.1 \times 10^{33}$

2. $Pb^{2+}(aq) + 4\ I^-(aq) \rightleftharpoons [PbI_4]^{2-}(aq)$ $K_f = \dfrac{[pbl_4^{2-}]}{[pb^{2+}][I^-]^4} = 3.0 \times 10^4$

3. $Ag^+(aq) + 2\ NO_2^-(aq) \rightleftharpoons [Ag(NO_2)_2]^-(aq)$ $K_f = \dfrac{[Ag(NO_2)_2^-]}{[Ag^+][NO_2^-]^2} = 7 \times 10^2$

4. $[Fe(CN)_6]^{3-}(aq) \rightleftharpoons Fe^{3+}(aq) + 6\ CN^-(aq)$ $K_d = \dfrac{[Fe^{3+}][CN^-]^6}{[Fe(CN)_6^{3-}]} = 1 \times 10^{-31}$

5. $[Zn(NH_3)_4]^{2+}(aq) \rightleftharpoons Zn^{2+}(aq) + 4\ NH_3(aq)$ $K_d = \dfrac{[Zn^{2+}][NH_3]^4}{[Zn(NH_3)_4^{2+}]} = 2.0 \times 10^{-9}$

28.3 Calculations

1. 7.58×10^{-10} M $Al^{3+}(aq)$ 2. 9.9×10^{-7} M $Cl^-(aq)$ 3. 1.1×10^{-10} M $Ba^{2+}(aq)$
4. 8.90×10^{-10} M $Ag^+(aq)$ 5. 2.3×10^{-6} M $Ag^+(aq)$ 6. 7.9×10^{-10} M $Ca^{2+}(aq)$
7. 3.8×10^{-3} M $Zn^{2+}(aq)$

29 Buffers and titrations

29.1 The common ion effect

1. 3.29 2. 7.60

29.2 Buffer solutions

1. 4.754 2. 2.99 3. 8.945

29.3   Calculating the pH of a buffer solution

1. 3.43   2. 3.35

29.4   Titration calculations

1. 7.000   2. 5.222   3. 8.840   4. 12.237   5. 8.644   6. 5.361   7. 0.903   8. 0.903   9. 13.097

30   Thermodynamics

30.1   Thermodynamic calculations

1. $-44\,\text{J mol}^{-1}\,\text{K}^{-1}$ 2. $-163\,\text{J mol}^{-1}\,\text{K}^{-1}$ 3. $-313\,\text{J mol}^{-1}\,\text{K}^{-1}$ 4. $-297\,\text{J mol}^{-1}\,\text{K}^{-1}$ 5. $14.6\,\text{kJ mol}^{-1}$
6. $-715.0\,\text{kJ mol}^{-1}$ 7. $96.8\,\text{J mol}^{-1}\,\text{K}^{-1}$ 8. Negative   9. Positive   10. 0.311

31   Electrochemistry

31.1   Review of redox reactions

1. $3\,Fe^{3+}(aq) + Al(s) \rightarrow 3\,Fe^{2+}(s) + Al^{3+}(aq)$

31.2   Balancing redox equations

1. Rule 1: Separate into two half-reactions.
   Rule 2: Balance all elements except O and H.
   Rule 3: Balance O.
   Rule 4: Balance H.
   Rule 5: Balance charge.
   Rule 6: Multiply half-reactions so electrons match.
   Rule 7: Add and cancel.
   Rule 8: Check.

   If you had to look, you are not ready to move on.

2. $2\,MnO_4^-(aq) + 5\,C_2O_4^{2-}(aq) + 16\,H^+(aq) \rightarrow 2\,Mn^{2+}(aq) + 10\,CO_2(g) + 8\,H_2O(l)$
3. $Pt(s) + 8\,H^+(aq) + 4\,NO_3^-(aq) + 6\,Cl^-(aq) \rightarrow PtCl_6^{2-}(aq) + 4\,NO_2(g) + 4\,H_2O(l)$
4. $C_2H_5OH(aq) + 2\,Cr_2O_7^{2-}(aq) \rightarrow 2\,CO_2(g) + 3\,H_2O(l) + 4\,CrO_2^-(aq)$

31.3   Electrochemical cells

1. $Ag(s) \mid Ag^+(aq) \parallel Ni^{2+}(aq) \mid Ni(s)$
2. $Pt\,(s), Cl_2(g) \mid Cl^-(aq) \parallel Mn^{2+}(aq), H^+(aq) \mid MnO_2, Pt$

31.4   Standard reduction potentials

1. 0.56 V   2. -0.129 V

31.5   Electrochemistry and thermodynamics

1. $-3.0146 \times 10^5\,\text{J mol}^{-1}$ 2. 1.100 V   3. 6.5×10^{52}

31.6   Nonstandard cells

1. 1.43 V   2. 1.09 V   3. $9.9 \times 10^{-4}\,\text{M}$

31.7   Electrolysis and calculations

1. $370.4\,\text{g}\,Cl_2(g)$ 2. $2.72 \times 10^4\,\text{s}$ 3. +3

32   Nuclear chemistry

32.2   Types of radiation

1. α, β, and γ 2. γ 3. A positron

32.3 Nuclear stability

1. Fluorine-19 because it has an even number of neutrons (even-odd nucleus), and fluorine-18 is an odd-odd nucleus.
2. Bromine-80 is an odd-odd nucleus, while both bromine-79 and bromine-81 are even odd nuclei.
3. Both are double magic nuclei. Calcium-48 has a mass much higher than the atomic mass of calcium (40.078), which indicates that the neutron–proton ratio is too high for it to be stable.

32.4 Balancing nuclear equations

1. $^{239}_{92}\text{U}$ 2. $^{14}_{7}\text{N}$

32.5 Kinetics of nuclear decay

1. $0.0510\ \text{y}^{-1}$ 2. $0.0862\ \text{d}^{-1}$ 3. $0.12\ \text{d}^{-1}$ 4. 34.2%   5. 1.697 y (October 12, 2025)   6. 4.467×10^9 y
7. Fraud. The sample is 3,000 years old, which means it is not old enough to date from the third dynasty.

32.6 Nuclear energy

1. $-1.756 \times 10^{13}\ \text{J mol}^{-1}$ 2. $-1.823 \times 10^{13}\ \text{J mol}^{-1}$ 3. $1.709 \times 10^{14}\ \text{J mol}^{-1}$